AF725638

Introduction to Electromagnetic Wave Propagation

For a complete listing of the *Artech House Remote Sensing Library*,
turn to the back of this book . . .

Introduction to Electromagnetic Wave Propagation

Paul Rohan

Artech House
Boston • London

Library of Congress Cataloging-in-Publication Data
Rohan, P. (Paul)
Introduction to electromagnetic wave propagation / Paul Rohan.
p. cm.
Includes bibliographical references and index.
ISBN 0-89006-545-4
1. Electromagnetic waves--Transmission. I. Title.
QC665.T7R64 1991 91-21545
539.2--dc20 CIP

British Library Cataloguing in Publication Data
Rohan, Paul
Introduction to electromagnetic wave propagation.
I. Title
530.141

ISBN 0-89006-545-4

© 1991 Artech House, Inc.
685 Canton Street
Norwood, MA 02062

All rights reserved. Printed and bound in the United States of America. No part of this publication may be reproduced or utilized in any form or by any means, electronic or mechanical, including photocopying, recording, or by any information storage and retrieval system, without permission in writing from the publisher.

International Standard Book Number: 0-89006-545-4
Library of Congress Catalog Card Number: 91-21545

10 9 8 7 6 5 4 3 2 1

To my children,
Sylvia Ingrid (Holčán)
and
Thomas Edward (Sun)

CONTENTS

PREFACE

The understanding of electromagnetic wave propagation and the phenomena influencing it are of paramount importance to all those concerned with radio communication, navigational aids, and radar.

This book derives from edited lecture notes for a course on electromagnetic wave propagation taught in one trimester to final year students in electrical and electronic engineering at the University of Adelaide.

The contents draw freely on available publications listed at the ends of chapters. The author is indebted to all the listed authors and we apologize to those whom we might inadvertently have omitted.

The author wishes to thank the following:

The Australian Journal of Physics, Melbourne

Professor K. G. Budden, Cambridge, England

The Clarendon Press, Oxford, England

Director, Comité Consultatif International des Radiocommunications, Genève

Director, Electronic Research Laboratory, Defence Research Centre, Salisbury, South Australia

Dover Publications, Inc., New York

Elsevier Publishing Company, Amsterdam

Krátké vlny, Prague

The Macmillan Press Ltd., London

McGraw-Hill Book Company, New York

National Institute of Standards and Technology, United States Department of Commerce, Gaithersburg, Maryland

Nature, London

P. Noordhoff N.V., Groningen, the Netherlands

Peter Peregrinus Ltd., London

Proceedings of the IEE, London

Proceedings of the IEEE, New York

Schulthess Polygraphischer Verlag, Zürich

VEB Verlag Technik, Berlin

Vsesojuznoje Agentstvo po Avtorskim Pravam, Moscow

We appreciate their kind permission to reproduce figures from their publications and to use material contained in some internal publications at the former Weapons Research Establishment (WRE), now the Defence Research Centre (DRCS).

Because most students at the advanced stage of their studies are conversant with Maxwell's equations, the first chapter of the book is meant to be only a brief refresher on this topic.

The second chapter is devoted to the study of basic phenomena applicable to electromagnetic waves of all frequency ranges. Their grouping in a separate chapter eases the continuity of the following chapters by avoiding digressions to the investigation of the phenomenon in question.

Ground wave propagation is treated in Chapter 3. It provides Rotheram's field strength-distance curves for various combinations of ground constants for propagation in an exponential reference atmosphere over a curved earth. This is followed, in Chapter 4, by the study of sky wave propagation, where a short investigation of ionospheric characteristics leads to the treatment of electromagnetic wave propagation in this medium.

The characteristics of waves at the various frequency ranges are described in Chapter 5, which includes a short look at anomalous propagation and satellite communication.

Chapter 6 treats noise, signal-to-noise ratio, and their limiting effects on propagation, and the concluding seventh chapter, looks at considerations to be given to link design, but is not meant to be a guide to this subject.

A few examples show the application of the material by either step-by-step solution or indicating the formulas to be employed, thus encouraging the serious reader to review the referenced sections. After all, *"Repetitio est mater studiorum."* All examples contain the sought solutions. Problem sets at the ends of chapters provide the reader with questions to ponder and to test his or her understanding of the subject.

A textbook should not go into minute detail, but it should encourage the reader to think in terms of the topic under study. Accordingly, the following treatise is not exhaustive, but it provides an introduction to electromagnetic wave propagation in adequate depth to make it suitable for course work, self-study, and due to the inclusion of graphs and nomographs, to become a handy reference work for practicing engineers and scientists.

Chapter 1
IN PLACE OF AN INTRODUCTION

1.1 MAXWELL'S EQUATIONS

Electromagnetic waves have existed since God allegedly said: "Let there be light," but humankind had to wait for Maxwell to recognize light to be electromagnetic radiation.

By extending the laws of induction from classical physics to displacement currents, Maxwell was able to express the magnitude of the electric and magnetic field at any point in space by a set of fundamental relations, known today as Maxwell's equations.

According to Faraday's law, the total electromotive force **E** induced in a closed circuit is equal to the negative of the time derivative of the magnetic flux through the circuit. This, in a mathematical formulation, yields

$$\oint_C \mathbf{E} \cdot d\mathbf{l} = -\int \frac{\partial \mathbf{B}}{\partial t} \cdot \mathbf{n} dS \tag{1.1}$$

with $d\mathbf{l}$ denoting an element of the contour C, dS is an element of the surface S bounded by C, and **B** is the magnetic flux. This equality must hold for any shape of the closed-circuit C, for any shape of the surface bounded by C, and therefore for any direction of the normal **n**.

Due to Stokes's theorem of vector analysis, the line integral of the vector **E** on the left is equal to the integral of the normal component of the curl of this vector over any surface S bounded by C, or

$$\oint_C \mathbf{E} \cdot d\mathbf{l} = \int_S \nabla \times \mathbf{E} \cdot d\mathbf{S} = -\int \frac{\partial \mathbf{B}}{\partial t} \cdot d\mathbf{S}$$

so that

$$\int_S \nabla \times \mathbf{E} \cdot d\mathbf{S} = -\int \frac{\partial \mathbf{B}}{\partial t} \cdot d\mathbf{S} \tag{1.2}$$

For this to be valid, the two integrands must be equal, and the differential form of Faraday's law in vector notation becomes

$$\nabla \times \mathbf{E} = -\frac{\partial \mathbf{B}}{\partial t} \tag{1.3}$$

which is one of Maxwell's equations.

Example 1.1

The electric field intensity at a point in free space is $\mathbf{E} = 300 \cos x \cos \omega t \mathbf{k}$.
Find the time varying intensity of the magnetic field.

Solution: Because

$$\nabla \times \mathbf{E} = -\frac{\partial E_z}{\partial x}\mathbf{j} = 300 \sin x \cos \omega t \mathbf{j}$$

then on the basis of Faraday's law

$$\mathbf{H} = -\frac{1}{\mu_0}\int \nabla \times \mathbf{E} dt,$$

where $\mu_0 = 4\pi 10^{-7}$, H/m is the permeability of free space. Inserting the computed values yields

$$\mathbf{H} = -\frac{3 \cdot 10^9}{4\pi\omega} \sin x \sin \omega t \mathbf{j}$$

On the basis of the rules of vector analysis, the divergence of 1.3 leads to

$$\text{div}\mathbf{B} = 0$$

which is another of Maxwell's equations.

Ampère's law states that the magnetomotive force along a closed path C equals the total current enclosed by the path, or

$$\oint_C \mathbf{H} \cdot d\mathbf{l} = \mathbf{I}_t \tag{1.4}$$

where $\mathbf{I}_t$ the total current may be expressed on the basis of the total current density $\mathbf{J}_t$ as

$$\mathbf{I}_t = \int_S \mathbf{J}_t \cdot d\mathbf{S} \tag{1.5}$$

Then,

$$\oint_C \mathbf{H} \cdot d\mathbf{l} = \int_S \mathbf{J}_t \cdot d\mathbf{S}$$

which, with the application of Stokes's theorem to the left-hand side of the equation, yields

$$\int_S (\nabla \times \mathbf{H}) \cdot d\mathbf{S} = \int_S \mathbf{J}_t \cdot d\mathbf{S}$$

which, on the basis of the previous reasoning, leads to

$$\nabla \times \mathbf{H} = \mathbf{J}_t \tag{1.6}$$

Example 1.2

A long, straight wire carries a current $i = 10 \cos 2\pi 50t$. Determine the inductive voltage drop across the two-turn square coil lying in the plane of the wire, as shown by Figure 1.1.

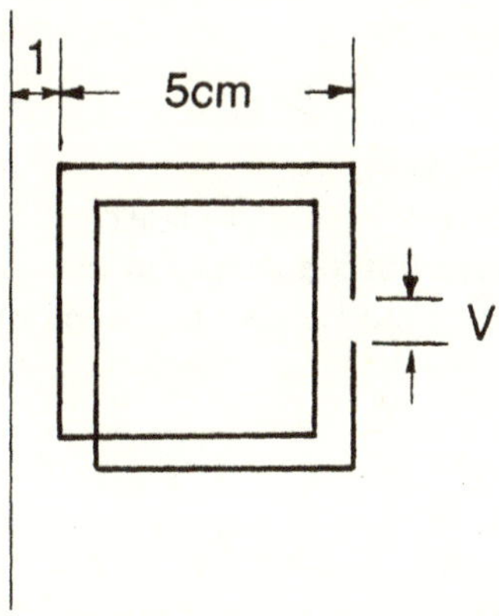

Figure 1.1 Example 1.2.

Solution: According to Ampère's law $H = i/l$, where l is the length of the magnetic path; the magnetic flux density is

$$B = \mu H = \mu \frac{i}{l}$$

The flux through an area A is

$$\phi = BA = \mu A \frac{i}{l}$$

and the voltage drop across the coil of N turns is

$$v = -N \frac{d\phi}{dt}$$

The average magnetic path length is $l = 2\pi(1 + 2.5) = 7\pi$ cm yielding, for a free-space permeability of $\mu_0 \doteq 4\pi 10^{-7}$ H/m,

$$\frac{d\phi}{dt} = -0.01 \sin 100\pi t$$

and, therefore,

$$v = 2 \cdot 10^{-2} \sin 100\pi t \text{ volts}$$

Taking the divergence of both sides of (1.6), one arrives at

$$\mathrm{div}\mathbf{J}_t = 0$$

which is true for charge-free cases only.

Assuming that the charge of a parallel-plate capacitor changes in time dt by dQ, and accepting that the current flows not only in the connecting leads, but also traverses the dielectric, the current in the circuit, called by Maxwell the *displacement current,* is

$$i = \frac{dQ}{dt}$$

and the displacement current density in the dielectric for capacitor plates of an area A, is

$$j = \frac{1}{A}\frac{dQ}{dt}$$

However, because the charge of a capacitor is

$$Q = CV$$

where V is the potential difference between the plates, the current density will be

$$j = \frac{C}{A}\frac{dV}{dt} \tag{1.7}$$

The capacitance of a parallel plate capacitor may be expressed by

$$C = \frac{\epsilon A}{d}$$

where ϵ is the dielectric constant of the material between the plates and d is their separation. The potential difference between the plates for a field **E** in the dielectric is $V = \mathbf{E}d$.

Inserting the preceding expressions into (1.7), one obtains

$$\mathbf{j} = \epsilon \frac{d\mathbf{E}}{dt}$$

which, with $\epsilon\mathbf{E} = \mathbf{D}$, yields

$$\mathbf{j} = \frac{\partial \mathbf{D}}{\partial t} \tag{1.8}$$

the variation of the displacement current density with time. **D** represents charge per unit area, which, when integrated over the area of the capacitor's plate:

$$\iint \mathbf{D} \cdot d\mathbf{A} = Q$$

leads to the total charge on the plate of the capacitor. This is a generalized form of Gauss's theorem. With Δv denoting a volume element and ρ the charge density, the preceding expression may be written as

$$\iint \mathbf{D} \cdot d\mathbf{A} = \rho \Delta v$$

By permitting Δv to approach zero, the limit:

$$\lim_{\Delta v \to 0} \frac{\iint \mathbf{D} \cdot \mathbf{dA}}{\Delta v} = \rho$$

becomes the divergence of the charge, so that

$$\mathrm{div}\mathbf{D} = \rho \tag{1.9}$$

Extending $\mathbf{I}_t$ to include the displacement current, the general form of Ampère's law becomes

$$\nabla \times \mathbf{H} = \mathbf{J} + \frac{\partial \mathbf{D}}{\partial t} \tag{1.10}$$

with **J** denoting the conduction current density.

Equations (1.9) and (1.10) are further two of Maxwell's equations. These four equations usually are extended by three constitutive equations:

$$\mathbf{B} = \mu\mathbf{H},\ \mathbf{D} = \epsilon\mathbf{E},\ \mathbf{J} = \sigma\mathbf{E}$$

which determine the dependence of the vector quantities on the medium in which the field is being considered. The parameters of these equations—namely, the permittivity ϵ, the permeability μ, and the conductivity σ—describe the medium.

Maxwell's equations:

$$\begin{aligned} \mathrm{curl}\ \mathbf{H} &= \sigma\mathbf{E} + \frac{\partial \mathbf{D}}{\partial t} \\ \mathbf{B} &= \mu\mathbf{H} \\ \mathrm{div}\mathbf{B} &= 0 \end{aligned} \tag{1.11}$$

$$\begin{aligned} \mathrm{curl}\ \mathbf{E} &= -\frac{\partial \mathbf{B}}{\partial t} \\ \mathbf{D} &= \epsilon\mathbf{E} \\ \mathrm{div}\mathbf{D} &= \rho \end{aligned} \tag{1.12}$$

constitute the starting relations of most theoretical electromagnetic wave propagation investigations.

Example 1.3

By using expressions (1.9) and (1.10) and the constitutive relation $\mathbf{J} = \sigma\mathbf{E}$, show that the charge density is $\rho = \rho_0 \exp[-(\sigma/\epsilon)t]$.

Solution: The divergence of (1.10) is

$$\nabla \cdot (\nabla \times \mathbf{H}) = \nabla \cdot \mathbf{J} + \nabla \cdot \frac{\partial \mathbf{D}}{\partial t}$$

the left-hand side of which is zero, so that

$$\nabla \cdot \mathbf{J} = -\nabla \cdot \left(\frac{\partial \mathbf{D}}{\partial t}\right)$$

Then, due to (1.9),

$$\nabla \cdot \mathbf{J} = -\frac{\partial}{\partial t}(\nabla \cdot \mathbf{D}) = -\frac{\partial \rho}{\partial t}$$

Further, as (1.9) may be written as $\rho = \epsilon(\nabla \cdot \mathbf{E})$ and from the constitutive relation $\nabla \cdot \mathbf{E} = 1/\sigma(\nabla \cdot \mathbf{J})$, the charge density becomes

$$\rho = -\frac{\epsilon}{\sigma}\frac{\mathrm{d}\rho}{\mathrm{d}t}$$

or

$$\frac{\sigma}{\epsilon}\rho + \frac{\mathrm{d}\rho}{\mathrm{d}t} = 0$$

This is a linear differential equation of the first order, the solution of which is

$$\rho = \rho_0 \exp\left(-\frac{\sigma}{\epsilon}t\right)$$

where ρ_0 may be determined from the initial conditions.

1.2 POTENTIAL FUNCTIONS

Although Maxwell's equations, by satisfying the required boundary conditions, lead to the solution of any electromagnetic field problem, their mathematics often is very involved. The differentials of potential functions that are scalar or vector point functions may yield results in a more manageable form. These functions need not have any physical significance.

To find the electric and magnetic fields produced by a system of charges, the field quantities must be expressed by potential functions, which may be determined by the wave equation when neither currents nor charges are present. Because the magnetic flux density **B** is solenoidal, its divergence is always zero. The magnetic field, according to Simonyi, then may be expressed by the curl of another vector **A** as

$$\mu \mathbf{H} = \nabla \times \mathbf{A} \tag{1.13}$$

The curl of the electric field vector **E** is not zero, and it cannot be expressed by the gradient of a scalar function.

Inserting (1.13) into Maxwell's equation for the curl of the electric field vector leads to

$$\nabla \times \mathbf{E} = -\frac{\partial}{\partial t}(\nabla \times \mathbf{A}) = -\left(\nabla \times \frac{\partial \mathbf{A}}{\partial t}\right) \tag{1.14}$$

from which

$$\nabla \times \left(\mathbf{E} + \frac{\partial \mathbf{A}}{\partial t}\right) = 0 \tag{1.15}$$

The curl of the vector in parentheses is zero, and the vector itself may be derived from a scalar potential:

$$\mathbf{E} + \frac{\partial \mathbf{A}}{\partial t} = -\nabla\varphi \tag{1.16}$$

so that the total electric field strength is

$$\mathbf{E} = -\frac{\partial \mathbf{A}}{\partial t} - \nabla\varphi \tag{1.17}$$

Both the electric and magnetic field strength for the dependent scalar potential ϕ may be determined. By inserting the values of **E** and **H** into Maxwell's first equation, we find

$$\nabla \times \nabla \times \mathbf{A} = \mu \mathbf{J} - \epsilon\mu \frac{\partial^2 \mathbf{A}}{\partial t^2} - \mu\epsilon\nabla \frac{\partial \varphi}{\partial t} \tag{1.18}$$

The vector **A** is uniquely defined by its arbitrarily chosen divergence:

$$\nabla \cdot \mathbf{A} = -\mu\epsilon \frac{\partial \varphi}{\partial t} \tag{1.19}$$

so that the vector potential **A**, with the vector identity:

$$\nabla \times (\nabla \times \mathbf{A}) = \nabla(\nabla \cdot \mathbf{A}) - \nabla^2 \mathbf{A}$$

becomes

$$\nabla^2 \mathbf{A} = \epsilon\mu \frac{\partial^2 \mathbf{A}}{\partial t^2} - \mu \mathbf{J} \tag{1.20}$$

Hertz expressed the vector **A** by the so-called Hertz vector **II** as

$$\mathbf{A} = \epsilon\mu \frac{\partial \mathbf{\Pi}}{\partial t} \tag{1.21}$$

which permits us to write

$$\nabla \cdot \mathbf{A} = -\epsilon\mu \frac{\partial}{\partial t} \nabla \cdot \mathbf{\Pi} \tag{1.22}$$

and, therefore,

$$\varphi = -\nabla \cdot \mathbf{\Pi} \tag{1.23}$$

if the time-independent constant is chosen to be zero.

The Hertz vector itself satisfies the wave equation in a current- and charge-free space, so that

$$\nabla^2 \mathbf{\Pi} - \epsilon\mu \frac{\partial^2 \mathbf{\Pi}}{\partial t^2} = 0 \tag{1.24}$$

Expressing then the electric and magnetic field strength in terms of the Hertz vector, we obtain

$$\mathbf{H} = \epsilon \frac{\partial}{\partial t} \nabla \times \mathbf{\Pi}$$
$$\mathbf{E} = -\epsilon\mu \frac{\partial^2 \mathbf{\Pi}}{\partial t^2} + \nabla(\nabla \cdot \mathbf{\Pi}) \tag{1.25}$$

By using the previously given vector identity, the electric field becomes

$$\mathbf{E} = \nabla \times \nabla \times \mathbf{\Pi} \tag{1.26}$$

If the time variation of the fields is given by $\exp(j\omega t)$ these two fields become

$$\mathbf{H} = j\omega\epsilon\nabla \times \mathbf{\Pi}$$
$$\mathbf{E} = \nabla \times \nabla \times \mathbf{\Pi} \tag{1.27}$$

Similarly to the derivation of expressions (1.27) for **H** from the vector potential (1.13), expressions for **E**, in a homogeneous charge-free medium also may be derived from the vector potential $\mathbf{E} = \nabla \times \mathbf{A}$.

The Hertz vector given by

$$\mathbf{A} = -\mu \frac{\partial \mathbf{\Pi}}{\partial t} \tag{1.28}$$

then yields

$$\mathbf{E} = -\mu\nabla \times \frac{\partial \mathbf{\Pi}}{\partial t}$$
$$\mathbf{H} = \nabla(\nabla \cdot \mathbf{\Pi}) - \nabla^2\mathbf{\Pi} \tag{1.29}$$

where the vector **Π** satisfies the wave equation.

If the time variation of the quantities of interest is given by $\exp(j\omega t)$, as before, Maxwell's equations for **E** and **H** in terms of the Hertz vector become

$$\mathbf{E} = \nabla \times \nabla \times \mathbf{\Pi}_e \quad \mathbf{E} = -j\omega\mu\nabla \times \mathbf{\Pi}_m$$
$$\mathbf{H} = j\omega\epsilon\nabla \times \mathbf{\Pi}_e \quad \mathbf{H} = \nabla \times \nabla \times \mathbf{\Pi}_m \tag{1.30}$$

where subscripts e and m indicate the basic vector potentials used for the derivation. The Hertz vectors of these expressions satisfy the wave equation:

$$\nabla^2 \mathbf{\Pi}_{e,m} + \epsilon\mu\omega^2 \mathbf{\Pi}_{e,m} = 0$$

Although expressions (1.13) and (1.17), which satisfy Maxwell's equation $\nabla \cdot \mathbf{B} = 0$, and the Maxwell-Faraday equation express **H** and **E** in terms of the potential functions **A** and φ, neither of these functions is uniquely defined because no relationship between them has been specified [Holt, 1963]. Restricting the investigation to linear, homogeneous, and isotropic media in which the permittivity ϵ and the permeability μ are constant and independent of position in space and time, although the fields depend on both the Maxwell-Ampère and Gauss expressions:

$$\nabla \times \mathbf{B} = \mu\epsilon \frac{\partial \mathbf{E}}{\partial t} + \mu \mathbf{J} \qquad (1.31)$$

$$\nabla \cdot \mathbf{E} = \frac{\rho}{\epsilon}$$

which, in terms of **A** and φ, read

$$\nabla \times \nabla \times \mathbf{A} = -\mu\epsilon \left(\nabla\varphi + \frac{\partial^2 \mathbf{A}}{\partial t^2} \right) + \mu \mathbf{J} \qquad (1.32)$$

$$\nabla \cdot \left(-\nabla\varphi - \frac{\partial \mathbf{A}}{\partial t} \right) = \frac{\rho}{\epsilon}$$

Use of the previously given vector equality for $\nabla \times \nabla \times \mathbf{A}$ leads to

$$\nabla^2 \mathbf{A} = \mu\epsilon \frac{\partial^2 \mathbf{A}}{\partial t^2} + \nabla(\nabla \cdot \mathbf{A} + \mu\epsilon\varphi) - \mu \mathbf{J} \qquad (1.33)$$

$$\nabla^2 \varphi = -\nabla \cdot \frac{\partial \mathbf{A}}{\partial t} - \frac{\rho}{\epsilon}$$

Setting

$$\nabla \cdot \mathbf{A} = -\mu\epsilon \frac{\partial \varphi}{\partial t} \qquad (1.34)$$

reduces expressions (1.33) to the wave equation for the potentials **A** and φ:

$$\nabla^2 \mathbf{A} = \mu\epsilon \frac{\partial^2 \mathbf{A}}{\partial t^2} - \mu \mathbf{J} \qquad (1.35a)$$

$$\nabla^2 \varphi = \mu\epsilon \frac{\partial^2 \varphi}{\partial t^2} - \frac{\rho}{\epsilon}$$

and establishes a relationship between the magnetic vector potential **A** (Wb/m) and the electric scalar potential φ (V).

Equations:

$$\mathbf{H} = \nabla \times \mathbf{A}$$

$$\mathbf{E} = -\nabla\varphi - \frac{\partial \mathbf{A}}{\partial t}$$

$$\nabla \cdot \mathbf{A} = -\mu\epsilon\varphi \tag{1.35b}$$

$$\nabla^2\mathbf{A} = \mu\epsilon\frac{\partial^2 \mathbf{A}}{\partial t^2} - \mu\mathbf{J}$$

$$\nabla^2\varphi = \mu\epsilon\frac{\partial^2 \varphi}{\partial t^2} - \frac{\rho}{\epsilon}$$

are Maxwell's equations in potential form. The functions $\mu\mathbf{J}$ and ρ/ϵ are the forcing functions of the wave equations or the sources of the waves of the potential functions **A** and φ, respectively.

Example 1.4

From the potential function:

$$\mathbf{A} = 10^{-8}\cos 3x \cos 4z \cos\omega t\mathbf{k}$$

in free space, determine the electric and magnetic field strength.

Solution: Expression (1.17) requires knowledge of the electric scalar potential φ for the determination of the electric field strength. It may be found from the wave equation (1.20), which, for free space, where the current density is zero, reads

$$\nabla^2\mathbf{A} = \mu\epsilon\frac{\partial^2 \mathbf{A}_z}{\partial t^2}$$

The second derivatives of A_z with respect to x and z (i.e., $-9A_z$ and $-16A_z$, respectively) yield

$$\nabla^2 A_z = -25A_z$$

which, when inserted into the wave equation, leads to

$$A_z = \mu_0\epsilon_0\frac{\partial^2 A_z}{\partial t^2} = -\omega^2\mu_0\epsilon_0 A_z$$

or

$$25 = \omega^2 \mu_0 \epsilon_0$$

from which the angular frequency, with $\epsilon = \epsilon_0 = 10^{-9}/36\pi$ F/m and $\mu = \mu_0 = 4\pi \cdot 10^{-7}$ H/m, the free-space permittivity and permeability, becomes

$$\omega = \sqrt{\frac{25}{\mu_0 \epsilon_0}} = 1.5 \times 10^9 \text{ rad/s}$$

The electric scalar potential, on the basis of (1.34), is

$$\varphi = -\frac{1}{\mu\epsilon} \int \nabla \cdot \mathbf{A} \, dt$$

Because $\nabla \cdot \mathbf{A} = -4 \times 10^{-8} \cos 3x \sin 4z \cos\omega t$, we arrive at

$$\varphi = 2.4 \cos 3x \sin 4z \sin\omega t$$

The electric field strength, from expression (1.17), is then

$$\mathbf{E} = 7.2 \sin 3x \sin 4z \sin\omega t\mathbf{i} + 15 \cos 3x \cos 4z \sin\omega t\mathbf{k}$$

The magnetic field strength follows from

$$\mathbf{H} = \frac{1}{\mu}(\nabla \times \mathbf{A}) = -\frac{1}{\mu}\frac{\partial A_z}{\partial x}\mathbf{j}$$

so that

$$\mathbf{H} = 0.024 \sin 3x \cos 4z \cos\omega t\mathbf{j}$$

1.3 RADIATION

The vector potential **A** due to a current flow in a closed circuit, as introduced in the previous section, is the sum of the vector potentials due to all elements in the circuit. In a more general form, it is the integral of the current density **J** over the volume in which it exists, or

$$\mathbf{A}(r) = \frac{\mu}{4\pi} \int \frac{\mathbf{J}(t - r/c)}{r} \, dV \qquad (1.36)$$

r is the distance of the point of observation from the source at the origin of spherical coordinates and $t - r/c$ accounts for the time delay due to the finite velocity c of propagation of electromagnetic waves [Jordan, 1953].

Expressing the current density **J** in terms of current element Idl (i.e., the filamentary currents I flowing along elemental length dl such that the current essentially is constant along their length) we obtain for the z-component of the vector potential

$$A_z = \frac{\mu}{4\pi} \frac{Idl \cos\omega(t - r/c)}{r} \tag{1.37}$$

Inserting the three spherical coordinate components of **A**; namely,

$$A_r = A_z \cos\theta, \quad A_\theta = -A_z \sin\theta, \quad A_\phi = 0 \tag{1.38}$$

into the components of $\mu\mathbf{H} = \nabla \times \mathbf{A}$, in the same directions, yields zero values for H_r and H_θ, leaving only

$$\begin{aligned} H_\phi &= \frac{1}{\mu r} (\nabla \times A)_\phi \\ &= \frac{1}{\mu r} \left[\frac{\partial}{\partial r} (rA_\theta) - \frac{\partial A_r}{\partial \theta} \right] \\ &= \frac{I_0 dl}{4\pi r} \left\{ \frac{\partial}{\partial r} [- \sin\theta \cos\omega(t - r/c)] - \frac{\partial}{\partial \theta} \left[\frac{\cos\theta}{r} \cos\omega(t - r/c) \right] \right\} \\ &= \frac{I_0 dl \sin\theta}{4\pi} \left[\frac{-\omega \sin\omega(t - r/c)}{rc} + \frac{\cos\omega(t - r/c)}{r^2} \right] \end{aligned} \tag{1.39}$$

The Maxwell-Ampère equation expresses the electric field strength by

$$\mathbf{E} = \frac{1}{\epsilon} \int (\nabla \times \mathbf{H}) \, dt \tag{1.40}$$

The curl of **H** yields only two components of **E**: one in the r-direction

$$E_r = \frac{1}{\epsilon r \sin\theta} \int \frac{\partial}{\partial \theta} (\sin\theta \mathbf{H}_\phi) \, dt \tag{1.41}$$

and the θ component

$$E_\theta = -\frac{1}{r} \int \frac{\partial (rH_\phi)}{\partial r} \, dt \tag{1.42}$$

Only the second one contributes to the power contained in the wave propagating in the direction of interest. After the indicated operations, this component is

$$E_\theta = \frac{I_0 dl \sin\theta}{4\pi\epsilon}\left[-\frac{\omega \sin\omega(t - r/c)}{rc^2} + \frac{\cos\omega(t - r/c)}{r^2 c} + \frac{\sin\omega(t - r/c)}{\omega r^3}\right] \tag{1.43}$$

We will show later that in a homogeneous isotropic dielectric the electric and magnetic fields are in phase. Analogous to the fundamental principles of physics, the product of the electric and magnetic fields yields power.

The first terms in the brackets of (1.39) and (1.43) are inversely proportional to the distance r from the transmitter, whereas the remaining ones to higher powers of this distance. The first terms account for the radiation field experienced at greater distances, and the other terms cause an induction field to exist in the vicinity of the source of radiation.

Restricting the following to the radiation field terms, the radiated power becomes

$$P(t) = H_\phi E_\theta = \frac{30\pi}{r^2} I_0^2 \left(\frac{dl}{\lambda}\right)^2 \sin^2\theta \sin^2\omega(t - r/c) \tag{1.44}$$

when the relations $\omega = 2\pi f$, $c = \lambda f = \omega\lambda/(2\pi)$, $2\epsilon c = 1/60\pi$ are used for simplifying the expressions for the amplitudes of the two fields to read

$$H_\phi = \frac{I_0 dl}{2\lambda r}\sin\theta$$

and

$$E_\theta = \frac{60\pi I_0 dl}{\lambda r}\sin\theta$$

respectively.

Surrounding the radiating current element by an imaginary, arbitrarily shaped closed surface and computing the power flow through the surface yields the total radiated power. Accepting, for simplicity's sake, the current element to be in the center of a closed spherical surface of radius r (Figure 1.2), divided into narrow ring-shaped strips of radius $r \sin\theta$ and width $r\, d\theta$, and therefore of surface area $2\pi r^2 \sin\theta\, d\theta$, lets us write (1.44) as

$$P(t) = 60\pi^2 I_0^2 \left(\frac{dl}{\lambda}\right)^2 \sin^3\theta \sin^2\omega(t - r/c)\, d\theta \tag{1.45}$$

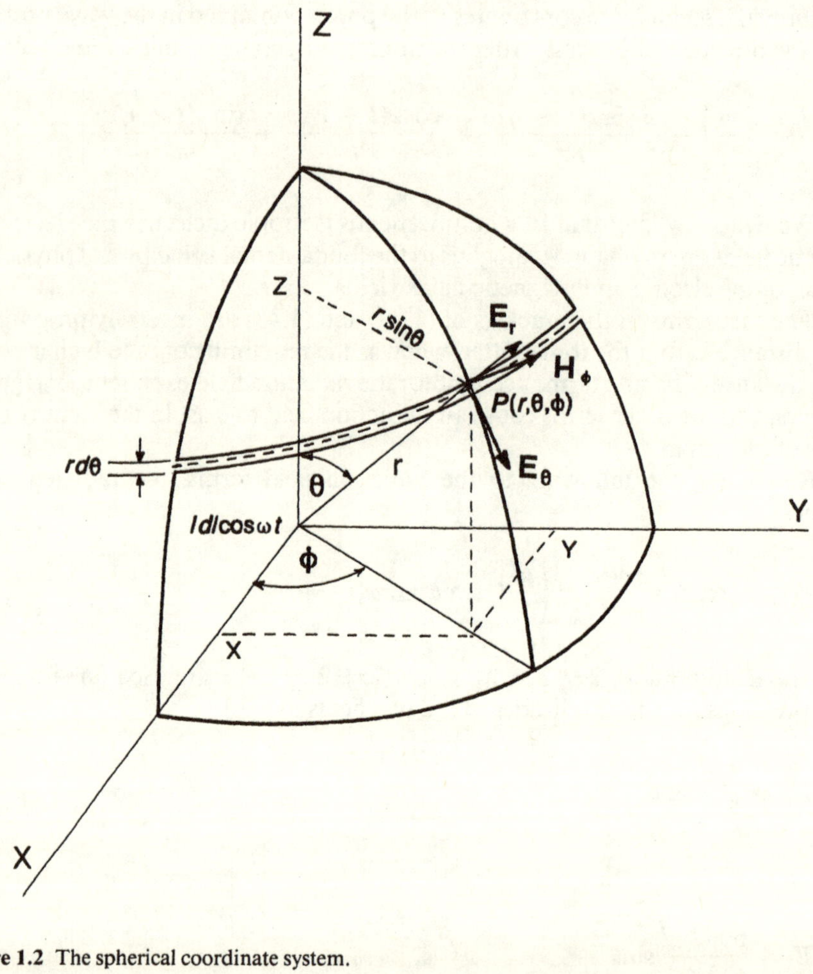

Figure 1.2 The spherical coordinate system.

that, integrated within the limits $0 < \theta < \pi$ and $0 < t < T$, yields the total radiated power.

Because

$$\int_0^\pi \sin^3\theta \, d\theta = \frac{4}{3}$$

and

$$\frac{1}{T}\int_0^T \sin^2\omega(t - r/c)\,dt = \frac{1}{2}$$

the average radiated power becomes

$$P_{av} = 40\pi^2 I_0^2 \left(\frac{dl}{\lambda}\right)^2 \tag{1.46}$$

or, because $I_0 = \sqrt{2}I_{eff}$,

$$P_{av} = 80\pi^2 I_{eff}^2 \left(\frac{dl}{\lambda}\right)^2 \tag{1.47}$$

Further, as $P = RI_{eff}^2$, the product

$$80\pi^2 \left(\frac{dl}{\lambda}\right)^2 \tag{1.48}$$

must have the dimension of resistance. It represents the radiation resistance of the current element, provided that $dl \ll \lambda$.

This value of resistance, with a current equal to that in the dipole, would dissipate as heat, a power equal to that radiated by the dipole.

Example 1.5

What is the value of the radiation resistance of a $\lambda/8$ long dipole?

Solution: From (1.48)

$$R = 80\pi^2/64 = 12.34\ \Omega$$

This example shows the inefficiency of very short dipole radiators. (Note: The radiation resistance of a half-wave dipole is 73.1 Ω. Its determination is beyond the scope of this book.)

The current as a function of the radiated power, on the basis of (1.47),

$$I_{eff} = \frac{\lambda}{\pi dl}\sqrt{\frac{P}{80}} \tag{1.49}$$

inserted into the expression for the amplitude of the electric field provides the effective electric field strength:

$$E_{\text{eff}} = 3 \sqrt{\frac{10P_{\text{av}}}{2} \frac{\sin\theta}{r}} \text{ V/m} \tag{1.50}$$

or, when expressed in the more often used units of mV/m, with the power inserted in kW and the distance r in km:

$$E_{\text{eff}} = \frac{300}{r_{\text{km}}\sqrt{2}} \sqrt{P_{\text{kW}}} \sin\theta \text{ mV/m} \tag{1.51}$$

The concept of the Poynting vector and its applications is relegated to Chapter 2.

1.4 FREQUENCIES AND DESIGNATIONS OF THE USED RANGE OF ELECTROMAGNETIC WAVES

The used range of electromagnetic waves extends in wavelength from about 100 km to less than a millimeter, as shown in Table 1.1. There is no strict dividing line between these categories although there is a difference in their behavior. Waves in these ranges find application in communication, navigation, and radar.

Table 1.1
The Range of the Used Electromagnetic Waves

Designation Frequencies	*Abbreviation*	*Frequency Range*	*Wavelength Range*
Extremely low	ELF	< 3 kHz	> 100 km
Very low	VLF	3–30 kHz	100–10 km
Low	LF	30–300	10–1 km
Medium	MF	300–3000	1 km–100 m
High	HF	3–30 Mhz	100–10 m
Very high	VHF	30–300	10–1 m
Ultra high	UHF	300–3000	1 m–10 cm
Super high	SHF	3–30 GHz	10–1 cm
Extremely high	EHF	30–300	1 cm–1 mm
Submillimeter waves		> 300 GHz	< 1 mm

PROBLEMS

1.1 A steady current I amp flows through a coil of length l m consisting of n turns that cross a square surface of side length $a > l$ placed so that one of its sides passes through the axis of the coil. Find the magnetomotive force along the circumference of the square surface.

1.2 A vertical 20-turn square frame antenna of 1 m^2 area rotates around its vertical axis in the earth's magnetic field of flux density 0.312 10^{-4} Wb/m^2 at a rate of 60 rpm (Figure 1.3). Determine the average induced electromotive force when the coil has rotated over a half-revolution from its initial position perpendicular to the magnetic field.

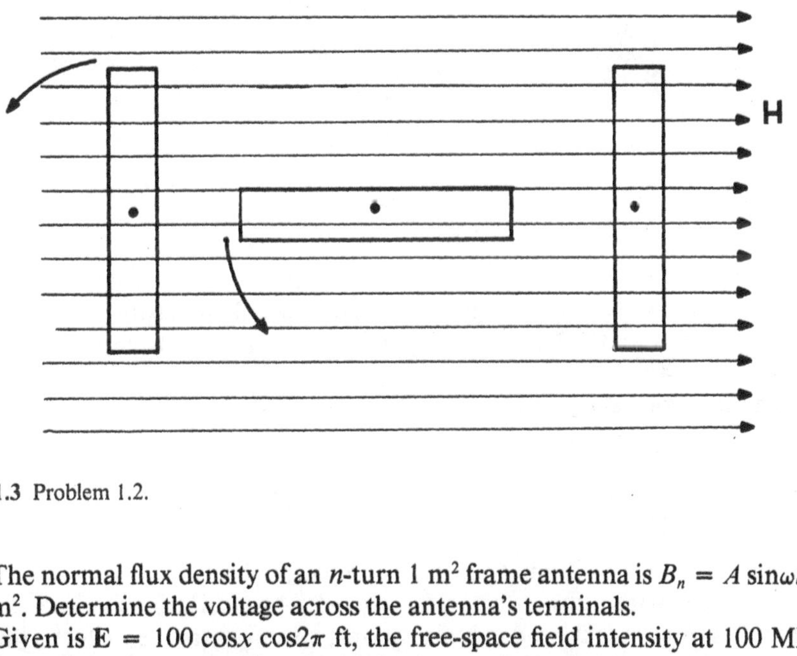

Figure 1.3 Problem 1.2.

1.3 The normal flux density of an n-turn 1 m^2 frame antenna is $B_n = A \sin\omega t$ Wb/m^2. Determine the voltage across the antenna's terminals.

1.4 Given is $\mathbf{E} = 100 \cos x \cos 2\pi ft$, the free-space field intensity at 100 MHz in the region of a square closed path in the xz plane of a rectangular coordinate system (Figure 1.4). Find the magnetomotive force around the path enclosing the area S and the total current over S.

1.5 Using the material of Chapter 1 derive the circuit laws

$$i = C\frac{dv}{dt} \text{ and } v = L\frac{di}{dt}$$

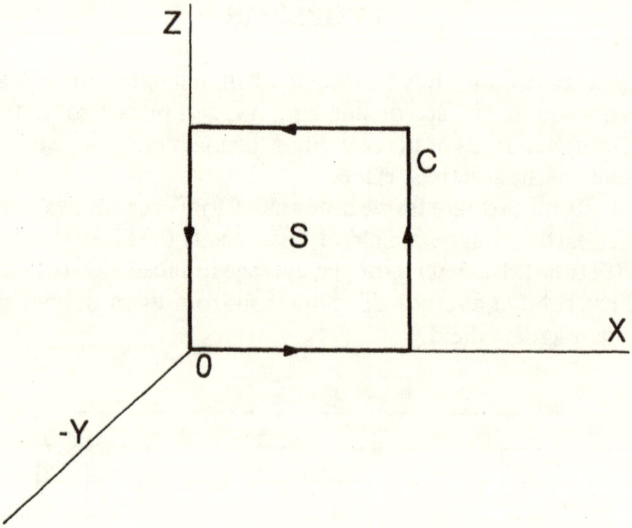

Figure 1.4 Problem 1.4.

1.6 A spherical conductor of radius r carries a charge Q. Determine (a) the free-space field intensity at a distance $d > r$, (b) the field intensity on the surface of a sphere concentric with the first one but having twice its radius, and (c) at a point in space at $d < r$.

1.7 The magnetic field in a nonferrous conductor is

$$\mathbf{H} = -\frac{10^{-6}\sqrt{2}}{\mu} e^{-10^3 x} \cos(10^5 t - 10^3 x + \pi/4)\mathbf{j}$$

Determine the conductivity σ of the conductor.

1.8 A charge Q within the volume V surrounded by a surface S causes a displacement current to leave the volume. Show that this current equals the time rate of change of the charge.

REFERENCES AND RECOMMENDED READING

Holt, C.A. (1963) *Introduction to Electromagnetic Waves and Fields,* John Wiley and Sons, New York.

Jordan, E.C. (1953) *Electromagnetic Waves and Radiating Systems,* Constable and Company, London.

Kong, J.A. (1975) *Theory of Electromagnetic Waves,* John Wiley and Sons, New York.

Simonyi, K. (1963) *Foundations of Electrical Engineering,* Pergamon Press, London.

Skilling, H.H. (1948) *Fundamentals of Electric Waves,* 2nd ed., John Wiley and Sons, New York.

Slater, J.C. (1942) *Microwave Transmission,* McGraw-Hill, New York.

Slater, J.C., and N. H. Frank (1947) *Electromagnetism,* McGraw-Hill, New York.

Stratton, J.A. (1941) *Electromagnetic Theory,* McGraw-Hill, New York.

Stuart, R.D. (1965) *Electromagnetic Field Theory,* Addison-Wesley, New York.

Chapter 2

GENERAL PRINCIPLES

A few general fundamental concepts are to be examined first and applied later to the more specific cases of electromagnetic wave propagation. The character of the medium in which propagation takes place is described by the appropriate values of the medium's permittivity ϵ, permeability μ, and conductivity σ. The media encountered in practice generally have properties between those ascribable to a dielectric and those of a semiconductor. The lowest charge-free layers of the atmosphere behave like a dielectric, whereas the ground and the sea may be considered semiconductors. The two extreme cases—namely, propagation in a homogeneous dielectric and in a homogeneous semiconductor—will be investigated in this chapter.

2.1 PROPAGATION IN AN IDEAL HOMOGENEOUS ISOTROPIC DIELECTRIC

Because the conductivity σ of a dielectric is zero, Maxwell's equations for this case reduce to [Stratton, 1941]

$$\text{curl}\,\mathbf{H} = \epsilon\frac{\partial \mathbf{E}}{\partial t} \qquad \text{div}\,\mathbf{H} = 0 \tag{2.1}$$

$$\text{curl}\,\mathbf{E} = \mu\frac{\partial \mathbf{H}}{\partial t} \qquad \text{div}\,\mathbf{E} = 0 \tag{2.2}$$

The aim of the following computations is the derivation of expressions describing the behavior of the propagating electromagnetic waves.

Differentiating the first expression of (2.1) with respect to time yields

$$\operatorname{curl} \frac{\partial \mathbf{H}}{\partial t} = \epsilon \frac{\partial^2 \mathbf{E}}{\partial t^2} \tag{2.3}$$

and the use of the curl operator in the first expression of (2.2) leads to

$$\operatorname{curl}(\operatorname{curl} \mathbf{E}) = -\mu \operatorname{curl} \frac{\partial \mathbf{H}}{\partial t} \tag{2.4}$$

Then inserting (2.3) into (2.4) yields

$$\frac{\partial^2 \mathbf{E}}{\partial t^2} = -\frac{1}{\mu\epsilon} \operatorname{curl} \operatorname{curl} \mathbf{E} \tag{2.5}$$

and similarly

$$\frac{\partial^2 \mathbf{H}}{\partial t^2} = -\frac{1}{\mu\epsilon} \operatorname{curl} \operatorname{curl} \mathbf{H} \tag{2.6}$$

Because curl curl = grad div $-$ ∇^2 and further because from (2.1) and (2.2) div **E** = 0 and div **H** = 0, expressions (2.5) and (2.6) become

$$\frac{\partial^2 \mathbf{E}}{\partial t^2} = \frac{1}{\mu\epsilon} \nabla^2 \mathbf{E} \tag{2.7}$$

$$\frac{\partial^2 \mathbf{H}}{\partial t^2} = \frac{1}{\mu\epsilon} \nabla^2 \mathbf{H} \tag{2.8}$$

where the Laplacian is given by the symbolic vector

$$\nabla^2 = \frac{\partial^2}{\partial x^2} + \frac{\partial^2}{\partial y^2} + \frac{\partial^2}{\partial z^2}$$

The orientation of the coordinate system is shown by Figure 2.1.

Expressions (2.7) and (2.8) show the relation between a space and time process, which may be referred to as a *wave;* expressions (2.7) and (2.8) are general wave equations. The solution of the wave equation may be restricted, without loss of generality, to a one-directional component as it is identical for all three components of a space wave [Jordan, 1953].

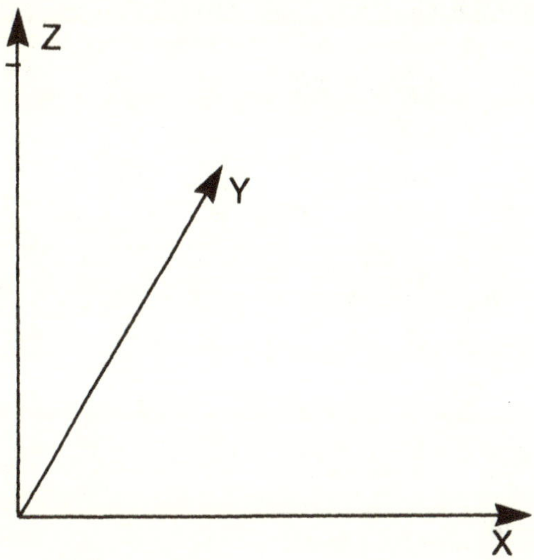

Figure 2.1 Orientation of the coordinate system.

Then, because $\partial \mathbf{E}/\partial y = \partial \mathbf{E}/\partial z = 0$ and, similarly, $\partial \mathbf{H}/\partial y = \partial \mathbf{H}/\partial z = 0$, we may write

$$\frac{\partial^2 E_y}{\partial x^2} = \mu\epsilon \frac{\partial^2 E_y}{\partial t^2}$$

This is a second-order partial differential equation that may be solved by substituting

$$u = x - v_0 t \quad v = x + v_0 t \tag{2.9}$$

so that $E_y(x,t)$ becomes $E_y(u,v)$ with

$$\frac{\partial E_y}{\partial x} = \frac{\partial E_y}{\partial u}\frac{\partial u}{\partial x} + \frac{\partial E_y}{\partial v}\frac{\partial v}{\partial x} = \frac{\partial E_y}{\partial u} + \frac{\partial E_y}{\partial v}$$

because

$$\frac{\partial u}{\partial x} = \frac{\partial v}{\partial x} = 1$$

as obtained by differentiating expressions (2.9) with respect to x. Further,

$$\frac{\partial^2 E_y}{\partial x^2} = \frac{\partial}{\partial x}\frac{\partial E_y}{\partial x} = \frac{\partial}{\partial x}\left[\frac{\partial E_y}{\partial u} + \frac{\partial E_y}{\partial v}\right]$$

$$= \frac{\partial}{\partial u}\left[\frac{\partial E_y}{\partial u} + \frac{\partial E_y}{\partial v}\right]\frac{\partial u}{\partial x} + \frac{\partial}{\partial v}\left[\frac{\partial E_y}{\partial u} + \frac{\partial E_y}{\partial v}\right]\frac{\partial v}{\partial x}$$

$$= \frac{\partial^2 E_y}{\partial u^2} + \frac{\partial^2 E_y}{\partial u \partial v} + \frac{\partial^2 E_y}{\partial u \partial v} + \frac{\partial^2 E_y}{\partial v^2}$$

$$= \frac{\partial^2 E_y}{\partial u^2} + 2\,\frac{\partial^2 E_y}{\partial u \partial v} + \frac{\partial^2 E_y}{\partial v^2}$$

Similarly, because

$$\frac{\partial u}{\partial t} = -v_0 \text{ and } \frac{\partial v}{\partial t} = v_0 \tag{2.10}$$

$$\frac{\epsilon E_y}{\partial t} = \frac{\partial E_y}{\partial u}\frac{\partial u}{\partial t} + \frac{\partial E_y}{\partial v}\frac{\partial v}{\partial t}$$

$$= -\frac{\partial E_y}{\partial u}v_0 + \frac{\partial E_y}{\partial v}v_0$$

and

$$\frac{\partial^2 E_y}{\partial t^2} = \frac{\partial}{\partial t}\frac{\partial E_y}{\partial t} = \frac{\partial}{\partial t}\left[-v_0\frac{\partial E_y}{\partial u} + v_0\frac{\partial E_y}{\partial v}\right]$$

$$= -v_0\frac{\partial}{\partial u}\left[\frac{\partial E_y}{\partial u} - \frac{\partial E_y}{\partial v}\right]\frac{\partial u}{\partial t}$$

$$- v_0\frac{\partial}{\partial v}\left[\frac{\partial E_y}{\partial u} - \frac{\partial E_y}{\partial v}\right]\frac{\partial v}{\partial t}$$

Then, because the fractions following the brackets in the foregoing expressions are equal to $-v_0$ and v_0, respectively, as shown by (2.10), these differential become

$$\frac{\partial^2 E_y}{\partial t^2} = v_0^2\left[\frac{\partial^2 E_y}{\partial u^2} - \frac{\partial^2 E_y}{\partial u \partial v} - \frac{\partial^2 E_y}{\partial u \partial v} + \frac{\partial^2 E_y}{\partial v^2}\right]$$

$$= v_0^2\left[\frac{\partial^2 E_y}{\partial u^2} - 2\,\frac{\partial^2 E_y}{\partial u \partial v} + \frac{\partial^2 E_y}{\partial v^2}\right]$$

Inserting the preceding two second derivatives into the wave equation and assuming that $v_0^2\mu\epsilon = 1$, an assumption to be proved later, we arrive at

$$\frac{\partial^2 E_y}{\partial u^2} + 2\frac{\partial^2 E_y}{\partial u \partial v} + \frac{\partial^2 E_y}{\partial v^2} = \frac{\partial^2 E_y}{\partial u^2} - 2\frac{\partial^2 E_y}{\partial u \partial v} + \frac{\partial^2 E_y}{\partial v^2}$$

from where, after canceling identical terms, we obtain

$$\frac{\partial^2 E_y}{\partial u \partial v} = 0 \tag{2.11}$$

Integrating (2.11) with respect to v yields $E_y = \int f_1(v)dv$ + a function not containing v; that is, $E_y = f_1(v) + f_2(u)$. Or, by using the definitions of u and v given by (2.9), we arrive at

$$E_y = f_1(x + v_0 t) + f_2(x - v_0 t) \tag{2.12}$$

which consists of two waves, one traveling away from the source and one traveling toward it. Both waves travel at the velocity v_0.

The correctness of the solution and the assumption of $v_0^2\epsilon\mu = 1$ may be ascertained by inserting

$$E_y = f(x \mp v_0 t) \tag{2.13}$$

into the one-dimensional wave equation. This yields

$$\begin{aligned} \mu\epsilon v_0^2 f''(x \mp v_0 t) &= f''(x \mp v_0 t) \\ \epsilon\mu v_0^2 &= 1 \end{aligned} \tag{2.14}$$

and therefore

$$v_0 = 1/\sqrt{\epsilon\mu} \tag{2.15}$$

is the velocity of propagation.

The permittivity ϵ and permeability μ of a medium usually is expressed by values relative to those of the corresponding parameters of free space:

$$\begin{aligned} \epsilon_0 &= 10^{-9}/36\pi \text{ F/m and} \\ \mu_0 &= 4\pi 10^{-7} \text{ H/m,} \end{aligned} \tag{2.16}$$

as $\epsilon' = \epsilon/\epsilon_0$ and $\mu' = \mu/\mu_0$ so that the velocity of propagation, when $\mu' = 1$, becomes

$$v_0 = \frac{1}{\sqrt{\epsilon'\epsilon_0\mu_0}} = \frac{1}{\sqrt{\epsilon'}\sqrt{\frac{4\pi}{36\pi}10^{-16}}} = \frac{3 \times 10^8}{\sqrt{\epsilon'}}\text{ m/s} \tag{2.17}$$

Further, because the relative dielectric constant of air $\epsilon' = 1$, the velocity of propagation in air equals the velocity of propagation of light.

Maxwell's second equation (2.2) may be developed to read

$$\begin{aligned} -\mu\frac{\partial H_x}{\partial t} &= \frac{\partial E_z}{\partial y} - \frac{\partial E_y}{\partial z} \\ -\mu\frac{\partial H_y}{\partial t} &= \frac{\partial E_x}{\partial z} - \frac{\partial E_z}{\partial x} \\ -\mu\frac{\partial H_z}{\partial t} &= \frac{\partial E_y}{\partial x} - \frac{\partial E_x}{\partial y} \end{aligned} \tag{2.18}$$

However, because

$$-\mu\frac{\partial H_x}{\partial t} = \frac{\partial H_z}{\partial t} = 0 \text{ and } \frac{\partial E_x}{\partial t} = 0$$

we are left with

$$-\mu\frac{\partial H_y}{\partial t} = -\frac{\partial E_z}{\partial x} \tag{2.19}$$

according to which the time variation of the magnetic field in the y-direction equals the variation of the z-component of the electric field propagating in the x-direction. The two fields are at right angles to each other. We may assume that the electric field varies with distance in the same way as determined earlier, in which case we may write

$$\mu\frac{\partial H_y}{\partial t} = \mp f'(x \mp v_0 t) \tag{2.20}$$

where the velocity of propagation, as derived earlier, is that given by (2.17).

Integrating expression (2.20) with respect to time

$$\int f'(x \mp v_0 t)\,\mathrm{d}t = \frac{1}{v_0} f(x \mp v_0 t)$$

and, accepting a zero integration constant, we arrive at

$$H_y = \mp \frac{1}{\mu v_0} f(x \mp v_0 t) \tag{2.21}$$

or, when inserting the previously obtained expression for v_0 and a unity relative permittivity μ',

$$H_y = \mp \sqrt{\epsilon'}\ \sqrt{\frac{\epsilon_0}{\mu_0}} f(x \mp v_0 t) \quad \text{A/m} \tag{2.22}$$

Using the values of ϵ_0 and μ_0 given by (2.16), we find that

$$\sqrt{\frac{\epsilon_0}{\mu_0}} = \sqrt{\frac{10^7}{36\pi 4\pi 10^8}} = \frac{1}{120\pi} \quad \text{S (siemens)}$$

or

$$H_y = \mp \frac{\sqrt{\epsilon'}}{120\pi} f(x \mp v_0 t) \quad \text{A/m} \tag{2.23}$$

Expressions (2.13) and (2.23) show that the electric and magnetic field components of the wave are in phase. The direction of the field vectors may be determined from the sign before the expression for H_y; the positive sign within the bracket denotes propagation in the positive x-direction. At an instant when the electric vector is in the positive z-direction, the magnetic vector is in the negative y-direction. When the wave propagates in the opposite direction, the direction of the H-vector reverses and that of the E-vector remains the same.

We observe that the dimension of the electric field is V/m, whereas that of the magnetic one is A/m so that their ratio (i.e., the constant 120π) must have the dimension of Ω; it is the impedance of free space.

The shape of the wave is determined by the initial conditions. Radio waves are radiated by antennas excited by high-frequency, usually, periodic currents of the argument $x/v_0 \mp t$. When the electric component is

$$E_z = E_0 e^{-j\omega(x/v_0 \mp t)} \tag{2.24}$$

the magnetic component becomes

$$H_y = \mp \frac{\sqrt{\epsilon'}}{120\pi} E_0 e^{-j\omega(x/v_0 \mp t)} \tag{2.25}$$

These two expressions let us determine the spatial relations of the two components of the propagating wave, as shown in Figure 2.2.

Example 2.1

Determine the amplitude of the magnetic component and the velocities of propagation of an electromagnetic wave propagating in air and the sea if the free-space amplitude of the electric component is 100 μV and the relative permittivities are $\epsilon'_{\text{air}} = 1$ and $\epsilon'_{\text{sea}} = 80$.

Solution:

$$H_{y-\text{air}} = \frac{1}{120\pi} 10^{-4} = 2.65 \times 10^{-7} \quad \text{A/m}$$

$$H_{y-\text{sea}} = \frac{\sqrt{80}}{120\pi} 10^{-4} = 2.37 \times 10^{-6} \quad \text{A/m}$$

$$v_{\text{air}} = 3 \times 10^8 \quad \text{m/s}$$

$$v_{\text{sea}} = \frac{3 \times 10^8}{\sqrt{80}} = 0.34 \times 10^8 \quad \text{m/s}$$

Electromagnetic waves are transverse waves as neither the electric nor the magnetic field has components in the direction of propagation. The electric and magnetic fields are at right angles to each other and to the direction of propagation. Both fields are in phase and propagate with the velocity of light.

We often are interested in determining the power supplied by an electromagnetic field to a unit volume in space. Rearranging the Maxwell-Ampère relation (1.10) to read

$$\mathbf{J} = \nabla \times \mathbf{H} - \frac{\partial \mathbf{D}}{\partial t}$$

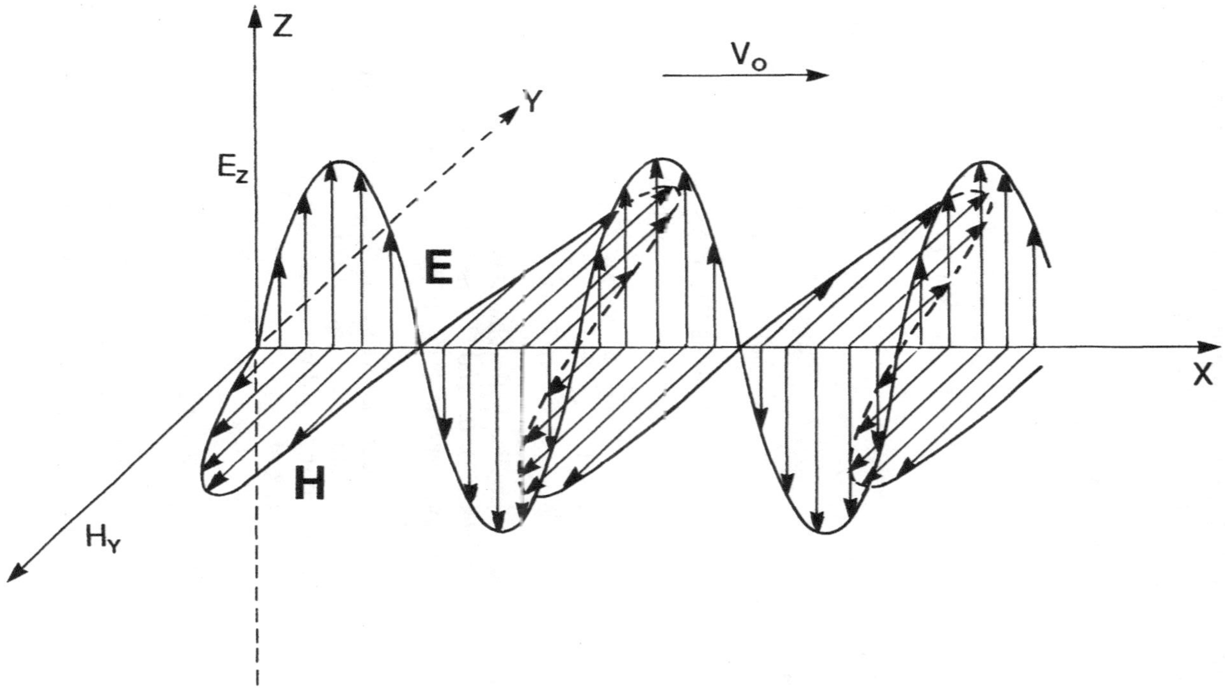

Figure 2.2 The electric and magnetic field components of a plane electromagnetic wave propagating in an ideal dielectric.

and writing, in analogy with general electric considerations, the power per unit volume supplied by an electric field to the charge carriers at a point as $\mathbf{E} \cdot \mathbf{J}$, we obtain by scalar multiplication of the preceding expression by $\mathbf{E}$,

$$\mathbf{E} \cdot \mathbf{J} = \mathbf{E} \cdot (\nabla \times \mathbf{H}) - \mathbf{E} \cdot \frac{\partial \mathbf{D}}{\partial t} \tag{2.26}$$

which, with the vector relation

$$\mathbf{E} \cdot (\nabla \times \mathbf{H}) = \mathbf{H} \cdot (\nabla \times \mathbf{E}) - \nabla \cdot (\mathbf{E} \times \mathbf{H}) \tag{2.27}$$

and the expression

$$\nabla \times \mathbf{E} = -\frac{\partial \mathbf{B}}{\partial t}$$

permits the expression of the total electromagnetic power per unit volume supplied to a point in space as

$$-\nabla \cdot (\mathbf{E} \times \mathbf{H}) = \mathbf{E} \cdot \mathbf{J} + \mathbf{E} \cdot \frac{\partial \mathbf{D}}{\partial t} + \mathbf{H} \cdot \frac{\partial \mathbf{B}}{\partial t} \tag{2.28}$$

This is the sum of the power supplied to the charge carriers as given by the first term and, due to $\mathbf{E} \cdot \partial\mathbf{D}/(\partial t) = \partial(\frac{1}{2}\epsilon E2)/(\partial t)$ and $\mathbf{H} \cdot \partial\mathbf{B}/(\partial t) = \partial(\frac{1}{2}\mu H2)/(\partial t)$, the power supplied to the electric and magnetic fields given by the second or third term on the right-hand side of this expression.

According to the divergence theorem of vector analysis the volume integral of the divergence of $\mathbf{E} \times \mathbf{H}$ over a volume V is equal to the surface integral of this product over a closed surface surrounding this volume. Then, integrating expression (2.28) over the volume, we arrive at

$$-\int_S (\mathbf{E} \times \mathbf{H}) \cdot d\mathbf{S} = \int_V \mathbf{E} \cdot \mathbf{J} dV + \int_V \left(\mathbf{E} \cdot \frac{\partial \mathbf{D}}{\partial t} + \mathbf{H} \cdot \frac{\partial \mathbf{B}}{\partial t}\right) dV \tag{2.29}$$

showing that the integral of $\mathbf{E} \times \mathbf{H}$ over a closed surface S yields the electromagnetic power flowing out of the volume V surrounded by S.

The vector

$$\mathbf{S} = \mathbf{E} \times \mathbf{H} \quad \text{W/m}^2 \tag{2.30}$$

the Poynting vector, yields the magnitude and direction of the energy flow per unit time per unit area at a point.

Example 2.2

A plane wave $E = 1000 \sin\omega t V$ passes through a material having a relative permittivity of 2.7. Its electric breakdown strength is 30 kV/m. Determine the maximum permissible rms power flow density and peak magnetizing force applicable to the material.

Solution:

$$H = \frac{\sqrt{\epsilon'}}{120\pi} 1000 \sin\omega t = 4.36 \sin\omega t$$

$$Z = \frac{E}{H} = \frac{120\pi}{\sqrt{\epsilon'}} = 229.43 \quad \Omega$$

$$S = E_{rms} \times H_{rms} = \frac{E^2_{rms}}{Z} = \frac{(3 \times 10^4)^2}{2 \times 229.43} = 1.96 \times 10^6 \quad \text{W/m}^2$$

$$E = 229.43H$$

$$S = 229.43H^2 \cdots H = \sqrt{\frac{S}{229.43}} \sqrt{2} = 130.76 \quad \text{A/m}$$

2.2 PROPAGATION IN A HOMOGENEOUS SEMICONDUCTOR

Because the conductivity σ of a semiconductor cannot be neglected, Maxwell's first equation in its complete form

$$\text{curl } \mathbf{H} = \sigma\mathbf{E} + \epsilon \frac{\partial \mathbf{E}}{\partial t}$$

becomes applicable [Doluchanow, 1956].

Assuming the previously adopted periodic variation of the exciting current given by expression (2.24), the instantaneous initial value of the electric field becomes

$$\mathbf{E} = \mathbf{E}_0 e^{j\omega t} \tag{2.31}$$

with $\mathbf{E}_0$ denoting its amplitude. Then, because

$$\frac{\partial \mathbf{E}}{\partial t} = j\omega \mathbf{E}_0 e^{j\omega t} = j\omega \mathbf{E}$$

the instantaneous value may be written as

$$\mathbf{E} = -\frac{j}{\omega}\frac{\partial \mathbf{E}}{\partial t}$$

which leads to

$$\text{curl } \mathbf{H} = \left(\epsilon - j\frac{\sigma}{\omega}\right)\frac{\partial \mathbf{E}}{\partial t} \tag{2.32}$$

The expression in parentheses is a complex dielectric constant:

$$\epsilon_c = \epsilon - j\frac{\sigma}{\omega} \tag{2.33}$$

which replaces the dielectric constant ϵ of the perfect dielectric in expressions applicable to the propagation of an electromagnetic wave of frequency ω in a semiconductor described by the dielectric constant ϵ and the conductivity σ.

The dielectric constant relative to that of free space is

$$\epsilon_c' = \frac{\epsilon_c}{\epsilon_0} = \epsilon' - j\frac{\sigma}{\omega\epsilon_0} = (n - j\kappa)^2 \tag{2.34}$$

with ϵ_0 given earlier by (2.16).

Expression (2.33) reduces to

$$\epsilon_c' = \epsilon' - j60\sigma\lambda \tag{2.35}$$

when $\omega = 2\pi c/\lambda$ and the velocity of light $c = 3.10^8$ m/s is inserted.

We will show later that n, the real term on the right-hand side of expression (2.34), is the refractive index of the medium in which the wave propagates.

Because the conduction current density is proportional to the conductivity σ, whereas the displacement current density to the relative dielectric constant ϵ', the ratio

$$\frac{i_{\text{displ.}}}{i_{\text{cond}}} = \frac{\epsilon'}{60\sigma\lambda}$$

permits the determination of the behavior of a medium at a given frequency. In a semiconductor where $\epsilon' \gg 60\sigma\lambda$, this ratio is larger than unity; whereas for media where the conduction current predominates, it is less than unity. The conduction current in an ideal dielectric is zero, hence the ratio becomes infinitely large.

Reverting to the general expression for the electric field as in (2.24), with

$$v_0 = \frac{c}{\sqrt{\epsilon_c'}} = \frac{c}{n - j\kappa} \tag{2.36}$$

we obtain

$$\begin{aligned} E_z &= E_0 e^{j\omega(x/v_0 \mp t)} \\ &= E_0 e^{-\omega\kappa x/c} e^{j\omega(t - \kappa n/c)} \end{aligned} \tag{2.37}$$

where

$$\delta = \frac{\omega\kappa}{c} = \frac{2\pi\kappa}{\lambda} \quad 1/\mathrm{m} \tag{2.38}$$

is an absorption coefficient and

$$v' = \frac{c}{n} \tag{2.39}$$

is the velocity of propagation in the semiconductor.

Comparing (2.37) and (2.24), we observe that the amplitude of the wave is distance dependent because absorption, determined by the coefficient δ, occurs and the velocity is reduced to that given by (2.39).

The magnetic field, on the basis of (2.25), (2.34), and (2.38), is

$$H_y = -\frac{\sqrt{\epsilon_c'}}{120\pi} E_0 e^{-\delta x} e^{j\omega(t - x/v')} \tag{2.40}$$

or, as due to (2.36),

$$\sqrt{\epsilon_c'} = n - j\kappa = \sqrt{n^2 + \kappa^2} e^{-j\tan^{-1}(\kappa/n)} \tag{2.41}$$

it reads

$$E_y = -\frac{\sqrt{n^2 + \kappa^2}}{120\pi} E_0 e^{-\delta x} \cos[(\omega(t - x/v') - \tan^{-1}(\kappa/n)] \quad (2.42)$$

The magnetic field in a semiconductor is shifted in phase with respect to the electric one by $\tan^{-1}(\kappa/n)$, as shown by Figure 2.3.

Whereas the wave propagating in a semiconductor remains transversal, the amplitude of both the electric and magnetic components diminishes with distance from the radiator, and the ratio of the magnetic to electric field amplitude at every point of their trajectory is

$$\frac{\sqrt{n^2 + \kappa^2}}{120\pi}$$

Squaring expression (2.41), when using expression (2.35),

$$\epsilon' - j60\sigma\lambda = n^2 - \kappa^2 - 2jn\kappa$$

permits the determination of n and κ by equating real and imaginary terms,

$$\epsilon' = n^2 - \kappa^2$$
$$60\sigma\lambda = 2n\kappa$$

From these

$$n = \pm\sqrt{\tfrac{1}{2}[\epsilon' + \sqrt{\epsilon'^2 + (60\sigma\lambda)^2}]}$$
$$\kappa = \pm\sqrt{\tfrac{1}{2}[-\epsilon' + \sqrt{\epsilon'^2 + (60\sigma\lambda)^2}} \quad (2.43)$$

Only the positive sign before the first square root has significance, as a negative κ would cause a rising amplitude of a wave propagating away from the source.

Example 2.3

Simplify the expressions for the refractive index n and the absorption coefficient δ for the case when the conduction current is much larger than the displacement one.
Solution:

$$\epsilon' \ll 60\sigma\lambda$$
$$n = \sqrt{30\sigma\lambda}$$

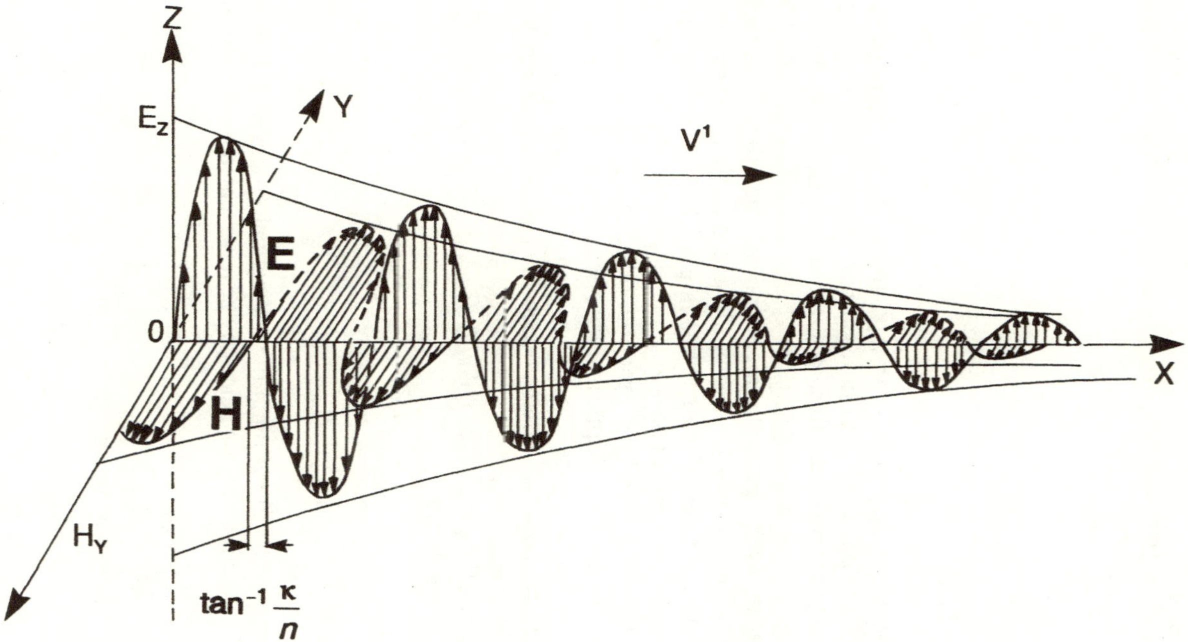

Figure 2.3 The electric and magnetic field components of a plane electromagnetic wave propagating in a semiconductor.

$$\kappa = \sqrt{30\sigma\lambda}$$

$$= \frac{2\pi}{\lambda}\sqrt{30\sigma\lambda} \quad 1/\text{m}$$

$$= 2\pi\sqrt{\frac{30\sigma}{\lambda}} \quad 1/\text{m}$$

2.3 POLARIZATION

Although light waves may be polarized or unpolarized, the electromagnetic waves of interest here always are polarized. The time varying behavior of the electric field vector of an electromagnetic wave at a fixed point in space determines the polarization of a plane wave.

Electromagnetic waves generally are elliptically polarized with their electric field vector varying periodically in magnitude and direction. Linear and circular polarizations are special cases of the general case of elliptical polarization. Linear polarization may be either vertical or horizontal, depending on whether the electric field vector, the magnitude of which varies periodically with time, is always in the vertical or horizontal plane. An elliptical and a circular polarization is termed *right-handed* if the electric vector rotates clockwise when viewed from the antenna in the direction of propagation and *left-handed* when the electric vector rotates counter-clockwise [Doluchanow, 1956].

Considering first the general case of *elliptic polarization* and assuming that the two time varying components

$$E_y = A\cos\omega t$$

$$E_z = -B\sin\omega t$$

of a wave traveling in the x-direction of the orthogonal system shown by Figure 2.1 differ in amplitude, the time varying field may be represented by the complex vector

$$\mathbf{E} = \mathbf{j}A\cos\omega t - \mathbf{k}B\sin\omega t \tag{2.44}$$

where $\mathbf{j}$ and $\mathbf{k}$ are unit vectors in the y or z direction.

Then, because

$$\frac{E_y^2}{A^2} + \frac{E_z^2}{B^2} = 1$$

defines an ellipse, the end point of the vector **E** will trace out an ellipse as shown by Figure 2.4.

In *linear polarization,* the vertical and horizontal components given by

$$E_y = E_{y0}\cos(\omega t - \varphi_1)$$
$$E_z = E_{z0}\cos(\omega t - \varphi_2) \tag{2.45}$$

may be in phase or have a phase difference of π rad. Accepting that $\varphi_1 = \varphi_2 = 0$, these components become

$$E_y = E_{y0}\cos\omega t$$
$$E_z = E_{z0}\cos\omega t$$

and their resultant is

$$E = \sqrt{E_y^2 + E_z^2} = \sqrt{E_{y0}^2 + E_{z0}^2}\cos\omega t \tag{2.46}$$

Because the ratio of thc two components given by (2.45)

$$\frac{E_z}{E_y} = \text{const} = \tan\alpha \tag{2.47}$$

which determines the orientation of the resultant, remains constant, it satisfies the definition of a linearly polarized wave. Only the horizontal component exists when $\alpha = 0$ and the polarization is said to be horizontal; whereas it is vertical for $\alpha = \pi/2$, when only the vertical component remains.

When both components have equal amplitudes but differ in phase by $\pi/2$ or $3\pi/2$ rad, *circular polarization* ensues.

Assigning the values $\varphi_1 = 0$ and $\varphi_2 = \pi/2$ rad to the two phase angles, the two field components when $E_{y0} = E_{z0} = E_0$ become

$$E_y = E_0\cos\omega t$$
$$E_z = E_0\sin\omega t \tag{2.48}$$

and the magnitude of the resultant becomes

$$E = \sqrt{E_y^2 + E_z^2} = E_0 = \text{const} \tag{2.49}$$

with its direction given by

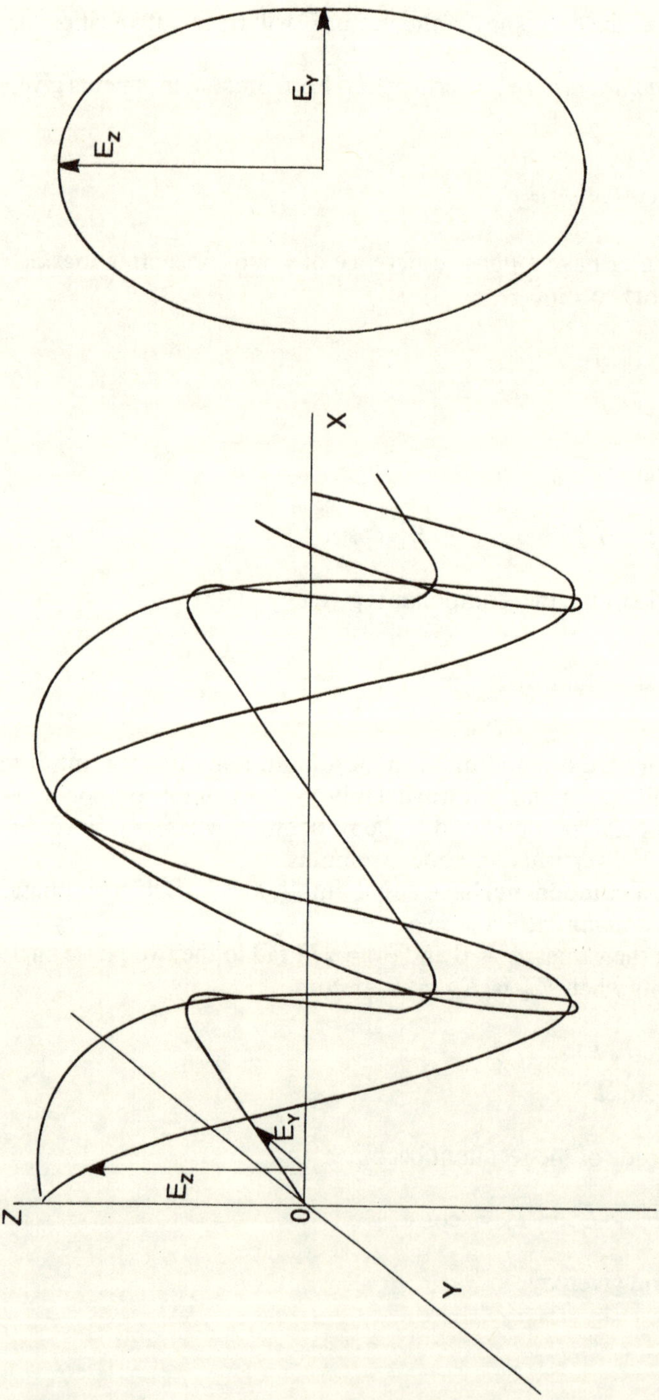

Figure 2.4 Elliptical polarization.

$$\tan\alpha = \frac{E_z}{E_y} = \tan\omega t \tag{2.50}$$

The resultant then has a constant magnitude but a periodic time varying orientation. It rotates with a constant angular frequency ω. The sense of rotation depends on whether the phase difference of the field components is $\pi/2$ or $3\pi/2$ rad. Circularly polarized waves of both senses of rotation are shown in Figure 2.5.

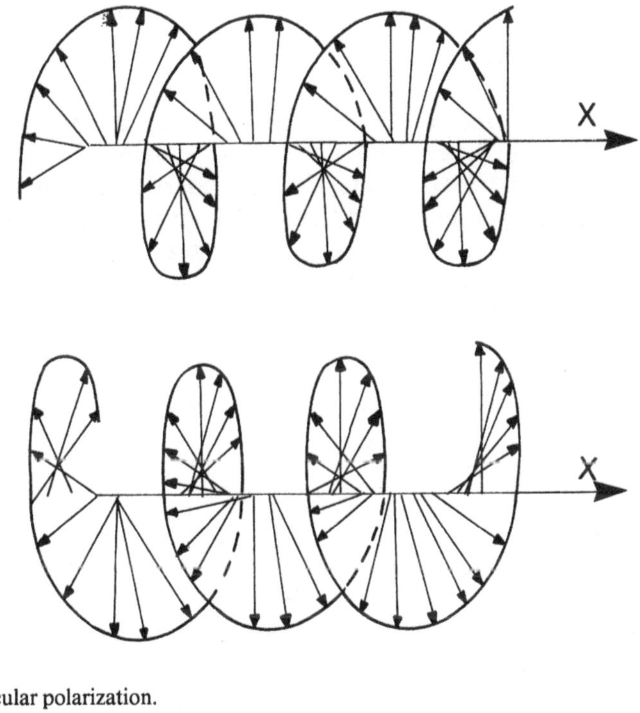

Figure 2.5 Circular polarization.

In linear polarization, we found that, although the orientation of the resultant is constant, its magnitude varies periodically with time; in circular polarization the opposite is true. Both, the magnitude and orientation of the resultant vary periodically with time at elliptical polarization.

We showed that electromagnetic waves propagating in homogeneous isotropic media are transverse; that is, they have no component in the direction of propagation. This need not prevail for propagation in inhomogeneous or anisotropic media, where a longitudinal component in the direction of propagation may develop. This component may be in phase with the perpendicular ones thus yielding a periodically varying magnitude of the resultant field vector inclined at a constant angle toward the direction of propagation. However, the longitudinal component may be shifted in phase with respect to the transverse ones causing the resultant electric field to vary continuously in direction thus producing an elliptically polarized wave inclined in the direction of propagation which also is described by the given expressions.

At ground constants encountered in practice, we find that the eccentricity of the ellipse is large, so that an assumption of linear polarization of the electric field in the direction of propagation and an appropriate tilt of the wavefront is acceptable.

2.4 BOUNDARY CONDITIONS

At the boundary of two media, the permittivity, permeability, and conductivity changes abruptly, causing the electric and magnetic fields to become discontinuous. The variation of the field from one side of the boundary to the other may be determined from Maxwell's equations. Usually, the behavior of the tangential and normal components of the fields are of interest because they represent the orthogonal components of a field reaching the boundary at any angle of incidence [Holt, 1963].

Erecting at the boundary of the two media, a rectangle of length Δl and differential height dh such that half of the height is on one side and the other half on the other side of the boundary and applying Maxwell's second law, we find that

$$\text{curl}\,\mathbf{E} = 0$$

because the area of the rectangle approaches zero when dh approaches zero, and therefore there can be no flux through it. By using Stokes's theorem of vector analysis, according to which the line integral of **E** around a closed path C equals the surface integral of the normal component of the same vector over the surface S bounded by C, the line integral around C, in this case the circumference of the rectangle, for dh approaching zero and Δl so small that **E** may be taken to be constant, is

$$E_{t1}\Delta l - E_{t2}\Delta l = 0 \tag{2.51}$$

showing that the tangential components of the electric field $\mathbf{E}_{t1}$ above and $\mathbf{E}_{t2}$ below the boundary are equal; hence the tangential component of the electric field is continuous across the boundary.

Considering the same rectangle and applying to it Maxwell's first equation, according to which the magnetomotive force around the contour C equals the total current through the surface S bounded by C, we find that, as neither of the two considered media is a perfect conductor, there can be no surface current density and as

S approaches zero as dh approaches zero, the total current over S approaches zero, too. Hence, provided that neither medium is a conductor,

$$\text{curl } \mathbf{H} = 0$$

or, using the same reasoning as earlier,

$$H_{t1} = H_{t2} \tag{2.52}$$

The tangential component of the magnetic field intensity is continuous across the boundary.

Applying Gauss's law to an infinitesimal parallelepiped of surface area ΔS and height dh enclosing a small portion of the surface between the two media, we find that the surface integral of the electric flux density **D** for dh, and hence the flux of **D** over the sides, approaching zero, is

$$D_{n1}\Delta S - D_{n2}\Delta S$$

The total charge within the volume of the parallelepiped, for a surface charge density ρ, is $\rho\Delta S$, so that

$$D_{n1} - D_{n2} = \rho \tag{2.53}$$

showing that the normal component of the electric flux density across the boundary between two media is not continuous when there is a surface charge.

Because no sources or sinks are within the parallelepiped introduced, the surface integral of the magnetic flux density **B** over the closed surface of the parallelepiped is zero. The flux of **B** through the sides, for the height dh of the parallelepiped approaching zero, is negligible. Then,

$$B_{n1}\Delta S - B_{n2}\Delta S = 0 \tag{2.54}$$

indicating that the normal component of the magnetic flux across the boundary of the two different media is continuous.

The relation between the tangential components of the electric flux density **D** and magnetic flux density **B** and the normal components of the electric, respectively, magnetic field **E** and **H** across the boundary may be obtained from the constitutive equations:

$$\mathbf{D} = \epsilon\mathbf{E} \text{ and } \mathbf{B} = \mu\mathbf{H}$$

Summarizing the findings for future reference:

$$\begin{aligned} &\text{(a)}\ E_{t1} = E_{t2} \\ &\text{(b)}\ H_{t1} = H_{t2} \\ &\text{(c)}\ D_{n1} - D_{n2} = \rho \\ &\text{(d)}\ B_{n1} = B_{n2} \end{aligned} \tag{2.55}$$

Example 2.4

Using (2.55), prove that the ratio of the tangents of the angle of incidence and angle of refraction equals the ratio of the permittivities of the media separated by a charge-free boundary.

Solution: From expression (2.55a), $E_{t1} = E_{t2} = E_t$.

From (2.55c), as the charge density on the given boundary is zero,

$$D_{n1} = D_{n2}$$

or

$$\epsilon_1 E_{n1} = \epsilon_2 E_{n2}$$

Further,

$$\frac{E_{n1}}{E_{n2}} = \frac{\epsilon_2}{\epsilon_1}$$

$$\tan\gamma = \frac{E_t}{E_{n1}} \text{ and } \tan\psi = \frac{E_t}{E_{n2}}$$

Then,

$$\frac{\tan\gamma}{\tan\psi} = \frac{\epsilon_1}{\epsilon_2}$$

2.5 REFLECTION

Antennas inadvertently radiate electromagnetic energy in directions other than the desired one. Some energy is directed toward the ground from whence it is reflected. The reflected radiation combines with the radiation traveling in space and the intensity of the resultant field at the point of their encounter is either enhanced or reduced depending on whether the two meet in phase or out of phase. This phenomenon is particularly pronounced at very high and ultra high frequencies.

Denoting by $\mathbf{E}_1$ that radiation that travels directly to the receiving antenna where it arrives with phase φ_1, and by $\mathbf{E}_2$ the other component arriving there with phase φ_2, the resultant field will be the vector sum [Rohan, 1983]

$$\mathbf{E} = \sqrt{\mathbf{E}_1^2 + \mathbf{E}_2^2 + 2\mathbf{E}_1\mathbf{E}_2 \cos(\varphi_1 - \varphi_2)}$$

Note that the magnitudes of these two vectors are

$$\mathbf{E}_1 = EG(\theta), \text{ and}$$

$$\mathbf{E}_2 = EG(\delta)\rho e^{-j\zeta}$$

respectively, where E is the amplitude of the antenna input energy, $G(\theta)$ and $G(\delta)$ are the antenna gains in the direction θ or δ, ρ is the magnitude of the composite reflection coefficient that expresses the effect reflection exerts on the incident radiation, and the total phase difference between the two field vectors is

$$\zeta = \varphi_1 - \varphi_2 = \frac{2\pi}{\lambda}\Delta R + \mu$$

with ΔR the difference between the reflected and direct path lengths, λ the wavelength of the radiation, and μ the phase shift suffered by the reflected ray on reflection. It is important, therefore, to know the magnitude and phase of the reflected radiation if we are to determine the field quantities at any point in space.

The coefficient of reflection for vertical polarization is the ratio of the reflected to the incident magnetic field or the ratio of the vertical components of the electric field. The ratio of the horizontal components yields the negative value of the coefficient of reflection. The coefficient of reflection for horizontal polarization is the ratio of the reflected to the incident electric field or the ratio of the corresponding vertical components of the magnetic field. Here again, the ratio of the incident and reflected horizontal components of the magnetic field yields the negative value of the coefficient of reflection as their direction differs by π rad.

Problems of reflection, refraction, and scattering often are treated by methods of geometrical optics that permit the representation of propagating waves by propagation along rays and use basic geometrical concepts for the solution. The accuracy of this approximation, which has been used since the early days of propagation studies for the solution of problems concerning large aperture antennas and very high frequencies, increases with an increasing frequency.

The basic relations justifying the geometrical optics approach and the limits of its validity are given in Section 5.4.1. Rays instead of waves will be considered in the following treatment of reflection.

2.5.1 Vertical Polarization

The configuration to be considered is depicted by Figure 2.6. The medium above the boundary at $z = 0$ is characterized by the constants of an ideal dielectric $\sigma = 0$, ϵ_0, μ_0 and hence velocity of propagation c, whereas the medium under the boundary is a semiconductor given by σ, ϵ, and μ and the velocity of propagation given by (2.17). Both media are taken to be isotropic and homogeneous so that the vectors of the electric and magnetic fields are perpendicular to the direction of propagation.

We showed that the polarization of the incident ray is the resultant of a vertical and horizontal component. The electric vector of the vertically polarized component oscillates in the z-x plane, whereas that of the horizontal one in a direction perpendicular to this, that is, in the x-y plane.

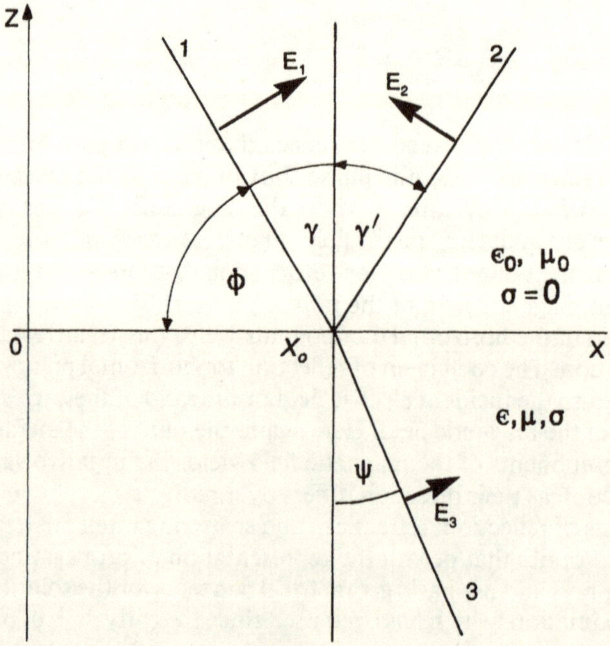

Figure 2.6 Reflection and refraction at vertical polarization.

The incident ray having an electric field instantaneous amplitude E_1 impinges the boundary at x_0 at an angle of incidence γ, or grazing angle ϕ, and yields a reflected ray of instantaneous amplitude E_2 at an angle γ' and a refracted ray of instantaneous amplitude E_3 at an angle ψ with respect to the normal to the boundary at x_0.

To satisfy the law of conservation of energy, we may write

$$E_1 \cos\gamma = E_2 \cos\gamma' + E_3 \cos\psi \tag{2.56}$$

for the field components that, according to (2.55), are continuous above and below the boundary.

The phase velocity of a wave in a medium described by the parameters ϵ and μ, as was shown earlier in (2.15), is

$$v_p = \frac{1}{\sqrt{\epsilon\mu}}$$

which in vacuo having the parameters ϵ_0 and μ_0 becomes the velocity of light c. Because $\epsilon = \epsilon'\epsilon_0$ and $\mu = \mu'\mu_0$, the phase velocity in terms of the relative parameters becomes

$$v_p = \frac{c}{\sqrt{\epsilon'\mu'}} \tag{2.57}$$

or, because the relative permeability of the media considered here is unity, the phase velocity becomes

$$v_p = \frac{c}{\sqrt{\epsilon'}} \tag{2.58}$$

The refractive index n of a medium is the ratio of the velocity of light in vacuo and the phase velocity in the considered medium; that is, for our case

$$n = \frac{c}{v_p} = \sqrt{\epsilon'} \tag{2.59}$$

as shown earlier by expressions (2.36) and (2.39).

Snell's law states that

$$n_1 \sin\gamma = n_1 \sin\gamma' = n_2 \sin\psi \tag{2.60}$$

where n_1 is the refractive index of the medium above the boundary and n_2 that of the medium below it. Because both the incident and reflected waves propagate in the same medium, according to Snell's law the angle of incidence and reflection must be equal. Further, because the velocity in the medium above the boundary is the velocity of light, yielding a unity refractive index, we find that

$$\sin\gamma = \sqrt{\epsilon_c'}\sin\psi \tag{2.61}$$

ϵ_c' the complex dielectric constant of the medium below the boundary.

The magnetic field vectors are perpendicular to the electric ones and to the direction of propagation; all three magnetic field vectors are parallel with the y-axis. By using then (2.23), we may express the boundary conditions (2.55) by

$$\frac{E_1 + E_2}{120\pi} = \frac{E_3\sqrt{\epsilon_c'}}{120\pi}$$

or

$$E_1 + E_2 = \sqrt{\epsilon_c'}E_3 \tag{2.62}$$

as the relative dielectric constant of the medium above the boundary is unity.

Expressions (2.56) and (2.62) constitute a system of equations with two unknowns from which we may determine Fresnel's reflection coefficient for vertical polarization as

$$\rho_v e^{-j\mu_v} = \frac{E_2}{E_1} = \frac{\sqrt{\epsilon_c'}\cos\gamma - \cos\psi}{\sqrt{\epsilon'}\cos\gamma + \cos\psi} \tag{2.63}$$

where μ_v is the phase shift on reflection. The refraction coefficient E_3/E_1 is of no interest in the present study.

The reflection coefficient, according to the definition given earlier in this section, is the ratio of the magnitudes of the reflected and incident fields. Both of these quantities are complex, thus indicating that not only the magnitude of the field but also its phase changes on reflection.

Expressing $\cos\psi$ on the basis of (2.61), we arrive at

$$\rho_v e^{-j\mu_v} = \frac{\epsilon_c'\cos\gamma - \sqrt{\epsilon_c' - \sin^2\gamma}}{\epsilon_c'\cos\gamma + \sqrt{\epsilon_c' - \sin^2\gamma}} \tag{2.64}$$

However, in high-frequency calculations, as a rule, we use the grazing angle (i.e., the complementary angle of the angle of incidence) so that the preceding expression for the reflection coefficient at vertical polarization becomes

$$\rho_v e^{-j\mu_v} = \frac{\epsilon_c' \sin\phi - \sqrt{\epsilon_c' - \cos^2\phi}}{\epsilon_c' \sin\phi + \sqrt{\epsilon_c' - \cos^2\phi}} \tag{2.65}$$

Example 2.5

Evaluate the coefficient of reflection and phase shift on reflection at a frequency of 10 GHz and a 75° grazing angle for a medium characterized by $\epsilon' = 30$ and $\sigma = 6.5$ mho/m.

Solution: The relative complex dielectric constant, as given by (2.35), is $\epsilon_c' = \epsilon' - j60\sigma\lambda$, which inserted into (2.65) yields under the square root sign an expression of the form $a - jb$, where $a = \epsilon' - \cos^2\phi$ and $b = 60\sigma\lambda$.

Applying De Moivre's formula lets us express the square root as

$$(a - jb)^{1/2} = (\sqrt{a^2 + b^2})^{1/2}(\cos\varphi/2 - j\sin\varphi/2)$$
$$\tan\varphi = b/a$$

The half-angle functions are

$$\cos\varphi/2 = \sqrt{\frac{1 + \cos\varphi}{2}} \quad \sin\varphi/2 = \sqrt{\frac{1 - \cos\varphi}{2}}$$

with $\cos\varphi = \frac{a}{\sqrt{a^2 + b^2}}$.

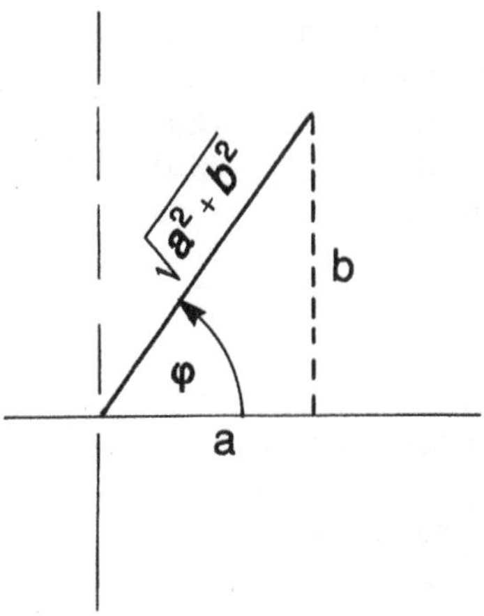

Then,

$$(a - jb)^{1/2} = (\sqrt{a^2 + b^2})^{1/2}(\sqrt{1/2 + 1/2 a/\sqrt{a^2 + b^2}} - j\sqrt{1/2 - 1/2 a/\sqrt{a^2 + b^2}})$$

Designating, for ease of writing, the expression:

$\epsilon' \sin\phi$ by A, $60\sigma\lambda \sin\phi$ by B, $(\sqrt{a^2 + b^2})^{1/2}\sqrt{1/2 + 1/2 a/\sqrt{a^2 + b^2}}$ by C, and $(\sqrt{a^2 + b^2})^{1/2}\sqrt{1/2 - 1/2 a/\sqrt{a^2 + b^2}}$ by D,

the coefficient of reflection becomes

$$\rho_v e^{-j\mu_v} = \frac{(A - C) - j(B - D)}{(A + C) - j(B + D)}$$

Inserting the given values and performing the outlined calculations leads to $|\rho_v| = 0.79$ and $\mu_v = -0.0127$ rad.

2.5.2 Horizontal Polarization

Reflection at horizontal polarization is shown by Figure 2.7. H_1, H_2, and H_3 show the incident, reflected, and refracted magnetic field vectors, and the crosses show the electric field vectors normal to the paper. On the basis of relation (2.55 a), we may write at $z = 0$ [Doluchanow, 1956]

$$E_1 + E_2 = E_3 \tag{2.66}$$

an expression to be used for the computation of the reflection coefficient.

The second expression follows from the boundary conditions of the magnetic field. Considering that the direction of the magnetic field vectors in Figure 2.7 is identical with the direction of the electric field vectors of Figure 2.6, with the application of equation (2.55 b), we arrive at

$$E_1 \cos\gamma - E_2 \cos\gamma = \sqrt{\epsilon_c'} E_3 \cos\psi \tag{2.67}$$

Equations (2.66) and (2.67) yield the ratio of the reflected to incident fields, or the reflection coefficient for horizontal polarization:

$$\rho_h e^{-j\mu_h} = \frac{\cos\gamma - \sqrt{\epsilon_c'}\cos\psi}{\cos\gamma + \sqrt{\epsilon_c'}\cos\psi} \tag{2.68}$$

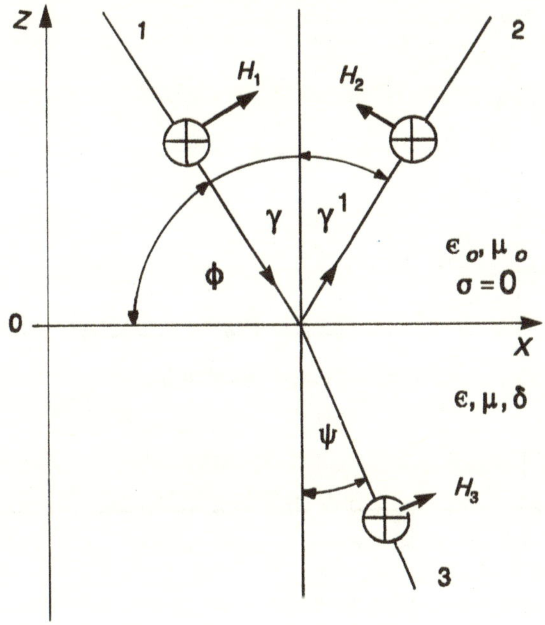

Figure 2.7 Reflection and refraction at horizontal polarization.

or, after expressing $\cos\psi$ on the basis of 2.61 and changing from the angle of incidence to the grazing angle:

$$\rho_h e^{-j\mu_h} = \frac{\sin\phi - \sqrt{\epsilon_c' - \cos^2\phi}}{\sin\phi + \sqrt{\epsilon_c' - \cos^2\phi}} \tag{2.69}$$

Figure 2.8 shows the grazing angle dependence of the magnitude and phase of the vertical and horizontal reflection coefficient at 3 GHz for reflection from a plane, smooth earth having a relative permittivity $\epsilon' = 10$ and conductivity $\sigma = 2$ mmho/m; Figure 2.9 shows the same for reflection from a smooth sea with $\epsilon' = 69$ and $\sigma = 6.5$ mho/m.

For most media encountered in practice one finds that $\epsilon' \gg 1$ so that expressions (2.65) and (2.69) simplify to read

$$\rho_v e^{-j\mu_v} = \frac{\sqrt{\epsilon_c'}\sin\phi - 1}{\sqrt{\epsilon_c'}\sin\phi + 1}$$

$$\rho_h e^{-j\mu_h} = \frac{\sin\phi - \sqrt{\epsilon_c'}}{\sin\phi + \sqrt{\epsilon_c'}}$$

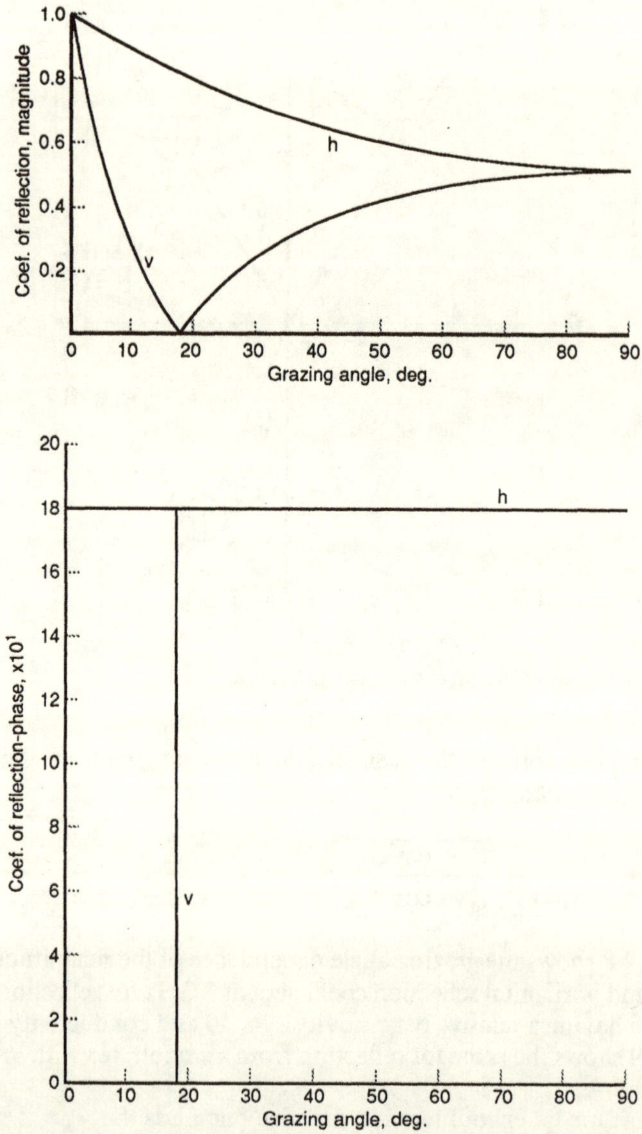

Figure 2.8 Reflection coefficient, magnitude and phase *versus* the grazing angle. Plane, smooth earth ($\epsilon_1' = 10$, $\sigma = 2 \times 10^{-3}$ S/m, $f = 3$ GHz); h = horizontal polarization, v = vertical polarization (after Rohan, 1983. © 1983 IEE).

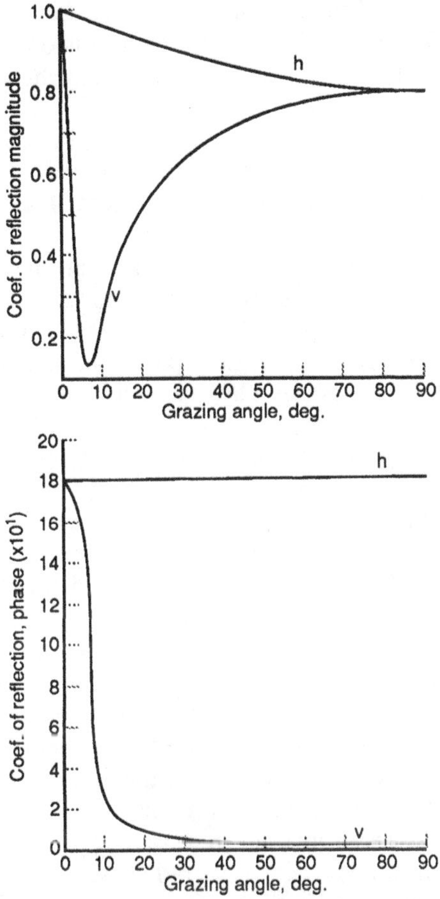

Figure 2.9 Reflection coefficient, magnitude and phase *versus* the grazing angle. Smooth sea ($\epsilon_1' = 69$, $\sigma = 6.5$ S/m, $f = 3$ GHz); h = horizontal polarization, v = vertical polarization (after Rohan 1983. © 1983 IEE).

These approximate expressions permit the observation of the variation of the magnitude and phase components with grazing angle variations [Kerr, 1951]. When the grazing angle ϕ tends to zero, the magnitude of the vertical and horizontal reflection coefficients tends to -1. The coefficient of reflection for horizontal polarization decreases from this value gradually when the grazing angle increases. The magnitude of the vertical coefficient of reflection falls sharply to a minimum when $\sin\phi = 1/\sqrt{\epsilon_c'}$ and rises again with an increasing grazing angle. The angle of incidence corresponding to the minimum value of ρ_v is a pseudo-Brewster angle occurring at vertical polarization at the surface of media having a complex dielectric constant. A Brewster angle at which the reflection coefficient falls to zero is observed in media having a real dielectric constant; that is, zero conductivity. This phenomenon does not occur at horizontal polarization.

On reflection from a perfect conductor, the magnitude of ρ_v falls to zero at the Brewster angle of incidence. It increases with rising values of the grazing angle to reach a value equal to that of the horizontal coefficient of reflection at a grazing angle of $\pi/2$ rad.

The dependence of the phase shift on reflection on the grazing angle also is more complicated at vertical polarization. At zero grazing angle the phase shift μ_v is π rad for both polarizations. Although the phase shift at horizontal polarization rises only slightly above this value, at vertical polarization it falls abruptly near the pseudo-Brewster angle to remain at a low value for the rest of the higher values of the grazing angle.

2.5.3 Circular Polarization

The constant amplitude electric field vector of a circularly polarized radiation rotates about an axis pointing in the direction of propagation, as shown in Section 2.3. Circularly polarized waves reflected by spherical objects are opposite sense circularly polarized; hence, circular polarization often is used in radar to reduce rain clutter because an antenna transmitting circularly polarized waves of one sense can receive, without excessive attenuation, only waves of the same sense. Although radar targets of interest might contain features that reverse the sense of polarization, they reflect some energy in a sense acceptable to the antenna.

We showed that a circularly polarized wave consists of equal magnitude vertical and horizontal components with a $\pi/2$ rad phase difference between them; each of these is reflected according to its appropriate coefficient of reflection. If the magnitude of the constituting components differs, the originally circularly polarized wave becomes elliptically polarized on reflection.

For same sense circularly polarized waves, the coefficient of reflection is the vector sum of the reflection coefficients of the constituting vertical and horizontal components, given by [Katz, 1963]:

$$\rho_{cs} = \tfrac{1}{2}\sqrt{\rho_v^2 + \rho_h^2 + 2\rho_v\rho_h \cos(\mu_h - \mu_v)} \tag{2.70}$$

while for an opposite sense circularly polarized wave it is

$$\rho_{co} = \tfrac{1}{2}\sqrt{\rho_v^2 + \rho_h^2 - 2\rho_v\rho_h \cos(\mu_h - \mu_v)} \tag{2.71}$$

The phase shift on reflection for same or opposite sense circularly polarized waves is

$$\mu_c = \tan^{-1} \frac{\rho_v \sin\mu_v \pm \rho_h \sin\mu_h}{\rho_v \cos\mu_v \pm \rho_h \cos\mu_h} \tag{2.72}$$

with the positive signs in the fraction applicable to same sense circularly polarized radiations.

Figure 2.10 shows the grazing angle dependence of the magnitude and phase of the two reflection coefficients for the same parameters as those used in Figure 2.8. The magnitude of same sense circularly polarized reflection coefficient for reflection

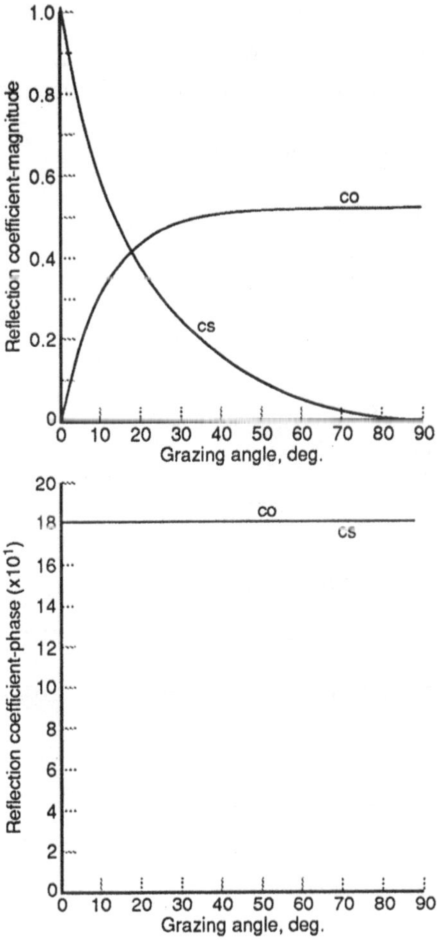

Figure 2.10 Reflection coefficient, magnitude and phase *versus* the grazing angle. Plane smooth earth ($\epsilon_1' = 10$, $\sigma = 2 \times 10^{-3}$ S/m, $f = 3$ GHz); *cs* = same circular polarization, *co* = opposite circular polarization (after Rohan, 1983. © 1983 IEE).

from a smooth surface near zero grazing angle incidence is close to those of linearly polarized waves; it falls off gradually to a very low value near normal incidence. The magnitude of opposite sense circularly polarized waves rises from zero at zero grazing angle to the value obtained for linearly polarized waves propagating under identical conditions.

Example 2.6

On the basis of Section 2.5.1 and 2.5.2, derive expressions for the dependence of the refractive index η on the angle of incidence at vertical and horizontal polarization.

Solution: (a) Vertical polarization. From expression (2.56), $E_1 \cos\gamma = E_2 \cos\gamma + E_3 \cos\psi$; therefore, $E_1 - E_2 = E_3 \cos\psi/\cos\gamma$. From expression (2.62), $E_1 + E_2 = \sqrt{\epsilon_c'}\, E_3$.

Adding the last two equations leads to

$$2E_1 = E_3\left(\sqrt{\epsilon_c'} + \frac{\cos\psi}{\cos\gamma}\right)$$

from where

$$\frac{E_3}{E_1} = \frac{2}{\sqrt{\epsilon_c'} + \dfrac{\cos\psi}{\cos\gamma}}$$

From expression (2.61), $\sin\gamma = \sqrt{\epsilon_c'} \sin\psi$. Then,

$$\frac{E_3}{E_1} = \rho_v e^{-j\nu_v} = \frac{2\sqrt{\epsilon_c'}\cos\gamma}{\sqrt{\epsilon_c' - \sin^2\gamma} + \epsilon_c' \cos\gamma}$$

(b) Horizontal polarization. From expression (2.66), $E_1 + E_2 = E_3$ and, from expression (2.67),

$$E_1 - E_2 = E_3 \frac{\sqrt{\epsilon_c'}\cos\psi}{\cos\gamma}$$

The sum of these equations yields

$$2E_1 = E_3\left(1 + \frac{\sqrt{\epsilon_c'}\cos\psi}{\cos\gamma}\right)$$

Then

$$\frac{E_3}{E_1} = \frac{2\cos\gamma}{\cos\gamma + \sqrt{\epsilon_c'}\cos\psi}$$

According to expression (2.61) $\sin\psi = \sin\gamma/\sqrt{\epsilon_c'}$ and then

$$\eta_h e^{-j\nu_h} = \frac{E_3}{E_1} = \frac{2\cos\gamma}{\cos\gamma + \sqrt{\epsilon_c' - \sin^2\gamma}}$$

The permittivity of the media is a complex, frequency dependent quantity, hence a phase shift $\nu_{v,h}$ will occur on refraction.

2.5.4 Diffuse Reflection

The previously derived reflection coefficients are applicable strictly to specular reflection. Specular reflection (i.e., reflection at which parallel incident rays remain parallel after reflection) occurs only from a smooth surface. If the rays are no longer parallel after reflection, reflection becomes diffuse.

In diffuse reflection, due to the roughness of the reflecting surface, energy is scattered in other directions as well as in the desired direction. This causes a reduction of the field strength of the reflected wave in the desired direction and thus of that of the resultant field mentioned earlier.

Roughness in a reflecting surface is defined in terms of the wavelength of the considered radiation. Figure 2.11 shows the highest and lowest level of surface protuberances of height h. The bottom ray incident at an angle γ suffers a phase shift

$$2\frac{2\pi h}{\lambda}\cos\gamma \tag{2.73}$$

with respect to the uppermost one as their path difference is

$$\Delta R = 2h\cos\gamma$$

Small phase differences caused by almost smooth surfaces are usually negligible. A π-rad phase difference causes a complete cancellation of the radiation in a given direction and its dispersion into other ones. A criterion dividing surfaces into smooth and rough ones was proposed by Rayleigh. Smooth surfaces are considered to be those that cause a phase difference below $\pi/2$ rad, and the phase shift caused

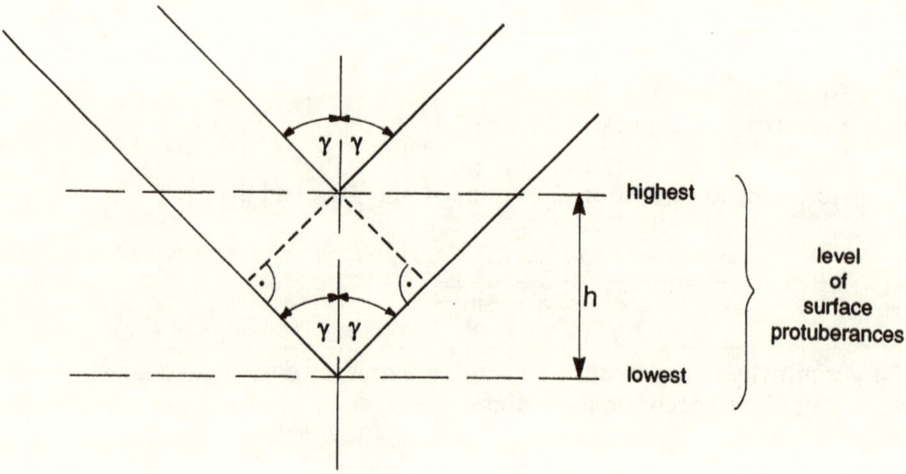

Figure 2.11 Criteria for surface roughness.

by rough ones exceeds this value. The critical height of surface proturberances, appropriate to this definition, is

$$h_c = \frac{\lambda}{8} \sec\gamma \tag{2.74}$$

so that for smooth surfaces $h < h_c$ and $h > h_c$ for rough ones.

To account then for the diminished reflected radiation in the desired direction, the reflection coefficients for specular reflection must be modified by a factor [Ament, 1953]

$$\rho_s = \exp\left[-2\,\frac{2\pi\sigma_h \cos\gamma}{\lambda}\right] \tag{2.75}$$

where σ_h is the standard deviation of the surface height about the mean height in the first Fresnel zone (to be defined later), or, when referring to the sea surface, about the significant sea wave height given by

$$\sigma_h = \frac{H_{1/3}}{4}$$

where $H_{1/3}$ is the average of the highest third of the waves in the wave record [Nathanson, 1969].

Figure 2.12 shows the dependence of the average scattering coefficient ρ_s on the grazing angle at 3 GHz with the height of surface irregularities as parameters. Other criteria of surface roughness, claiming to be more realistic, have been proposed; however, Rayleigh's criterion has been generally accepted.

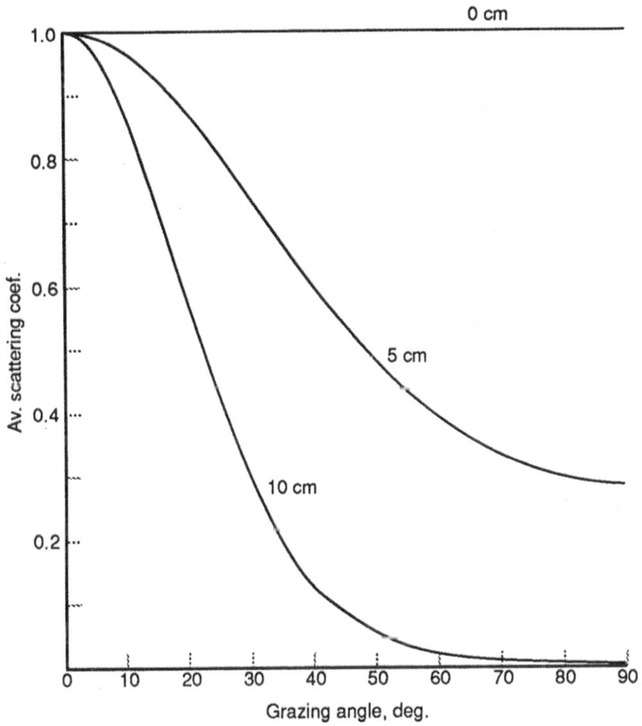

Figure 2.12 The average scattering coefficient as a function of the grazing angle with heights of surface irregularities as a parameter, $f = 3$ GHz (after Rohan, 1983. © 1983 IEE).

2.5.5 Divergence Factor

The surface of the earth, even when neglecting its irregularities, is not plane but convex. In accordance with Snell's law, which shows that the angles of incidence and reflection are equal, a conical bunch of rays illuminating the earth's surface diverges on reflection thus diminishing the field strength proportional to the square root of the ray density (Figure 2.13). A divergence factor

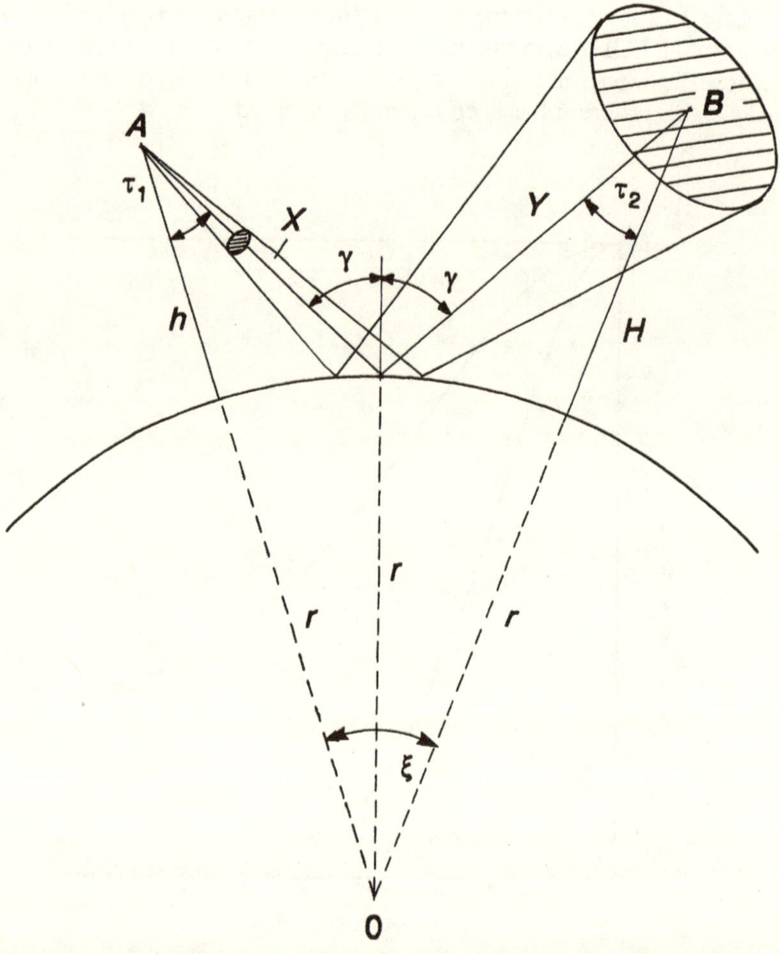

Figure 2.13 The divergence factor.

$$D = \frac{r(X + Y)\sqrt{\sin\gamma \cos\gamma}}{\sqrt{[(r + H)X \cos\tau_2 + (r + h)Y \cos\tau_1](r + h)(r + H) \sin\xi}} \qquad (2.76)$$

the square root of the ratio of the cross section of a bunch of rays reflected from a plane surface to that reflected from a convex one takes account of this phenomenon [Van der Pol and Bremmer, 1939].

The meaning of symbols in 2.76 is as follows: X and Y are the slant ranges of the transmitting resp. receiving antenna from the point of reflection; r is the earth radius modified for the appropriate atmospheric index of refraction (to be dealt with later); γ is the angle of incidence and reflection in the specular direction; h and H are the transmitting resp. receiving antenna height; τ_1 and τ_2 are the angles between X and Y and the lines connecting the center of the earth with the transmitting resp. receiving antenna; and ξ is the angle between the above two lines.

In theoretical considerations, when the earth is taken to be spherical, the divergence factor reduces to

$$D = \frac{1}{\sqrt{1 + \frac{2XY}{r(X+Y)}\left[1 + \frac{1 + \frac{2XY}{r(X+Y)}}{\cos\gamma}\right]}} \tag{2.77}$$

2.5.6 The Composite Reflection Coefficient

A composite reflection coefficient that considers all the preceding influences on the reflected wave is obtained by modifying the specular reflection coefficient given by (2.65), (2.69), (2.70), or (2.71), depending on polarization, by the expressions obtained in Section 2.5.4 and 2.5.5. It reads

$$\rho = \rho_{v,h,c}\rho_s D \tag{2.78}$$

where $\rho_{v,h,c}$ is the specular reflection coefficient for vertical, horizontal, or circular polarization, while ρ_s is the average scattering coefficient and D the divergence factor.

Figures 2.14 and 2.15 show the dependence of the composite reflection coefficient on the grazing angle with the height of surface irregularities as parameter. Comparing these figures with Figures 2.8 and 2.9, we find that the divergence factor has greatest influence at very low values of grazing angle. This causes the composite reflection coefficient to rise sharply from its initial zero value at grazing incidence to a value given by the product of the plane earth reflection coefficient and the scattering coefficient, whereas the shape of the curves showing the dependence $\rho = \rho(\phi)$, where ϕ is the grazing angle, differs little from those obtained for specular reflection when the surface irregularities are small. They differ markedly for surface irregularities exceeding the wavelength. At a given grazing angle, the composite reflection coefficient is lower than the specular one.

2.6 FRESNEL ZONES

We mentioned earlier (Section 2.5) that the resultant field at a point in space, for theoretical considerations, may be assumed to be due to at least two fields, one propagating directly from the source of radiation to the considered location and one arriv-

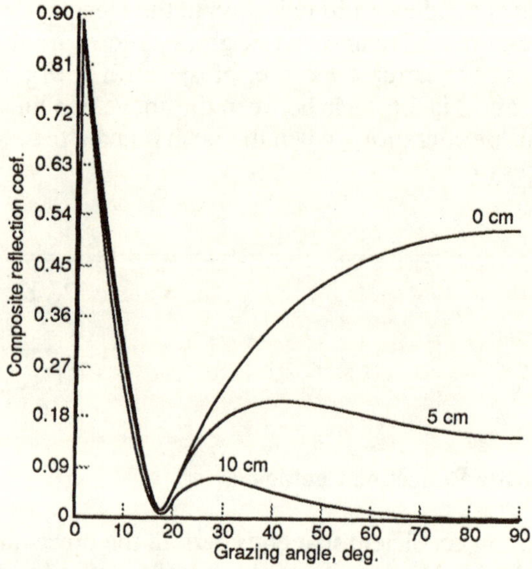

Figure 2.14 The vertical polarization composite reflection coefficient *versus* grazing angle with the height of surface irregularities as parameters; $\lambda = 10$ cm (after Rohan, 1983. © 1983 IEE).

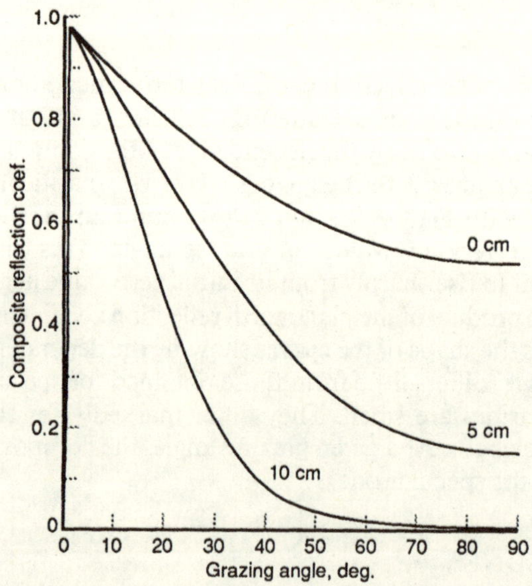

Figure 2.15 The horizontal polarization composite reflection coefficient versus grazing angle with the height of surface irregularities as a parameter; $\lambda = 10$ cm (after Rohan, 1983. © 1983 IEE).

ing there after reflection from a point on the surface of the earth. In reality, the energy source illuminates a large area of the earth. There it induces currents radiating in all directions with such phase relations that most of the radiated energy cancels out and only components having a desired phase relation with respect to the component propagating directly (i.e., without reflection) contribute to the final field.

The path lengths of the direct and reflected rays differ. If the path length of the direct ray is designated by R, and that of the reflected one, the sum of branches X and Y shown by Figure 2.16 for a plane earth, by R_1, then their difference

$$R_1 - R = \delta$$

The region around the location from which the reflected radiation has a phase associated with a cooperative contribution to the final field and extended roughly to points for which the relative phase deviation amounts to π rad or a half-wavelength λ, is called the *first Fresnel zone.* It is characterized by

$$R_1 - R < \lambda/2$$

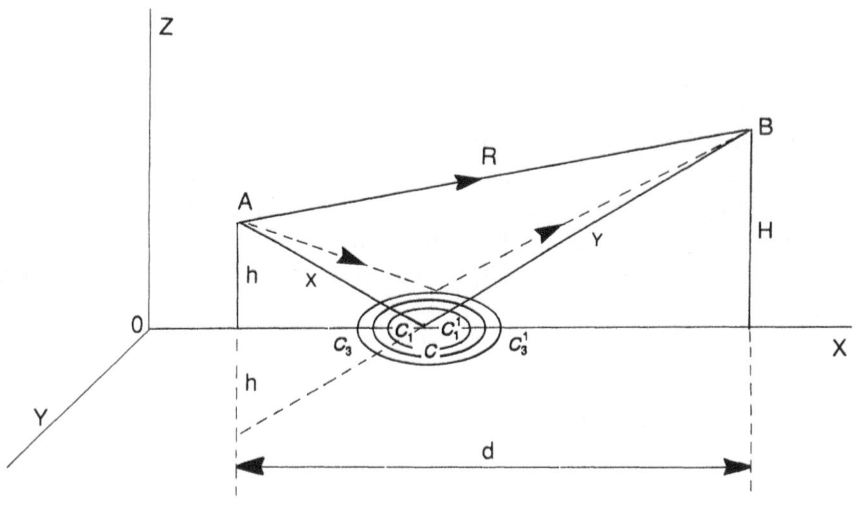

Figure 2.16 Fresnel zones in the x-y plane.

The expression

$$R_1 = X + Y = R + \delta = \text{const}$$

where $\delta = \delta_0 + n\lambda/2$, with $n = 1, 2, 3, \ldots$ the Fresnel zone number, is the equation of ellipsoids of revolution with major axes incorporating the slant range R joining the foci at A and B. Their intersections with the earth's surface are ellipses with limits (Figure 2.16):

$$\overline{AC_n} + \overline{C_nB} = \delta_0 + n\lambda/2$$
$$\overline{AC'_n} + \overline{C'_nB} = \delta_0 + n\lambda/2$$

Here, $\delta_0 = 2hH/d$, with h and H the height of the transmitting resp. receiving antenna and d their separation, is the approximate path length difference between the direct and once-reflected ray traveling between points A and B. It is obtained from the difference of

$$R_1 = \sqrt{(H + h)^2 + d^2}$$

and

$$R = \sqrt{(H - h)^2 + d^2}$$

by the binomial expansion of the expressions under the square root signs [Kerr, 1951].

The contribution of individual Fresnel zones is directly dependent on the amplitude of excitation. Because zones bounded by adjacent ellipses reflect energy in phase opposition, their contributions tend to cancel out each other. However, because the amplitudes of excitation decrease gradually, no complete cancellation occurs. At an oblique incidence, where the zones are inclined at large angles to the direction of propagation, the amplitude of excitation varies greatly even over a single zone. Thus, the reradiation area is decreased approximately to half of the first Fresnel zone.

The center of the Fresnel zones is at a distance of

$$x_{on} = \frac{d}{2}\left\{1 - \frac{\dfrac{H^2 - h^2}{d^2}}{\left[\dfrac{\delta}{d} + \sec\theta\right]^2 - 1}\right\} \tag{2.79}$$

from the transmitting antenna. The intercepts of the minor axes are

$$y_{1,2} = \pm \frac{d}{2} \sqrt{\left(\frac{\delta}{d}\right)^2 + \frac{2\delta}{d}\sec\theta} \sqrt{1 - \frac{\left(\dfrac{H+h}{d}\right)^2}{\left[\dfrac{\delta}{d} + \sec\theta\right]^2 - 1}} \tag{2.80}$$

while those of the major axes are at

$$x_{1,2} = x_0 \pm y_{1,2} \sqrt{1 + \frac{1}{\left[\dfrac{\delta}{d} + \sec\theta\right]^2 - 1}} \tag{2.81}$$

The area of the individual Fresnel zones is given by

$$A = \frac{\pi d^2}{4} \frac{(\delta/d + \sec\theta)[(\delta/d)^2 + 2\delta/d \sec\theta]}{\sqrt{(\delta/d + \sec\theta)^2 - 1}} \left[1 - \frac{\left(\dfrac{H-h}{d}\right)^2}{(\delta/d + \sec\theta)^2 - 1}\right] \tag{2.82}$$

The slant angle $\theta = \tan^{-1}(H - h)/d$ is the inclination of the ellipsoids' major axis with respect to the earth's surface.

Although the first, the innermost Fresnel zone is the most important contributor to the development of reradiated energy of desirable phase, important contributions also come from areas in front of the transmitting and receiving antennas. A rigorous analysis shows that the field can be calculated to within a few percent by retaining only two or three Fresnel zones and sometimes only a single one. The elongation of the ellipsoids grows larger the larger is the ratio $2\pi R/\lambda$. With waves of very short wavelength, the ellipsoids become very narrow and ray propagation akin to optical ray propagation arises.

Rough surfaces cause diffuse reflection reducing the reflected field to values inadequate for developing a distinct lobing pattern of the antenna's vertical coverage diagram (see Section 3.5), thus reducing the signal strength at greater distances. Mirror reflection occurs only on perfectly flat surfaces. For this reason the ground near the base of the transmitting antennas is leveled.

2.7 DEPOLARIZATION

Depolarization is the change of polarization from that of the incident wave to that of the reflected one. The meaning of the terms of linear, circular, and elliptic polarization were given earlier. Linear and circular polarizations, from a mathematical point of view, are only special cases of elliptical polarization.

Beckmann (1968) shows that an electric field **E** is the sum of its space components:

$$\mathbf{E} = E_h\mathbf{e}_h + E_v\mathbf{e}_v = E_h\left(\mathbf{e}_h + \frac{E_v}{E_h}\mathbf{e}_v\right) \tag{2.83}$$

and the ratio of the two complex field magnitudes E_h and E_v, denoted by p, is the complex polarization factor uniquely defining any polarization of an electromagnetic wave. The amplitude ratio of the components is the magnitude of the polarization factor, whereas the difference of their phases is its argument.

If p_1 defines the polarization of the incident and p_2 that of the scattered wave, then

$$q = p_2/p_1 \tag{2.84}$$

is the depolarization factor. It equals the ratio of the generalized scattering coefficients

$$\chi_v = E_{v_2}/E_{v_1} \quad \chi_h = E_{h_2}/E_{h_1} \tag{2.85}$$

where the subscripts 1 and 2 refer to the incident resp. scattered values of the waves. The generalized scattering coefficients reduce to the earlier derived reflection coefficients when scattering from a properly oriented smooth plane occurs.

The degree of polarization is

$$P = \frac{M^2 - N^2}{M^2 + N^2} \leq 1 \tag{2.86}$$

with M and N the major or minor axis of the polarization ellipse.

The depolarization ratio is

$$D = \frac{|E_c|^2}{|E_p|^2} \tag{2.87}$$

with c denoting the cross-polarized component, that is, the component perpendicular to the direction of the incident linear polarization; and p the parallel polarized one, parallel to the above direction.

If the direction of interest in this expression is at an angle γ to the axes of the polarization ellipse, the depolarization ratio becomes

$$D = \frac{N^2 + M^2 \tan^2\gamma}{N^2 \tan^2\gamma + M^2} \tag{2.88}$$

which, for the case of $\gamma = 0$, yields the cross-polarization ratio

$$D = N^2/M^2 \tag{2.89}$$

For scattering by a perfectly conducting smooth plane in the specular direction (i.e., where the laws of geometrical optics apply), the direction of the depolarization factor

$$q = \rho_v/\rho_h = -1 \tag{2.90}$$

This intimates that a smooth plane, in general, depolarizes all incident polarizations except a horizontal one, the complex polarization factor p of which is zero, and a vertical one with $p = \infty$.

At near grazing incidence onto a finitely conducting plane, the vertical and also the horizontal reflection coefficient are equal to -1, yielding a unity depolarization factor indicating that no depolarization occurs. In general, a large, smooth conducting plane surface having a -1 depolarization factor, depolarizes incident radiation. It reverses the sense of a rotational polarization and the orientation of the polarization ellipse; however, it does not affect the ellipse's shape.

Beckmann shows that expression (2.90) applies as well to rough, perfectly conducting surfaces, provided that the radii of curvature of the individual surface irregularities are much larger than a wavelength of the incident radiation. If the surface is not perfectly conducting, the depolarization factor q will be random because the reflection coefficients will vary over the surface in accordance with the local angle of incidence and, in case of a nonhomogeneous scattering surface, also with the local electrical properties of the surface.

If the radii of curvature of the surface irregularities are smaller than a wavelength, the local diffraction effects complicate matters and the depolarization factor must be determined on the basis of its definition given by expression (2.84).

The absence of depolarization of purely horizontally ($p_1 = 0$) and purely vertically ($p_1 = \infty$) polarized radiation also applies to rough surfaces, regardless of the conductivity and distribution of the surface irregularities.

If the surface is oriented so that its normal wave at the point of incidence does not lie in the plane of incidence, an incident wave will be scattered out of the plane of incidence thus also causing the depolarization of purely horizontally and purely vertically polarized waves.

2.8 DIFFRACTION

Diffraction occurs when edges, corners, or vertices of boundary surfaces are hit or grazed by an electromagnetic radiation at wavelengths much shorter than the linear dimensions of these features. The exact analytical solution of problems of diffraction relies on solutions of the wave equation for the electric and magnetic fields that satisfy the differential equation for the region within the diffracting body or obstacle and simultaneously the conditions of continuity of the tangential components at the boundary surfaces [Joos, 1934]. Such a solution determines the field intensity, polarization, *et cetera* behind the obstacle. The mathematical difficulties restrict this approach to the simple cases of diffraction of a plane wave from a sphere, cylinder, or a screen, the surfaces of which coincide with those of an orthogonal curvilinear coordinate system.

The solution of more complicated diffraction problems relies on the methods of an extended version of geometrical optics, known as the *geometric theory of diffraction.* Geometrical optics uses straight and curved lines or rays to treat reflection and refraction, but it fails to account for the phenomenon of diffraction [Keller, 1962]. The geometric theory of diffraction introduces additional, so-called diffracted rays produced by the incident radiation at the previously mentioned features. Some of these rays extend the radiated field into the shadow region. The laws of diffraction follow from the appropriately modified Fermat's principle of least time, according to which the actual path of propagation between points A and B is such that the time taken to traverse it is minimal or, more correctly, the variation of the line integral

$$\int_A^B n(x, y, z)\, \mathrm{d}s$$

with $n(x, y, z)$ denoting the relative index of refraction, is zero.

The geometric theory of diffraction improves on classical physical approximations and extends them by considering phase differences and polarization. Its methods are based on the assumption that processes causing refraction and diffraction depend only on the incident field and the properties of the medium in the immediate vicinity of the individual points of the mentioned surfaces. Asymptotic expansions of boundary value problems for Maxwell's equations at rising frequencies of radiation determine the local fields.

The inclusion of higher-order diffraction effects, in some cases, may extend the applicability of the geometric theory of diffraction to structures having dimensions comparable to or even shorter than a wavelength [James, 1976].

A more detailed treatment of the geometric theory of diffraction is beyond the scope of this book. Interested readers are referred to the publications of Keller, James, and Jones listed at the end of this chapter.

2.9 THE RECIPROCITY THEOREM

Similar to the reciprocity theorem of linear network theory, according to which the electromotive force in the kth mesh due to a unit current in the mth one is the same as the electromotive force in this mesh due to a unit current in the kth one, there exists a reciprocity theorem applicable to antennas and electromagnetic fields [Simonyi, 1963].

If the fields $\mathbf{E}_a$ and $\mathbf{H}_a$ are due to an applied field $\mathbf{E}'_a$ and fields $\mathbf{E}_b$ and $\mathbf{H}_b$ due to an applied field $\mathbf{E}'_b$, then the requirement of reciprocity is satisfied when

$$\mathbf{E}'_a\mathbf{E}_b = \mathbf{E}'_b\mathbf{E}_a \tag{2.91}$$

In an isotropic medium Maxwell's equations for these two systems become

$$\begin{aligned} \text{curl } \mathbf{H}_a &= (\sigma + j\omega\epsilon)\mathbf{E}_a + \sigma\mathbf{E}'_a \\ \text{curl } \mathbf{E}_a &= -j\omega\mu\mathbf{H}_a \\ \text{curl } \mathbf{H}_b &= (\sigma + j\omega\epsilon)\mathbf{E}_b + \sigma\mathbf{E}'_b \\ \text{curl } \mathbf{E}_b &= -j\omega\mu\mathbf{H}_b \end{aligned} \tag{2.92}$$

with $\sigma\mathbf{E}'_i$ the conduction current density due the applied fields $\mathbf{E}'_i$.

Dot multiplying these equations respectively by $\mathbf{E}_b$, $\mathbf{H}_b$, $-\mathbf{E}_a$, $-\mathbf{H}_a$, and adding them, we obtain

$$\begin{aligned} \mathbf{E}_b \cdot \text{curl } \mathbf{H}_a + \mathbf{H}_b \cdot \text{curl } \mathbf{E}_a - \mathbf{E}_a \cdot \text{curl } \mathbf{H}_b \\ - \mathbf{H}_a \cdot \text{curl } \mathbf{E}_b = \sigma(\mathbf{E}'_a\mathbf{E}_b - \mathbf{E}_a\mathbf{E}'_b) \end{aligned} \tag{2.93}$$

and because

$$\mathbf{E}_b \cdot \text{curl } \mathbf{H}_a - \mathbf{H}_a \cdot \text{curl } \mathbf{E}_b = \text{div}(\mathbf{E}_b \times \mathbf{H}_a)$$

and

$$\mathbf{E}_a \cdot \text{curl } \mathbf{H}_b - \mathbf{H}_b \cdot \text{curl } \mathbf{E}_a = \text{div}(\mathbf{E}_a \times \mathbf{H}_b)$$

expression (2.93) becomes

$$\operatorname{div}(\mathbf{E}_a \times \mathbf{H}_b) - \operatorname{div}(\mathbf{E}_b \times \mathbf{H}_a) = \sigma(\mathbf{E}_a'\mathbf{E}_b - \mathbf{E}_a\mathbf{E}_b') \tag{2.94}$$

Taking the volume integral over the whole isotropic space and applying Gauss's divergence theorem, according to which the integral of the divergence of a vector over a given volume equals the surface integral of that vector over the surface bounding the volume, we find that the integral over the surface at infinity vanishes yielding

$$\mathbf{E}_a'\mathbf{E}_b = \mathbf{E}_a\mathbf{E}_b'$$

as in (2.91), thus proving that the two systems are reciprocal.

The expression

$$\int_S \mathbf{n} \cdot (\mathbf{E}_a \times \mathbf{H}_b - \mathbf{E}_b \times \mathbf{H}_a)\, dS = 0 \tag{2.95}$$

on which the validity of (2.91) depends is Lorentz's reciprocity theorem.

As an example of reciprocity one may consider the reciprocal behavior of an antenna coverage diagram for transmission and reception. The distribution of the electromagnetic field produced by an antenna as a radiator is proportional to the directional gain $G(\theta, \phi)$. The power received by the same antenna depends on its aperture, which, given by

$$A(\theta, \phi) = G(\theta, \phi)\frac{\lambda^2}{4\pi} \tag{2.96}$$

depends on the directional gain given earlier. Hence, the performance of an antenna as a radiator and receiver is reciprocal.

The reciprocity theorem does not apply to certain materials (e.g., ferrites) used in some microwave devices; similarly, the ionosphere that, being subjected to the effects of the earth's magnetic field, is nonisotropic and hence does not support reciprocity [Wait, 1964].

Example 2.7

Show that interchanging the location of the transmitter and receiver in Figure 2.17 will not affect the received signal level.

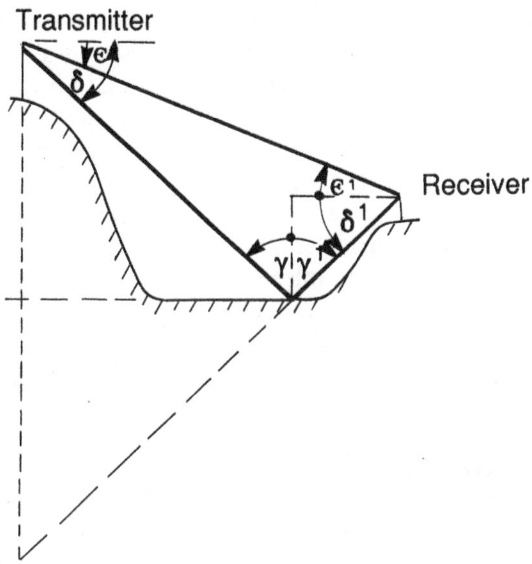

Figure 2.17 Example 2.7.

Solution: The signal level radiated or received by an antenna depends on the antenna's directional gain; that is, the gain at the angles at which the direct and reflected rays are transmitted resp. received by the antenna.

Because the angles of incidence and reflection γ and γ' must be equal, the angles of radiation resp. reception of the preceding two rays, as may be shown by simple geometrical considerations, also must be equal, or $\epsilon = \epsilon'$ and $\delta = \delta'$. Hence, interchanging the transmitter and receiver locations in Figure 2.17 will not affect the received signal level.

PROBLEMS

2.1 A source of voltage E is applied to a straight wire of diameter d and length l made of a material of resistivity ρ. Express the power dissipated in the wire by using Poynting's vector and compare it with the dissipated power computed from Ohm's law.

2.2 At a given point in space, the field due to a vertical antenna is $H_\varphi = 1/(r) \sin\theta \cos(\omega t - \beta r)$ when the electric field is $E_\theta = 377 H_\varphi$. Determine the energy flow, in Watts, from the volume surrounded by a spherical surface S of radius r with its center at the origin.

2.3 An ideal coaxial cable, having neither resistance, capacitance, nor conductance, has its inner and outer conductors connected to a battery so that the power flows outward from the inner conductor and into the outer one. Determine the power carried by the cable if the radius of the inner conductor is a, that of the outer one is b, and the direct current in the cable is I_{amp}. The uniform charge distribution on the inner conductor is q C/m.

2.4 An elliptically polarized wave with the amplitude relation of its components given by $|B| = 3|A|$, propagates in the $x =$ direction. Plot the curve described by the end of the resulting complex vector.

2.5 State the criterion determining the sense of circularly polarized waves.

2.6 Determine the magnitude of the horizontal and vertical components of the incident, reflected, and refracted rays of a vertically polarized wave at the boundary of air and sea water for a grazing angle of 60°.

2.7 Evaluate, under the given conditions, the quantities specified in Example 2.5, for horizontal polarization.

2.8 Determine the refractive index and absorption coefficient of sea water at a frequency of 10 GHz. How does the velocity of propagation compare with that in free space? Compare the obtained values with corresponding quantities for soil having $\epsilon' = 10$ and $\sigma = 2 \times 10^{-3}$ mho/m.

2.9 Derive expressions for the magnitude and phase of the refractive index at horizontal polarization and plot their dependence on the grazing angle for a medium having $\epsilon' = 10$ and $\sigma = 2$ mmho/m. (See Example 2.5 for a hint on evaluation.)

2.10 Determine the refractive index of seawater if the velocities of propagation obtained in Example 2.1 apply.

2.11 Show that the results of Problem 2.6 satisfy the relations between the tangential resp. normal components at the boundary of two dissimilar media.

2.12 Using the data of Figure 2.8 determine the magnitude and phase of the reflection coefficient for same and opposite sense circular polarization at a grazing angle of 10°. Compare the results with values read from Figure 2.10.

2.13 Determine the acceptable height of terrain unevenness at the point of reflection permitting the development of the reflected field in adequate strength to yield a distinct lobing of the 20-m high radar antenna's vertical coverage diagram at 1300 MHz when observing an aircraft flying at a height of 200 m over a plane earth at a range of 6 km.

2.14 Simplify the formula for the divergence factor for communication between two antennas of equal height separated by a distance R on the surface of a spherical earth.

2.15 What area of the ground near a 100-m high transmitting antenna should be leveled for improving radiation to a 10-m high receiving antenna at a distance of 50 km (plane earth) at a frequency of 1 MHz?

REFERENCES AND RECOMMENDED READING

Ament, W.S. (1953) "Toward a Theory of Reflection by a Rough Surface," *Proc. IRE,* Vol. 41, No. 1, January, pp. 142–146.

Beckmann, P., and A. Spizzichino (1963) *The Scattering of Electromagnetic Waves from Rough Surfaces,* Pergamon Press, New York.

Beckmann, P. (1968) *The Depolarization of Electromagnetic Waves,* Golem Press, Boulder, CO.

Bremmer, H. (1949) *Terrestrial Radio Waves,* Elsevier Publishing Co., Amsterdam.

Bremmer, H. (1958) "Propagation of Electromagnetic Waves," in the *Encyclopedia of Physics,* Vol. 16, *Electric Fields and Waves,* ed. S. Flugge, Springer-Verlag, Berlin, Gottingen, and Heidelberg.

Doluchanow, M.P. (1956) *Die Ausbreitung von Funkwellen,* VEB Verlag Technik, Berlin.

Holt, C.A. (1963) *Introduction to Electromagnetic Waves and Fields,* John Wiley & Sons, New York.

James, G.L. (1976) *Geometrical Theory of Diffraction for Electromagnetic Waves,* IEE Electromagnetic Waves Series, Vol. 1, Peter Peregrinus Ltd., London.

Jones, D.S. (1979) *Methods in Electromagnetic Wave Propagation,* Clarendon Press, Oxford.

Joos, G. (1934) *Theoretical Physics,* Blackie & Son Limited, London and Glasgow.

Jordan, E.C. (1953) *Electromagnetic Waves and Radiating Systems,* Constable and Company Ltd., London.

Katz, I. (1963) In C.I. Beard et al., "Radar Reflectivity of the Ocean Surface for Circular Polarization," *IEEE Trans. on Antennas and Propagation,* Vol. AP-11, No. 4, July, pp. 451–453.

Keller, J.B. (1962) "Geometrical Theory of Diffraction," *J. Optics Society of America,* Vol. 52, No. 2, February, pp. 116–130.

Kerr, D.E. (1951) *Propagation of Short Radio Waves,* M.I.T. Radiation Lab. Series, Vol. 13, McGraw-Hill, New York.

Millington, G. (1971) "The Reciprocity Principle in Radio Propagation," *Marconi Rev.,* Vol. 34, No. 183, pp. 233–252.

Monteath, G.D. (1973) *Applications of the Electromagnetic Reciprocity Principle,* Pergamon Press, Oxford.

Nathanson, F.E. (1969) *Radar Design Principles—Signal Processing and the Environment,* McGraw-Hill, New York.

Price, G.H. (1986) "On the Relationship between the Transmitting and Receiving Properties of an Antenna," *IEEE Trans. on Antennas and Propagation,* Vol. AP-34, No. 11, November.

Rodimov, A.P. (1966) "Degree of Polarization of Radio Signals," *Radio Engineering Electronic Physics,* Vol. 11, No. 11, November, pp. 1728–1734.

Rohan, P. (1983) *Surveillance Radar Performance Prediction,* Peter Peregrinus Ltd., London.

Schelkunoff, S.A. (1943) *Electromagnetic Waves,* D. Van Nostrand Co., New York.

Simonyi, K. (1963) *Foundations of Electrical Engineering,* Pergamon Press, London.

Stratton, J.A. (1941) *Electromagnetic Theory,* McGraw-Hill, New York.

Van der Pol, B., and H. Bremmer (1939) "Further Note on the Propagation of Radio Waves over a Finitely Conducting Spherical Earth," *Phil. Magazine,* Vol. 27, Series 7, No. 182, March, pp. 261–275.

Wait, J.R. (1964) "Some Remarks Concerning Nonreciprocity in Radio Propagation," *IEEE Trans. on Antennas and Propagation,* Vol. AP-12, May, pp. 372–373.

Weeks, W.L. (1964) *Electromagnetic Theory for Engineering Applications,* John Wiley & Sons, New York.

Chapter 3
ELECTROMAGNETIC WAVE PROPAGATION

3.1 INTRODUCTION

Problems of electromagnetic wave propagation have occupied scientists and engineers ever since electromagnetic radiation was produced intentionally. In a homogeneous medium electromagnetic waves propagate at a constant velocity along straight paths. The presence of the ground and ionized atmospheric layers causes their absorption and the deviation of their paths from the straight one. Obstacles in the propagation path, having radii of curvature on the order of a wavelength, bend the path. The hump due to the curvature of the earth lying between the transmitting and receiving antennas located beyond the transmitting antenna's horizon constitutes such an obstacle at broadcast and lower frequencies at which a wave hugging the earth's surface due to bending may reach distances of 3000 to 4000 km at usable signal levels. At higher frequencies absorption is so high that bending does not contribute to long-distance propagation.

Waves traveling along these paths are termed *ground waves* in contrast to space or sky waves, which are not radiated along the ground's surface and travel through space until they reach an obstacle, such as the boundary between air and the ground or sea, where they are reflected, or the ionized regions of the upper atmosphere, where their path may be bent enough for their return to ground. Long-distance communication at high frequencies relies on the waveguide-like propagation of waves within the spherical waveguide constituted on the one side by the earth and on the other by the ionized layers.

The first method of counteracting the effect of the finite conductivity of the ground on the radiation from a short, vertical antenna was presented by Sommerfeld in 1909. His analysis describes the complete field and defines the capabilities of the earth's surface for supporting ground-wave propagation, but his results, in addition to leading to such anomalies as negative attenuation and zero field intensity at finite distances from the transmitting antenna, were too complex for general engineering applications.

Norton, in 1935, remedied the shortcomings of Sommerfeld's approach by changing a single sign in the computations. Norton also reduced the complexity of Sommerfeld's results to formulas and graphs applicable to straightforward calculations of the field intensity.

Waves launched by an isotropic radiator exhibit spherical equiphase surfaces; however, the study of ground-wave propagation may be simplified by considering the ground to be plane and the radiator, so far that the radius of curvature of the equiphase surfaces is large enough for the wave, to be considered a plane wave, too. The solution of Maxwell's equations for an inhomogeneous plane wave over a flat surface having finite losses was obtained by Zenneck [1907].

Figure 3.1 shows various propagation paths. The ground wave in path 1 hugs the earth's surface due to bending. Path 2 is an ionospheric path occurring at high frequencies. Paths 3 and 4 depict the direct and reflected paths between the two elevated, mutually "visible" antennas mentioned in Section 2.5.

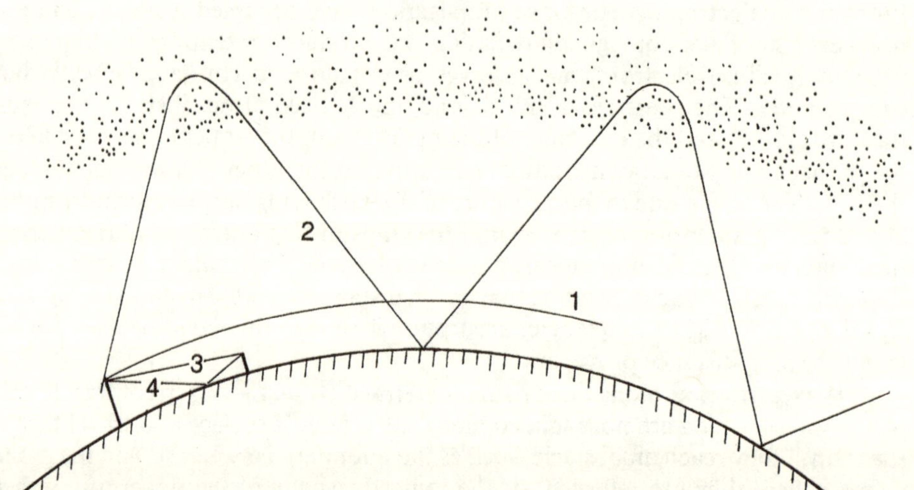

Figure 3.1 Various propagation paths.

3.2 PROPAGATION OVER A PLANE SEMICONDUCTING EARTH

Based on the fundamental concepts of Chapter 2, we can derive expressions for the components of a propagating wave. The surface of the ground is taken to be the boundary between medium 1 above and medium 2 below it. Medium 1 is described by ϵ_0, μ_0, and $\sigma = 0$; and medium 2 is characterized by ϵ_2, μ_2, and σ_2.

The vertical component of the time dependent electric field in medium 1 above the ground may be expressed on the basis of (2.24) by

$$E_{1z} = E_{01_z} e^{j\omega t} \tag{3.1}$$

where E_{01_z} is the amplitude of the z-component in that medium.

Waves traveling along the surface of the ground set up electric currents in the earth itself. These attenuate the waves because energy is absorbed from them to maintain the currents that are changed into heat. The parts of the waves in contact with the earth's surface loose energy, which is replaced by energy taken from the parts of the waves immediately above it.

The velocity of propagation given by (2.17) shows that it differs for the medium above the boundary from that below it due to the difference in their permittivity. Hence, because the complex dielectric constant of the ground is larger than that of air, the wavelength in the ground will be shorter.

Due to the difference in the velocities of propagation in the two adjacent media, a longitudinal component of the field, E_{1x}, develops in the direction of propagation. Accordingly, the electric lines of force no longer are perpendicular to the boundary but have a forward tilt (Figure 3.2). This accounts for the directional properties of tilted antennas.

The longitudinal and vertical components are not in phase, so the resultant electric field vector is elliptically polarized. The wave then generally is not plane; however, as its radius of curvature is large when compared with the wavelength and depth of penetration near the boundary, it may be considered to be a plane wave.

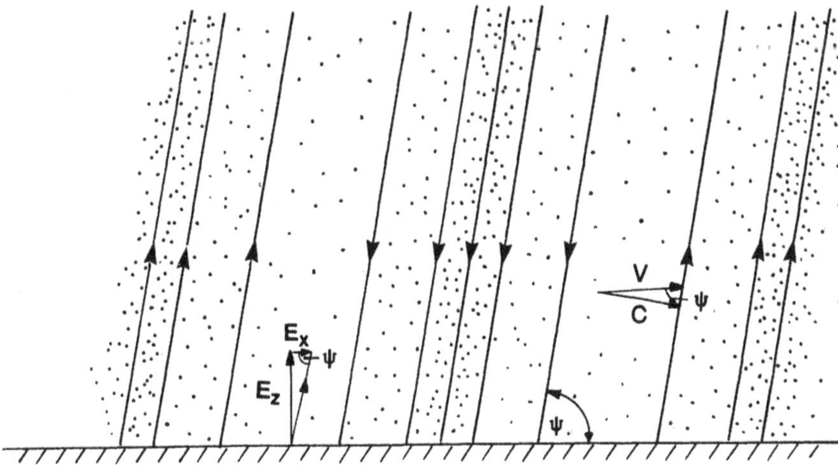

Figure 3.2 Tilt of the wavefront.

The magnetic component perpendicular to the vertical component of the electric one and to the direction of propagation is given in accordance with the boundary conditions and expression (2.22) by

$$H_{1y} = -\sqrt{\frac{\epsilon_c}{\mu_0}}\,E_{1x} \tag{3.2}$$

Both H_{1y} and E_{1x} are tangential to the boundary.

The horizontal field components, in terms of the ground constants, may be determined from these expressions. However, because the usual ground constants do not change the wavelength appreciably, we can use expressions (2.24) and (2.25) to express the field components as

$$E_{1z} = E_0 e^{-j\omega(x/v_0 - t)}$$

$$H_{1y} = -\sqrt{\frac{\epsilon_0}{\mu_0}}\,E_{1z} = -\frac{\sqrt{\epsilon'}}{120\pi}\,E_0 e^{-j\omega(x/v_0 - t)}$$

Further, in accordance with (2.55 b), the tangential components in the two media must be equal, so that at $z = 0$

$$H_{2y} = H_{1y} \tag{3.3}$$

From expressions (2.25) and (3.2), we obtain

$$E_{1x} = -\sqrt{\frac{\mu_0}{\epsilon_c}}\;H_{1y} = \sqrt{\frac{\epsilon_0}{\epsilon_c}}\,E_{1z} \tag{3.4}$$

and, due to the boundary condition (2.55 a), the relation

$$E_{2x} = E_{1x} \tag{3.5}$$

By using then the complex dielectric constant of (2.33) and relations (2.34) [Doluchanow, 1956], we arrive at

$$E_{1x} = \frac{E_{1z}}{\sqrt{\epsilon' - j60\sigma\lambda}} \tag{3.6}$$

determining the longitudinal component of the electromagnetic field in the air just above the boundary. It is shifted in phase and its amplitude is very much diminished when compared with that of the vertical component.

Writing

$$\sqrt{\epsilon' - j60\sigma\lambda} = r\,e^{-j\varphi}$$

where

$$r^2 = \sqrt{\epsilon'^2 + (60\sigma\lambda)^2}$$

and using the previously derived expressions, the longitudinal component of the electric field becomes

$$E_{1x} = \frac{E_{01z}}{\sqrt[4]{\epsilon'^2 + (60\delta\lambda)^2}} \cos(\omega t + \varphi/2) \tag{3.7}$$

where

$$\varphi = \tan^{-1}\frac{60\sigma\lambda}{\epsilon'} \tag{3.8}$$

Expression (3.7) shows that the amplitude of the longitudinal component of the electric field depends on the ground constants and the used wavelength.

The relation of the normal components of the electric field above and below the boundary follows from (2.55 c). For a charge-free case then,

$$\epsilon_0 E_{1z} = \epsilon_c E_{2z} \tag{3.9}$$

which, with (2.35), yields

$$\begin{aligned} E_{2z} &= \frac{E_{1z}}{\epsilon' - j(60\sigma\lambda)} \\ &= \frac{E_{01z}}{\sqrt{\epsilon'^2 + (60\sigma\lambda)^2}} \cos(\omega t + \varphi) \end{aligned} \tag{3.10}$$

which is a good approximation of conditions for a plane earth of either a high conductivity or high relative permittivity.

Due to the phase difference between the longitudinal and vertical components of the electric field, the field becomes elliptically polarized in the direction of propagation. Expression (3.8) indicates that the phase shift between the longitudinal and vertical components of the electric field in medium 1 above the boundary tends to zero when $\epsilon' \gg 60\sigma\lambda$ and to $\pi/2$ for the opposite case.

The eccentricity of the ellipse with the usual ground constants is very large, thus permitting the assumption of the electric field's coincidence with the major axis of the ellipse in the direction of propagation. The resulting field is tilted at an angle

$$\psi = \tan^{-1}\frac{E_{1z}}{E_{1x}} = \tan^{-1}\sqrt[4]{\epsilon'^2 + (60\sigma\lambda)^2} \tag{3.11}$$

toward the direction of propagation.

The previously mentioned loss of energy at the boundary of two media may be determined by considering a parallelepiped of unit width, height h, and length dx erected on the boundary [Franke, 1937].

The energy entering into its left-hand side is

$$\mathbf{S} = \mathbf{E}_{1z}\mathbf{H}_{1y}h$$

which with (3.4) becomes

$$\mathbf{S} = -\sqrt{\frac{\epsilon_0}{\mu_0}}\,\mathbf{E}_{1z}^2 h$$

The energy lost through the parallelepiped's base is

$$\mathbf{W} = \mathbf{E}_{1x}\mathbf{H}_{1y}dx$$

which, again with (3.4), may be written as

$$\mathbf{W} = -\sqrt{\frac{\epsilon_0}{\mu_0}}\,\mathbf{E}_{1x}E_{1z}dx$$

The energy loss then is

$$\mathbf{W} = -\mathrm{d}\mathbf{S}$$

or, when performing the indicated differentiation,

$$\frac{\mathrm{d}\mathbf{E}_{1z}}{\mathbf{E}_{1z}} = -\frac{1}{2h}\frac{\mathbf{E}_{1x}}{\mathbf{E}_{1z}}\,dx$$

Then, by using the ratio $\mathbf{E}_{1x}/\mathbf{E}_{1z}$ obtained in (3.11), we obtain the expression

$$E_{1z} = E_0 \exp\left(-\frac{1}{\sqrt[4]{\epsilon'^2 + (60\sigma\lambda)^2}}\frac{d}{2h}\right) \tag{3.12}$$

for the variation of the electric field with distance. The symbol d for distance was substituted for x and h stands for the height of the ionosphere.

This expression is similar to the Austin-Cohen formula, to be given later. It is interesting to note that the first fraction in the exponent, with $\epsilon' = 81$ and $\sigma = 4.0$ mho/m, as applicable to seawater, at a frequency of approximately 290 kHz, has a value of 0.002, a value very close to 0.0015 obtained by Sommerfeld for the attenuation constant α for propagation over the sea, as used in the Austin-Cohen formula.

The exact investigation of ground-wave propagation is rather involved, therefore in practice we resort to the use of published curves and data for field strength computations.

Example 3.1

Determine the tilt of the wavefront at the boundary of air and seawater ($\epsilon' = 81$, $\sigma = 4$ S/m) at a frequency of 300 kHz and the velocity vector component parallel to the boundary.

Solution: From expression (3.11),

$$\begin{aligned}\varphi &= \tan^{-1}\sqrt[4]{\epsilon'^2 + (60\sigma\lambda)^2}\\ &= \tan^{-1}\sqrt[4]{81^2 + (60 \times 4 \times 1000)^2}\\ &= 89.88°\end{aligned}$$

From Figure 3.2,

$$\begin{aligned}v &= \frac{c}{\sin\varphi}\\ &= 3 \times 10^5 + 0.6249 \text{ km/s}\end{aligned}$$

Hence, the velocity of propagation along the surface of the sea is larger than that in a direction perpendicular to the wavefront.

3.2.1 Plane Earth, Both Antennas at the Earth's Surface

An antenna raised to a height of less than a wavelength is said to be at the earth's surface. The daytime coverage of broadcasting stations, which operate mainly in the frequency range of about 530–1620 kHz, relies on surface waves providing field strengths adequate for disturbance-free reception. In the early days of radio, this was the topic of many extensive studies. The results of such studies and their supporting measurements supplied data to verify computations and construct curves permitting the determination of the field-strength of broadcast transmitters radiating waves propagating over various terrains. Due to the high attenuation of these waves, they provide adequate field strength for disturbance-free reception only in the vicinity of the transmitter, at distances over which the earth may be considered a plane. An empirical formula shows this distance to be

$$R_{\text{km}} \approx \frac{80}{\sqrt[3]{f_{\text{MHz}}}} \tag{3.13}$$

Assuming free-space conditions where the relative permittivity ϵ' of the medium in which propagation takes place is unity, the energy density of the field, on the basis of (2.24) and (2.25), as given by the Poynting vector (2.30)

$$\mathbf{S} = \mathbf{E}_z \times \mathbf{H}_y \tag{3.14}$$

becomes

$$W = E_{\text{eff}}^2/120\pi \quad \text{W/m}^2 \tag{3.15}$$

Further, an isotropic point source of power P at the center of a sphere of radius R produces on the sphere's surface a power density

$$W = \frac{P}{4\pi R^2} \quad \text{W/m}^2 \tag{3.16}$$

Practical antennas, however, are not isotropic, but exhibit a directional gain G due to which the above power density in the direction of the increased gain becomes

$$W = \frac{PG}{4\pi R^2} \quad \text{W/m}^2 \tag{3.17}$$

This, together with (3.15), shows that the effective field strength at a distance R from the source of radiation is

$$E_{\text{eff}} = \frac{\sqrt{30PG}}{R} \quad \text{V/m} \tag{3.18}$$

or, when expressing it in the more often used units of mV/m,

$$E_{\text{eff}} = \frac{173\sqrt{P_{\text{kW}}G}}{R_{\text{km}}} \quad \text{mV/m} \tag{3.19}$$

with the source power in kW and the distance in km.

If the radiating element is at the surface of an ideally conducting plane, a condition applicable in the first approximation to the earth's surface, the radiated energy propagates only over a half-sphere, and expression (3.19) is to be multiplied by $\sqrt{2}$ to obtain the required effective value of the electric field intensity. Multiplying (3.19) by the preceding factor and inserting the gain of an elementary dipole, $G = 1.5$ equates the numerical terms of expressions (3.19) and (1.51).

The peak value, due to $E_{\text{eff}} = E_0/\sqrt{2}$, becomes

$$E_0 = \frac{245\sqrt{P_{\text{kW}}G}}{R_{\text{km}}} \quad \text{mV/m} \tag{3.20}$$

In reality, as was mentioned earlier, part of the radiated energy is dissipated as heat in the ground over which propagation takes place. Designating the attenuation factor of the field by the symbol A, the peak value of the field strength in the direction of increased antenna gain is

$$E_0 = \frac{245\sqrt{P_{\text{kW}}G}}{R_{\text{km}}} A \quad \text{mV/m} \tag{3.21}$$

Factor A represents the ground's influence on the surface wave. Many attempts have been made to obtain a usable expression for A.

Sommerfeld's investigations of the field of a vertical dipole located at the surface of a plane, finitely conducting earth show that it consists of two parts represented by the z-components of the Hertz vectors:

$$\begin{aligned} \Pi_1 &= \frac{e^{jk_1R}}{R} + \int_0^\infty f_1(\lambda)J_0(\lambda r)e^{-\sqrt{\lambda^2-k_1^2}z}\lambda d\lambda \\ \Pi_2 &= \frac{e^{jk_2R}}{R} + \int_0^\infty f_2(\lambda)J_0(\lambda r)e^{-\sqrt{\lambda^2-k_2^2}z}\lambda d\lambda \end{aligned} \tag{3.22}$$

having zero radial and azimuthal components. Here, R is the slant range of the point of measurement from the dipole resp. from its image, and k_1 and k_2 are the propagation constants in the air and ground respectively; that is, $k_n^2 = (2\pi/\lambda)^2(\epsilon_n' + j60\sigma_n\lambda)$.

Each of these vectors consists of a space wave representing that part of the radiation of the dipole or image that reaches the receiving antenna directly and a surface wave, given by the integrals considering the medium's effect on the resultant field.

The application of Sommerfeld's results met with difficulties and many attempts were made to reduce them to a more amenable form. Van der Pol [1931] obtained comparatively simple expressions for the vector electric field intensity in the neighborhood and at great distances from an antenna located near a plane earth of finite conductivity and carrying a current of arbitrary distribution. For vertical polarization, these read

$$\mathbf{E} = j\frac{2\pi}{\lambda}\left[\Pi + \frac{\lambda^2}{4\pi^2}\nabla\nabla\Pi \right] \mathrm{e}^{-j\omega t}$$
$$\Pi_z^v = \frac{\mathrm{e}^{j2\pi R_1/\lambda}}{R_1} - \frac{\mathrm{e}^{j2\pi R_2/\lambda}}{R_2} + V \tag{3.23}$$

where R_1 is the slant range above the earth and R_2 the range to the antenna's image. The intensity of the surface wave exceeds that of the space wave near the ground's surface, the opposite applies above the ground.

The term V was expressed by Norton [1937] as

$$V = [(1 - \rho_v)F + 1 + \rho_v]\frac{\mathrm{e}^{j2\pi R_2/\lambda}}{R_2} \tag{3.24}$$

with ρ_v the vertical polarization reflection coefficient of the ground as given by expression (2.65).

The magnitude of

$$F = 1 + j\sqrt{\pi p_1}\,\mathrm{e}^{-p_1}\,\mathrm{erfc}(-j\sqrt{p_1}) \tag{3.25}$$

evaluated at the surface of the ground (i.e., at grazing incidence) when

$$p_1 = p\,\mathrm{e}^{jb}$$
$$p = \frac{\pi R}{\lambda}\frac{1}{\sqrt{(\epsilon' + 1)^2 + (60\sigma\lambda)^2}}$$
$$b \cong \tan^{-1}\frac{\epsilon' + 1}{60\sigma\lambda} \tag{3.26}$$
$$\mathrm{erfc}(-j\sqrt{p_1}) = \frac{2}{\sqrt{\pi}}\int_{-j\sqrt{p_1}}^{\infty}\mathrm{e}^{-x^2}dx = j\frac{2}{\sqrt{\pi}}\int_{+j\infty}^{\sqrt{p_1}}\mathrm{e}^{x^2}dx$$
$$x = 60\sigma\lambda$$

with ϵ' the relative dielectric constant of the ground, σ its conductivity in S/m, and λ the wavelength in m, yields the ground wave attenuation factor

$$A = |F| \tag{3.27}$$

applicable to antennas having enough top loading to cause a uniform current distribution over the vertical portion of the antenna.

The ground-wave attenuation factor, valid only for antennas at ground level, is a function of the frequency and polarization of the radiation, the distance traveled by the wave and the electric characteristics of the ground over which propagation takes place. At short distances from the transmitter, the value of A is very close to unity, a value it reaches at zero distance. This value decreases with increasing distance and frequency.

Van der Pol approximated Norton's attenuation factor by the empirical formula

$$A_{10} \cong \frac{2 + 0.3p}{2 + p + 0.6p^2} \text{ for } b < 5° \tag{3.28}$$

which may be corrected by the empirical extension to read

$$A \cong A_{10} - \sin b \sqrt{p/2}\, e^{-5p/8} \tag{3.29}$$

and allegedly become applicable for all values of b.

The attenuation factor for horizontal polarization may be determined from (3.29) or Figure 3.3 when using the values of

$$\begin{aligned} p &\cong \frac{\pi R}{\lambda} \frac{x}{\cos b'} \\ b' &\cong \tan^{-1} \frac{\epsilon' - 1}{x} \\ b &= 180° - b' \end{aligned} \tag{3.30}$$

Knight and Robson [1984] found that, at some frequencies and values of ground conductivity, the attenuation factor, given by expression (3.28), at distances exceeding 1 km, is between 10 and 20% greater than that obtained from the CCIR curves. Although not seeking a reason for the discrepancy, they propose to compute the attenuation factor for $b = 0°$ from the expression

$$A_0 = \frac{2 + 0.33p}{2 + p + 0.6p^2}$$

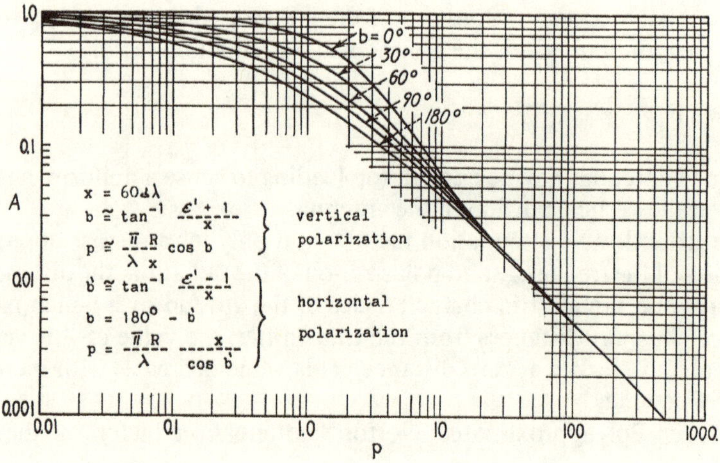

Figure 3.3 Factor A, appearing in (3.21), which takes into account the effect of ground losses of the surface wave (after Terman, © McGraw-Hill, 1943).

and for $b = 90°$ from

$$A_{90} = \frac{2 + 170p}{2 + 210p + 310p^2}$$

Their final value of the factor, determined by

$$A = A_0 - (A_0 - A_{90}) \sin b$$

was found to agree closely with values recommended by the CCIR.

Parameter b is a phase constant, which depends on the power factor of the ground represented as a conductance shunted by a dielectric. At low frequencies, at which the ground appears to be resistive, b is small and the numerical distance becomes independent of the permittivity. At high frequencies, at which the ground has a capacitive character, the phase constant tends to $\pi/2$ rad and the numerical distance is determined mainly by the permittivity. The changeover in the ground's character occurs at a frequency of about 2 MHz.

The value of x at the lower frequencies of the used frequency range is rather large so that the horizontal polarization numerical distance, and hence the attenuation at a given distance R, is very much larger than the corresponding values for vertical polarization. Only vertically polarized surface waves provide usable signal levels at these frequencies; antennas capable of transmitting and receiving vertically polarized waves therefore are used at lower frequencies. Higher frequency surface waves, due to their great attenuation, cannot be used for communication to greater distances. Elevated antennas capable of launching space waves of either polarization are used here exclusively.

Little attenuation occurs while $p < 1$, but it increases rapidly when p exceeds this value. At large numerical distances its value approaches the value of $1/(2p)$ when it becomes independent of b. Beyond about 100 km, it is inversely proportional to distance so that the field strength of the ground wave is inversely proportional to the square of the distance.

Examination of the formulas shows that for long-distance propagation of ground waves at broadcast frequencies high-ground conductivity and longer wavelengths are preferred. At high frequencies the attenuation is reduced by high ground permittivity and by lowering the used frequency.

By redefining Sommerfeld's numerical distance, Norton extended the applicability of the given formulas for the whole frequency range used in radio communication. Note, however, that the formulas are valid only for the determination of the field intensity at the surface of the earth as they are based on the wave function of short antennas at the earth's surface. They yield erroneous results even at short distances when used for elevated antennas. Because Sommerfeld neglected considering the convexity of the earth, field strengths determined by his formulas, at great distances, are much higher than those experienced in practice.

Norton [1936] presented curves for the ground-wave attenuation factor as a function of the "numerical distance" p and the parameter b. The curves are reproduced above in Figure 3.3.

3.3 PHASE AND GROUP VELOCITY

The phase and group velocities, though of little importance in communication, are rather important in the operation of various navigational aids. An expression for the phase velocity may be obtained from the expression for the attenuation factor written as

$$A = |A|e^{-j\varphi} \tag{3.31}$$

where the magnitude $|A|$ is a function of the "numerical distance" given earlier, while its phase is the argument.

The instantaneous value of the phase is [Doluchanow, 1956]

$$\phi = \omega(t - R/c) - \varphi \tag{3.32}$$

with φ, which depends on the ground's parameters, not constant. The total differential of the instantaneous phase with respect to time and distance leads to an expression for the phase velocity. Writing then

$$d\phi(t, R) = \omega\, dt - \frac{\omega}{c}\, dR - \frac{\delta\varphi}{\delta R}\, dR = 0$$

we obtain the phase velocity v_p as

$$\frac{\omega}{\omega/c + \delta\varphi/\delta R} = \frac{dR}{dt} = v_p \tag{3.33}$$

Because $\omega = 2\pi c/\lambda$, and $\delta\varphi/\delta R = (\delta\varphi/\delta p)(\delta p/\delta R)$, this expression becomes

$$v_p = \frac{c}{1 + \dfrac{\delta\varphi}{\delta p}\dfrac{\delta p}{\delta R}\dfrac{\lambda}{2\pi}} \tag{3.34}$$

Differentiating then the expressions for the "numerical distance" p with respect to range, we arrive at

$$\frac{\lambda}{2\pi}\frac{\delta p}{\delta R} = \frac{1}{2}\frac{1}{\sqrt{(\epsilon' + 1)^2 + (60\sigma\lambda)^2}} \tag{3.35}$$

for vertical polarization, and

$$\frac{\lambda}{2\pi}\frac{\delta p}{\delta R} = \frac{1}{2}\sqrt{(\epsilon' - 1)^2 + (60\sigma\lambda)^2} \tag{3.36}$$

for horizontal polarization.

The term $\delta\varphi/\delta p$, the slope of the tangent to the appropriate curve (Figure 3.4) at a point determined by p and b, together with expression (3.35) or (3.36) inserted into (3.34), yields the phase velocity for a known dependence of the argument φ of the complex attenuation factor A on distance. At vertical polarization, for low values of the numerical distance p, the phase velocity is lower than the free-space velocity c, increasing with increasing values of p and tending to the free-space value for $p >$ 10.

Due to the frequency dependence of the phase velocity, the group velocity generally will be different from the phase velocity. It may be computed by considering that maximal energy at a range R will occur when the phase difference of the constituent frequencies is minimal because the energies then add in phase.

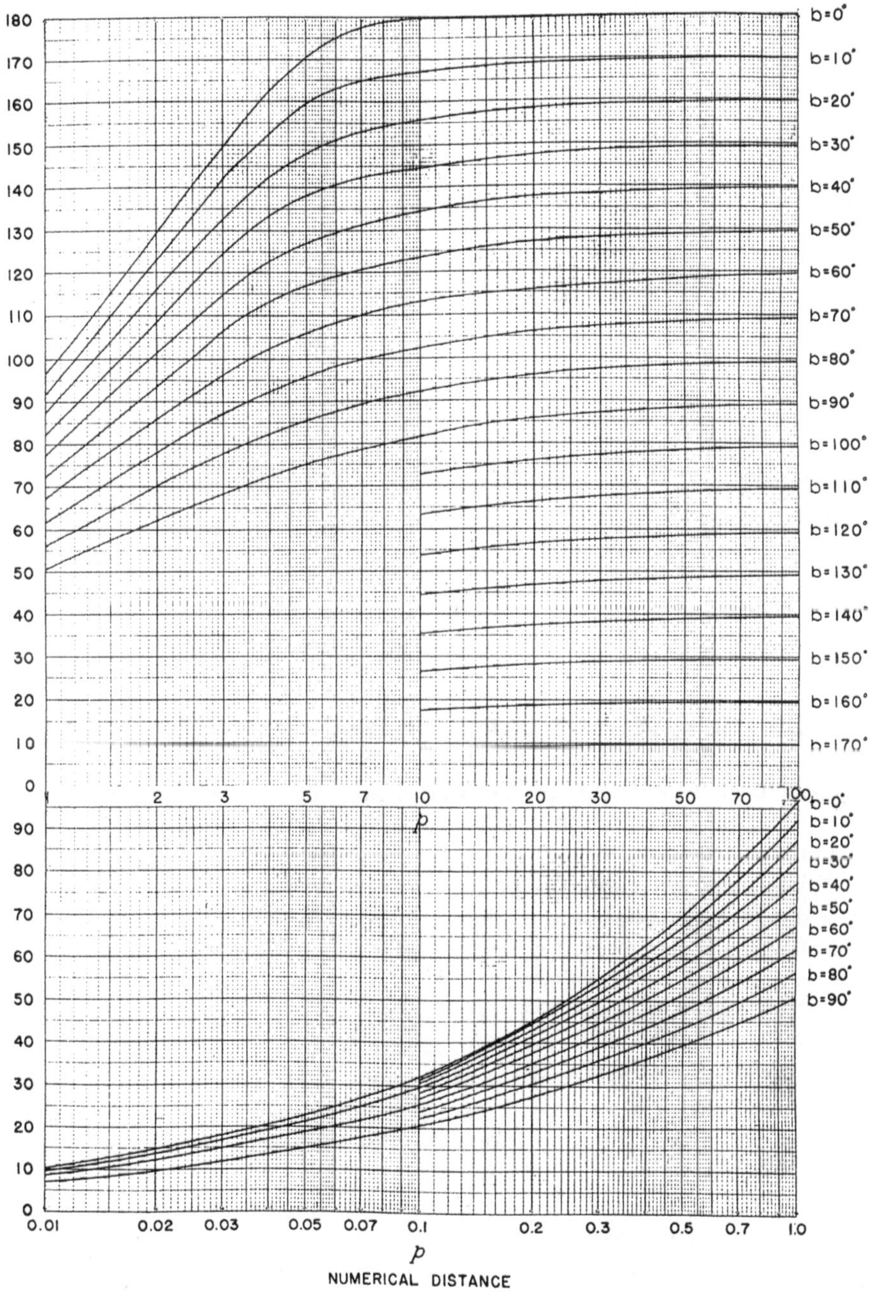

Figure 3.4 The phase of the surface wave attenuation function (after Norton, 1941, © 1941 IRE).

Writing the instantaneous phases of the waves at frequencies ω_1 and ω_2 as

$$\phi_1 = \omega_1(t - R/v_1) - \varphi_1$$
$$\phi_2 = \omega_2(t - R/v_2) - \varphi_2 \tag{3.37}$$

which at the transmitter, where R and t tend to zero, yields for $\varphi_1 = \varphi_2$ a maximum when

$$\phi_1 - \phi_2 = (\omega_1 - \omega_2)t - R(\omega_1/v_1 - \omega_2/v_2) = 0$$

or

$$R = \frac{\omega_1 - \omega_2}{\omega_1/v_1 - \omega_2 v_2}\, t \tag{3.38}$$

This distance, differentiated with respect to time, leads to the group velocity:

$$v_g = \mathrm{d}R/\mathrm{d}t = \frac{\omega_1 - \omega_2}{\omega_1/v_1 - \omega_2/v_2} = \mathrm{d}\omega/\mathrm{d}(\omega/v) \tag{3.39}$$

3.4 SURFACE WAVE OVER GROUND OF VARYING CONSTANTS

The dielectric constant and conductivity of the ground is not constant but varies from place to place. The problem of determining field strength in case of propagation over an inhomogeneous earth has attracted the attention of a number of researchers.

When the total propagation path is over land where the variation of these quantities is not excessive, we compute the attenuation for average values of the ground constants, usually determined by measurement. In composite paths, part of which is over land and part over sea, we compute the field strength at the boundary of the two path parts and adjust the field of the second part at that range by making it equal to the field caused there by propagation over the first part [Eckersley, 1930].

This is illustrated in Figure 3.5, where curve $\overline{AB}$ shows the decay of the field of a transmitter if propagation took place over ground having constants ϵ_1' and σ_1, and $\overline{CD}$ that of the field of the same transmitter for propagation over ground given by ϵ_2' and σ_2. Because the boundary between the two types of ground is at a distance d_1 from the transmitter, curve CD, corresponding to the path over the second type of ground, is displaced so as to intersect curve $\overline{AB}$ at E, at this boundary. The fully drawn curve is an approximation of the transmitter's field decay over the whole path of interest.

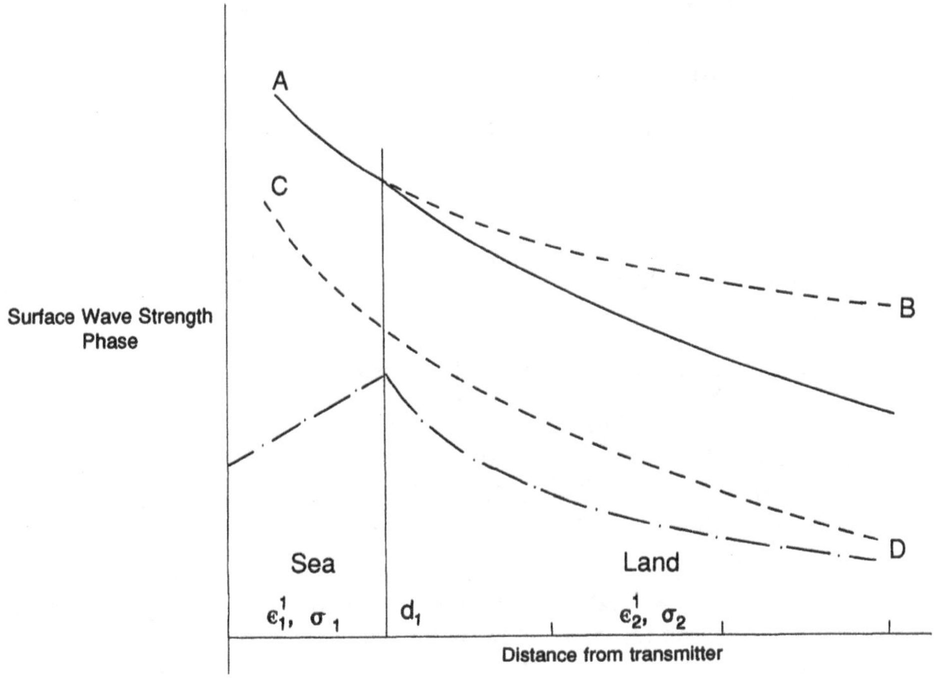

Figure 3.5 Surface wave strength and phase for propagation over two different types of ground.

A closer approximation to values obtained by measurements may be obtained by taking the geometric mean of field strengths computed once for propagation in one direction between the ends of the path of interest and then for propagation in the opposite direction. This method satisfies the reciprocity principle for the propagation of the ground wave over the considered path [Millington, 1949].

Example 3.2

Determine the phase velocity of a 750 kHz vertically polarized transmission over

(a) dry soil ($\epsilon' = 4$, $\sigma = 0.01$ mho/m)
(b) seawater ($\epsilon' = 80$, $\sigma = 4$ mho/m)

at a location 1000 km from the transmitter.

Solution: Expressions (3.34) and (3.35), with $c = 3 \times 10^8$ m/s and $\lambda = 400$ m, will be used. Parameters x, b, and p are from expressions in Figure 3.3.

(a) $x = 60 \times 0.01 \times 400 = 240$

$$b = \tan^{-1}\frac{5}{240} = 1.19°$$

$p = 32.72$

(b) $x = 60 \times 4 \times 400 = 96000$

$$b = \tan^{-1}\frac{81}{96000} = 0.048°$$

$p = 0.0818$

By drawing tangents to the appropriate curves of Figure 3.4 at points given by these parameters we obtain

(a) $$\frac{\partial\varphi}{\partial p} = \frac{180 - 179}{99} = 0.01 \text{ and } \frac{\lambda}{2\pi}\frac{\partial p}{\partial R} = 2.083 \times 10^{-3}$$

yields $v_p = 299.994$ km/s

(b) $$\frac{\partial\varphi}{\partial p} = \frac{38 - 1}{0.99} = 37.37 \text{ and } \frac{\lambda}{2\pi}\frac{\partial p}{\partial R} = 5.208 \times 10^{-6}$$

yields $v_p = 299.942$ km/s

Extending Eckersley's method of determining field strength to the boundary between land and sea, Millington found that a considerable amount of field strength can be recovered before the curve showing the dependence of the field strength on distance from the transmitter acquires the shape characteristic of propagation over sea. He attributes this phenomenon to the large redistribution of energy in the transition region, where apparently energy is fed from higher levels down to the ground thus causing an increase in field strength for some way beyond the boundary.

When propagation is in the opposite direction, that is, from sea to land, a rapid drop instead of an increase in the field strength is observed beyond the boundary before the variation of the field strength with distance assumes the characteristic shape for land propagation. Figure 3.6 illustrates these findings. The sea curve applies only to propagation over a sea, and the land curve is only for propagation over land. The land-sea boundaries for propagation in this direction are at the intersections of the land curve with the rising curves. The rapidly falling curves intersect the sea curve at the sea-land boundary for a sea-to-land propagation direction. It is interesting to note that the reciprocity requirements, a condition for the occurrence of field-strength recovery, are fulfilled at the boundary. Experimental evidence shows that the recovery is more pronounced at lower frequencies.

Wait proposed an analytical approach to the problem by computing the mutual impedance between vertical electric dipoles, one on either side of the boundary, located on the surface of a smooth spherical earth having a position-dependent surface impedance. He reduced a two-dimensional integral equation for the sought field strength to a line integral to the path attenuation function, which in addition to the ground and sea constants depends on the antenna height and uses it in the com-

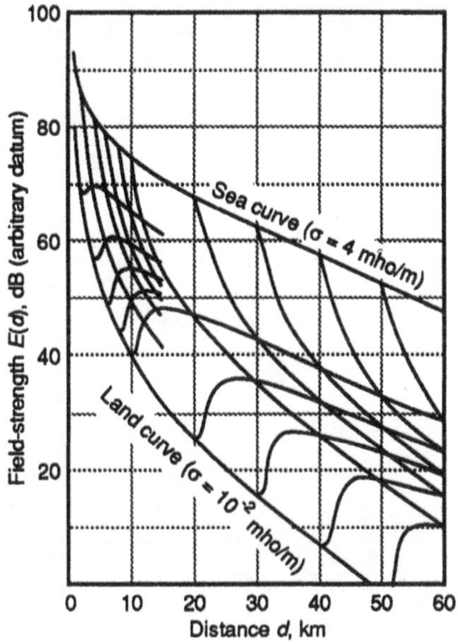

Figure 3.6 Field strength curves for various positions of a land-sea or sea-land boundary, obtained by the reciprocal empirical method; f = 3 MHz (after Millington, 1949. © IEE).

putation of the field strength. The results of the computations are in close agreement with experimental data even as far as the recovery effect beyond the boundary is concerned.

The sudden change of ground constants at propagation over a sea-land boundary causes abrupt disturbances of the wave's phase (Figure 3.5, dot-dashed line) that depends on them. The magnitude of these disturbances varies with frequency and the ground constants on either side of the boundary. The phase changes cause wavefront distortions that in turn cause incorrect bearing indications of some direction finding measurements [Clemmow, 1953; Pressey and Ashwell, 1956; Pressey, Ashwell, and Fowler, 1956]. This phenomenon often is called the *shore effect* or coastal refraction.

The shore effect is most pronounced when the path over land is short in comparison with the total path length. In an idealized case, where the shoreline is cut off sharply below the sea, which is considered to be a perfect conductor, the refraction angle, or the deviation from the true direction, is [Al'pert, 1973]

$$\sigma = \frac{\tan\theta}{2\pi\sqrt{R_L/\lambda}} \operatorname{Im}\left[\frac{j\sqrt{-j}}{\sqrt{\epsilon' - j4\pi\sigma/\omega}}\right]$$

where θ is the angle of incidence of the wave at the sea-land boundary, R_L is the path length over land, and ϵ' and σ are the permittivity and conductivity of the land surface (Figure 3.7).

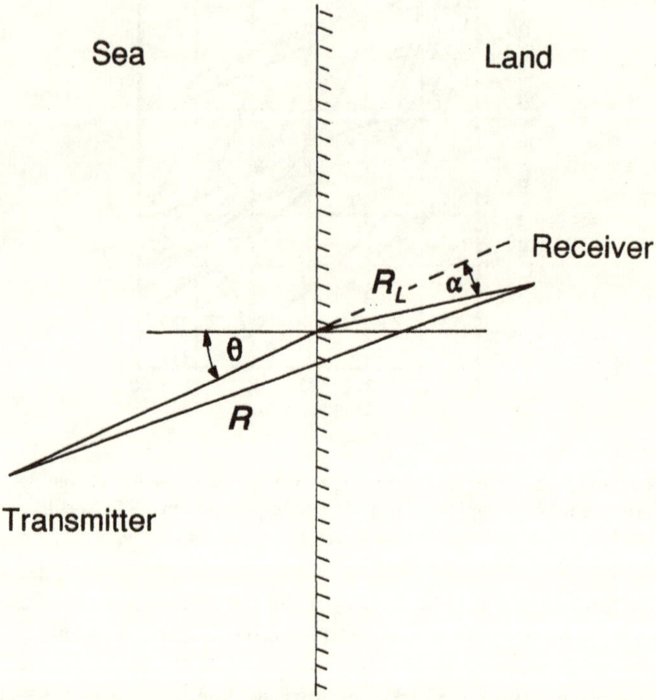

Figure 3.7 Apparent bearing caused by coastal refraction.

3.5 PROPAGATION OF ELECTROMAGNETIC WAVES OVER A SEMICONDUCTING GROUND WHEN BOTH ANTENNAS ARE RAISED

With both antennas elevated, the energy received by the raised receiving antenna is due to rays arriving there from various paths. This usually is referred to as *multipath propagation.*

Assuming that the receiving antenna receives a ray propagating directly from the transmitter and additional rays that arrive after reflection from the ground at point C, and possibly further rays reflected by other objects, like buildings or hills, the resultant field will be the vector sum of all constituent fields (Figure 3.8).

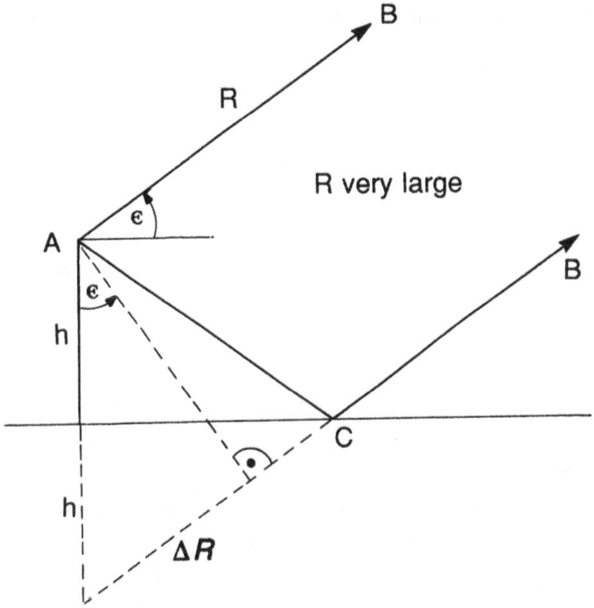

Figure 3.8 Propagation with both antennas elevated.

It is obvious that the fields of the reflected rays will be affected not only in magnitude but also in phase. The resulting field may be expressed then on the basis of (3.19) as

$$\mathbf{E} = \frac{173\sqrt{P_{\mathrm{kW}}G}}{R_{\mathrm{km}}}[1 + \rho e^{j(2\pi/\lambda)\Delta R} + (1 - \rho)Ae^{j(2\pi/\lambda)\Delta R} + \cdots] \tag{3.40}$$

where the first term in the bracket represents the direct ray's contribution, the second one that of the ray reflected from the ground, and the third one, the part due to the rest of rays. The range ΔR is the difference of the path lengths between that of the direct ray and those of the indirect ones; ρ is the magnitude of the complex reflection coefficient given by (2.65), (2.69), (2.70), and (2.71) and arguments μ_v, μ_h, and μ_c. A is the previously defined ground influence factor.

In engineering practice we usually consider only the first two terms, as indicated in Section 2.5 on reflection. The path difference, on the basis of Figure 3.8, is

$$\Delta R = 2h \sin\epsilon \tag{3.41}$$

where ϵ is the ray elevation angle at the transmitting antenna and h is the antenna's height.

The corresponding phase shift between the direct and reflected rays becomes then

$$\psi = \frac{2\pi}{\lambda} \Delta R \tag{3.42}$$

The phase shift μ occurring on reflection can be computed from the complex coefficient of reflection. It must be added to ψ, the phase shift due to path length difference, so that the total phase shift is

$$\xi = \mu + \frac{2\pi}{\lambda} \Delta R \tag{3.43}$$

We showed in Section 2.5.2 that the phase shift on reflection of a horizontally polarized radiation remains very close to π rad for the whole range of grazing angles of interest. For the earlier-mentioned two rays to meet in phase, the phase shift due to path length difference, at horizontal polarization, must equal an odd multiple of π rad. Maxima of the resultant field then occur for

$$2h \sin\epsilon = (2n - 1)\pi \tag{3.44}$$

while minima, at horizontal polarization, occur when the total phase difference is an odd multiple of π rad or the phase shift due to the path length difference added to that occurring on reflection must be an even multiple of π rad; that is, when

$$2h \sin\epsilon = 2n\pi \tag{3.45}$$

Then, as $\lambda/2 = \pi$, the elevation angles ϵ for maxima of the field are given by

$$\epsilon = \sin^{-1}(2n - 1)\frac{\lambda}{4h} \qquad n = 1, 2, \ldots, k \tag{3.46}$$

whereas those for minima by

$$\epsilon = \sin^{-1}\frac{n\lambda}{2h} \qquad n = 1, 2, \ldots, k \tag{3.47}$$

When the elevation angle ϵ varies from 0 to $\pi/2$ rad a lobing pattern is created in the vertical plane containing the antenna.

Evaluating n from (3.47) for $\epsilon = \pi/2$ yields the number of minima as

$$n = 2h/\lambda$$

and the number of maxima, obtained in the same way from (3.46), is the nearest smaller integer to

$$n = \frac{4h + \lambda}{2\lambda}$$

The number of lobes and minima increases while the angles of their directions decrease with an increasing antenna height and frequency. Numbering of the lobes and minima starts usually at the bottom with the first lobe being at an angle ϵ, obtained from expression (3.46) for $n = 1$.

These expressions for the angles of the maxima and minima are not applicable for vertical and circular polarization, because the phase shift on reflection at those polarizations changes with grazing angle variations to a greater extent than that for horizontal one, as shown in Section 2.5. The phase difference of the two rays required for lobe or minimum formation at these polarizations has to be computed for the appropriate phase shifts on reflection.

Figure 3.9 shows the free-space coverage diagram of a rectangular aperture antenna far removed from the ground. The minor lobes are sidelobes due to the sin$x$/x radiation pattern of such an aperture [Rohan, 1983]. Figure 3.10 depicts the lobing pattern caused by the combination of the direct and reflected rays at horizontal polarization, where the magnitude and phase of the reflection coefficient retain a fairly constant value for all grazing angles; the lobing pattern is fully developed. At vertical polarization, where both the magnitude and the phase of the reflection coefficient vary to a great extent, the field due to the reflected ray being not as intense as that at horizontal polarization causes an incomplete development of the lobes and thus gaps under the main lobe and the sidelobes. Further note that the maximum range of the free-space coverage diagram is only about half that of those with lobing as in this case, the two rays meeting in phase reinforce one another.

Example 3.3

Plot the dependence of the elevation angle of the first lobe of the vertical coverage diagram on the antenna height when the 100 MHz radiation is horizontally polarized and the antenna height varies from 1 to 100 m.

Solution: Setting $n = 1$ in (3.4) yields the required data for drawing Figure 3.11.

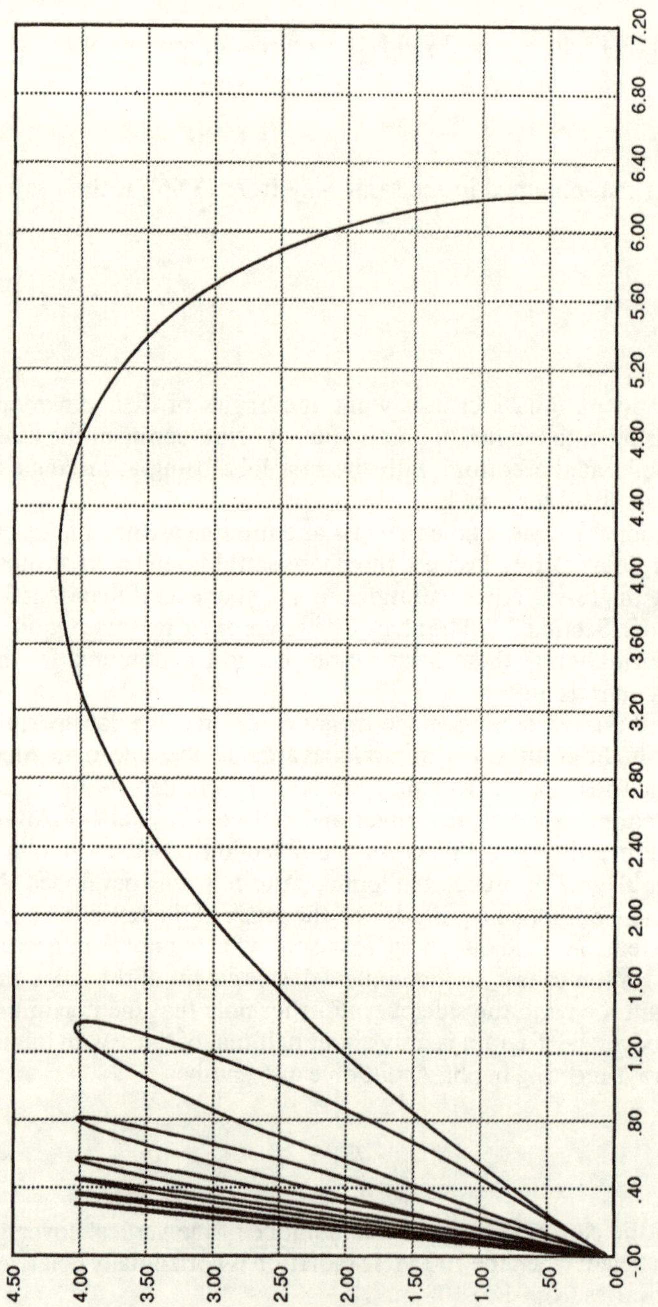

Figure 3.9 Free-space vertical coverage diagram of a rectangular aperture antenna (after Rohan, 1983, © IEE).

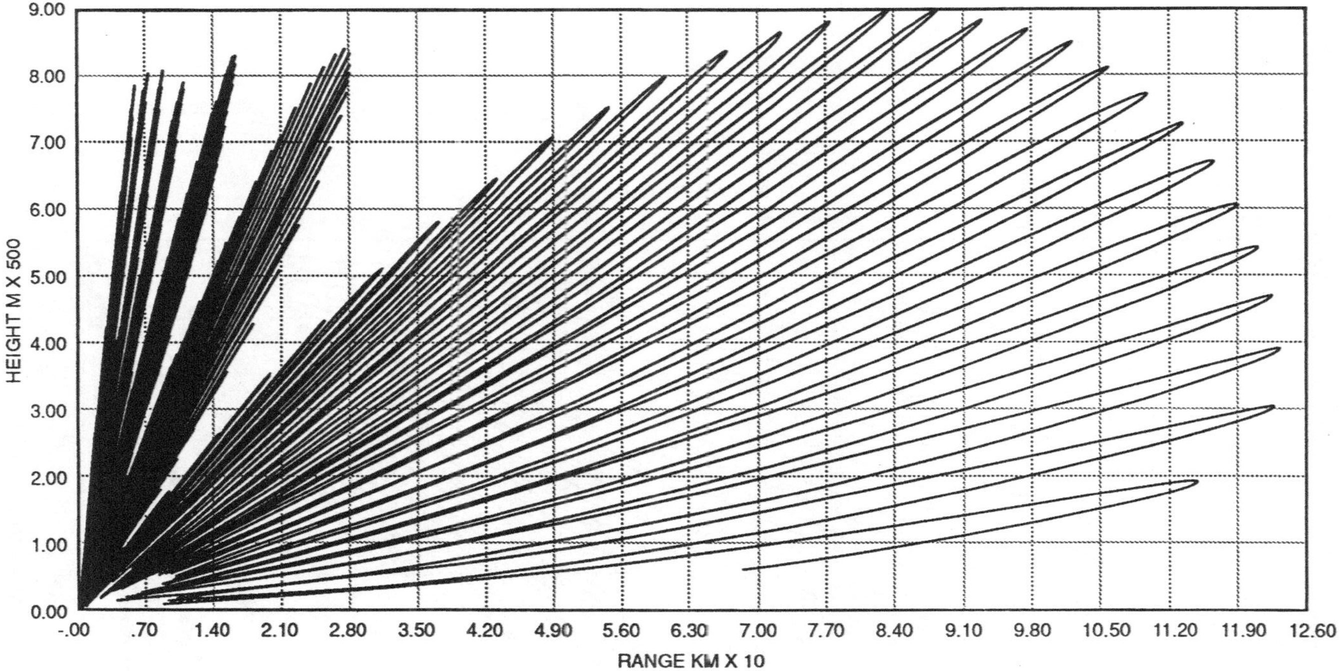

Figure 3.10 The lobing pattern of a rectangular aperture antenna: (a) horizontal polarization (after Rohan, 1983, © IEE).

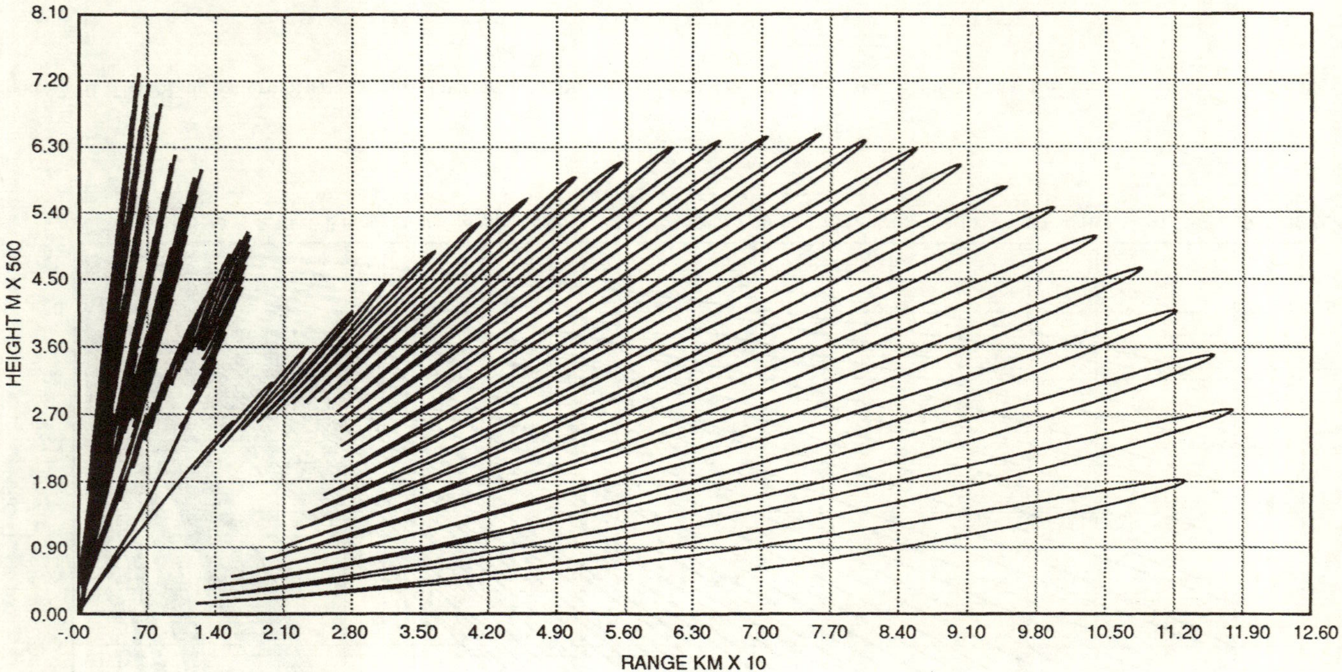

Figure 3.10 (b) vertical polarization (after Rohan, 1983, © IEE).

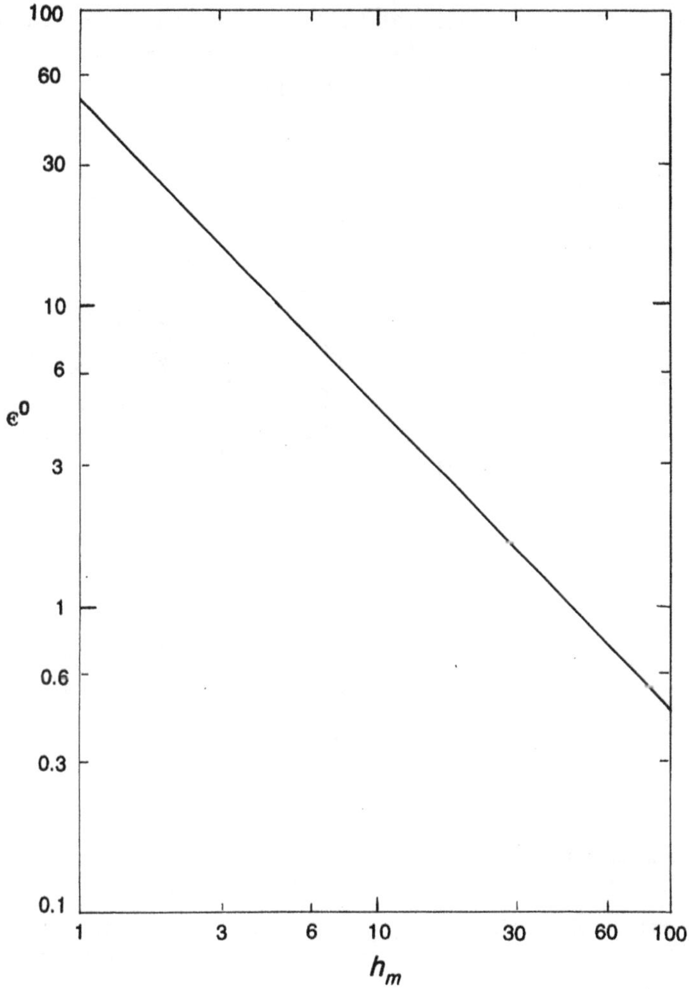

Figure 3.11 Exercise 3.3.

3.6 GROUND-WAVE PROPAGATION OVER A SPHERICAL EARTH

The formulas given so far are applicable strictly to propagation over a plane earth, or to a spherical earth to distances at which the earth may be considered to be plane. Their validity becomes questionable near the optical horizon (Figure 3.12) given by the expression

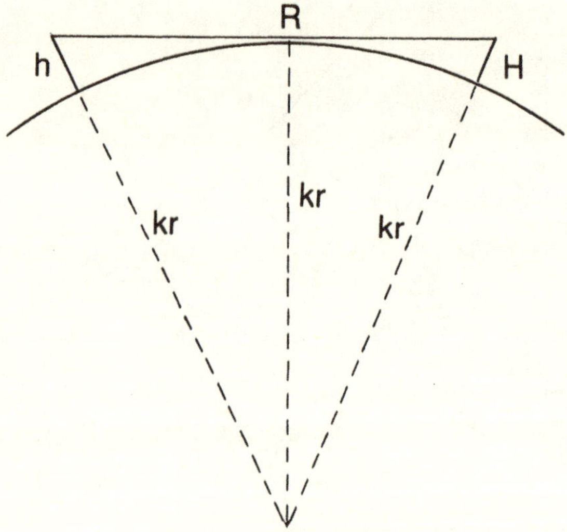

Figure 3.12 The optical horizon.

$$R = \sqrt{2kr}(\sqrt{h} + \sqrt{H}) \tag{3.48}$$

where h and H are the antenna heights, r the radius of the earth, and k is an earth radius correction factor, given by (5.44), the value of which depends on the refractive index of the atmosphere (see Section 5.4.2). For the hitherto most often used atmospheric profile defining a fictitious, so-called standard atmosphere that provides a good approximation over the first few kilometers of the true atmosphere, the value of k is $\frac{4}{3}$, so that the optical horizon for an average earth radius of 6378 km becomes

$$R_{km} = 4.12(\sqrt{h} + \sqrt{H}) \tag{3.49}$$

when the antenna heights are measured in meters.

Thorough investigation of propagation of electromagnetic waves over a spherical earth is based on the pioneering work of Watson [1918], who provided a rigorous theory of the problem, though restricted to an earth of infinite conductivity. Others presented general solutions for other values of the earth's constants and for both horizontal and vertical polarizations, so that problems of propagation over ground of any values of parameters may be solved. Generally, the computations are tedious and too complicated for practical applications.

Van der Pol and Bremmer [1937] extended the plane-earth field intensity formulas by terms considering the shadowing of the antennas by the earth's hump and their elevation to heights exceeding a wavelength. Burrows and Gray [1941] presented formulas given by these authors in a series of graphs and have shown that the field of an antenna for waves propagating over a spherical earth may be expressed by

$$E = 2E_0AF_sG_1G_2 \tag{3.50}$$

where E_0 is the free-space field (3.20) that doubles in magnitude for propagation over a perfectly conducting plane earth at vertical polarization, A is the plane-earth ground influence factor given by (3.28) or (3.29), F_s is the shadow factor accounting for the influence of the earth's hump, and G_1 and G_2 are the height factors considering the influence of raising the antennas. The height factors approach unity for antenna heights below a wavelength.

The shadow factor concept was introduced by Van der Pol and Bremmer [1937]. Its value, given by the ratio of the received field for propagation over a spherical earth to that for propagation over a plane one of the same constants, in addition to the polarization of the wave, depends on the ground constants and hence on the frequency of the propagating wave. Burrows and Gray produced a universal curve of the shadow factor (Figure 3.13 (a), curve 0) as a function of

$$\eta = \zeta_r f(\delta)$$

where

$$\zeta_r = R\sqrt[3]{\frac{2\pi}{\lambda(kr)^2}} \tag{3.51}$$

and, $f(\delta)$, unity for horizontal polarization, for vertical polarization, assumes the values read from the solid lines of Figure 3.13 (b) when

$$\delta = \frac{1}{\epsilon_c'}\left(\frac{2\pi}{\lambda}kr\right)^{2/3} \tag{3.52}$$

with kr as the modified earth radius depending on the atmospheric refractivity (see Section 5.4.2.2).

Because variation in the ground constants spreads the shadow factor curves, those due to the extreme values of the constants are designated by ± 10 in Figure 3.13 (a). Linear interpolation between the curves yields the shadow factor appropriate to the used ground constants.

Figures 3.13 (c) and (d) yield in heavy lines the modulus and phase angle of δ for the following ground constant combinations and a "standard" atmosphere described in Section 5.4.2:

Ground	*Permittivity* ϵ'	*Conductivity* σ *mho/m*
Sea water	80	4
Fresh water	80	5×10^{-3}
Moist soil	30	0.02
Fertile land	15	5×10^{-3}
Rocky ground	7	10^{-3}
Dry soil	4	0.01
Very dry soil	4	10^{-3}

The thin lines of Figure 3.13 (c) show the modulus of δ for some additional values of the ground constants.

Example 3.4

A 2 mV/m field-strength is required at a place 350 km from a 1 MHz transmitter. Propagation for the first 60 km is over sea ($\epsilon' = 80$, $\sigma = 5$ mho/m), then over ground ($\epsilon' = 4$, $\sigma = 10^{-2}$ mho/m). What radiated power will satisfy the requirement?

Solution: According to expression (3.13), a spherical earth is to be considered for the path over ground that is 290 km long. Because both antenna heights are less than a wavelength, the height factors of expression (3.50) are $G_1 = G_2 = 1$.

Expression (3.51) yields $\zeta_r = 1.94$, for which we read $F_s = 0.55$ from Figure 3.13 (a). From (3.25), $\tan b = 2.778 \times 10^{-2}$ and therefore $b = 1.59°$. From (3.24), p

= 16.87, which leads to A = 0.04, as obtained from (3.28). Then, expression (3.50) yields 45.45 mV/m free-space field strength at the boundary of the two path parts.

Expression (3.13) intimates that a plane-earth approximation is acceptable for the path over sea. Here $\tan b = 9 \times 10^{-4}$, $b = 0.05°$, $p = 0.00698$, and $A = 0.99755$. These values inserted into (3.21) yield P = 125 kW for the required radiated power.

The effect of raising either antenna off the ground, at moderate antenna heights, is independent of the distance, curvature of the earth, and height of the other antenna. The height factors G_1 and G_2 become then

$$G = 1 + j\chi \tag{3.53}$$

a

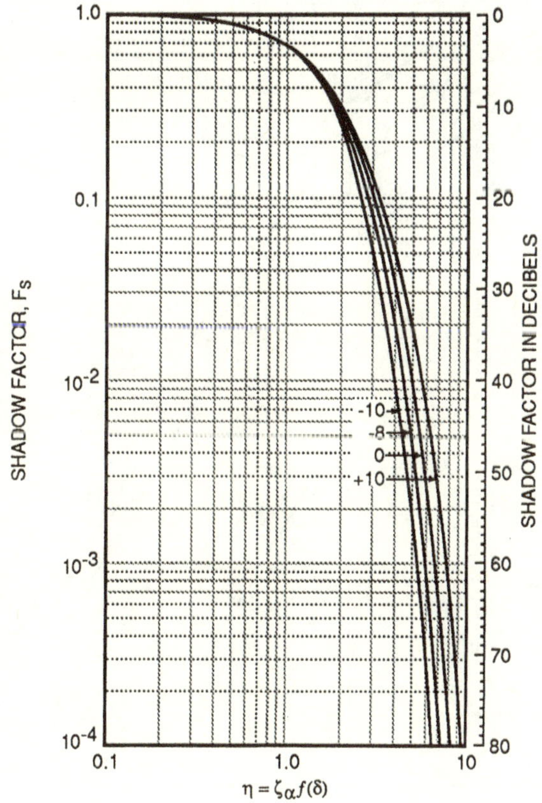

Figure 3.13 (a) The shadow factor as a function of η (after Burrows and Gray, 1941, © 1941 IRE).

b

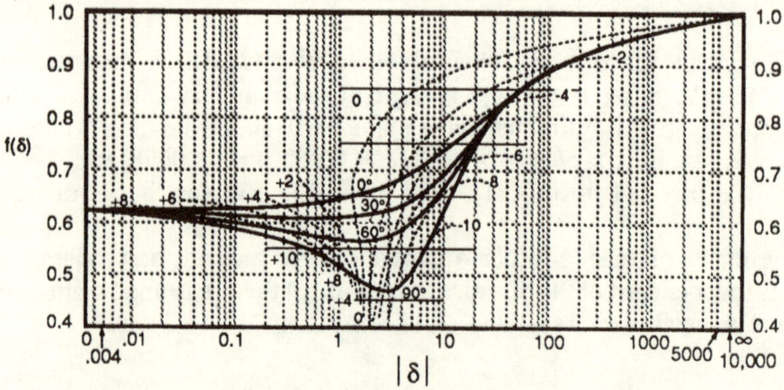

c

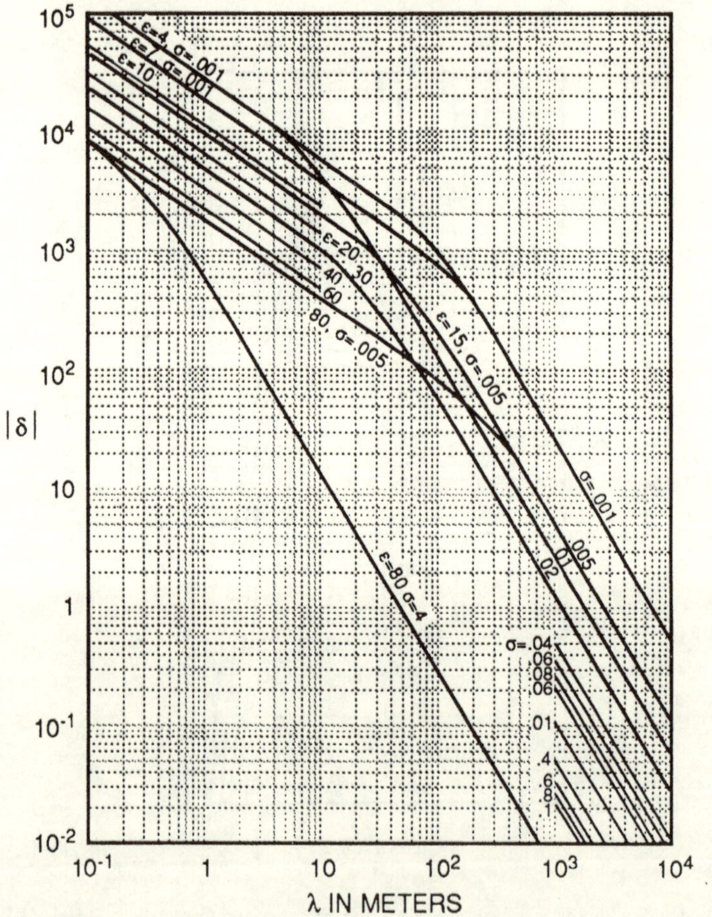

Figure 3.13 (b) the factor $f(\delta)$; (c) the modulus of δ for various ground constants and a "standard" atmosphere (after Burrows and Gray, 1941, © 1941 IRE).

d

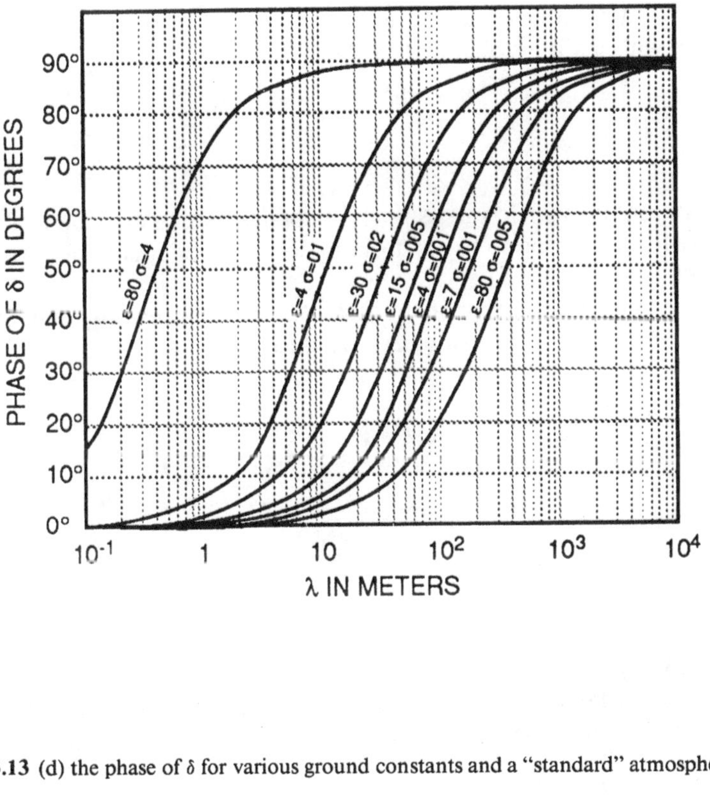

Figure 3.13 (d) the phase of δ for various ground constants and a "standard" atmosphere

where χ for vertical polarization is

$$\chi = \frac{2\pi h}{\lambda} \frac{\sqrt[4]{(\epsilon' - 1)^2 + (60\sigma\lambda)^2}}{\sqrt{\epsilon'^2 + (60\sigma\lambda)^2}} \tag{3.54}$$

$$\chi = \frac{2\pi h}{\lambda} \sqrt[4]{(\epsilon' - 1)^2 + (60\sigma\lambda)^2} \tag{3.55}$$

for the horizontal one. The antenna height h stands for either antenna (h or H).

At medium antenna heights, where the value of the height factor exceeds 20, good agreement with reality is obtained by using

$$E = \frac{2\pi hH}{\lambda R} E_0 F_s \tag{3.56}$$

for the computation of the field intensity. This applies particularly to ultrahigh frequencies, where the field is found to be substantially independent of polarization and of the ground constants.

Raising the antenna has a distinct effect when communicating over long distances. Here, expression (3.56) must be extended to read

$$E = \frac{4\pi hH}{\lambda R} E_0 F_s g_1 g_2 \tag{3.57}$$

with

$$g_{1,2} = \frac{G_{1,2}}{\chi} \tag{3.58}$$

and χ given by (3.54) or (3.55), depending on the polarization.

If one antenna is kept low while the other one is raised so that it approaches the line of sight of the low antenna, the field is computed by using the formula

$$E = 2E_0 G_1 L(\delta) \left(\frac{\chi_0}{\chi_r}\right)^{1/4} F_L \tag{3.59}$$

with G_1 the height factor of the low antenna,

$$\chi_0 = \frac{\zeta_r^2}{2} \tag{3.60}$$

$$\chi_r = \sqrt[3]{\left(\frac{2\pi}{\lambda}\right)^2 \frac{h^3}{kr}} \tag{3.61}$$

The ground parameter $L(\delta)$ for vertical polarization is given by Figure 3.14, and that for horizontal polarization may be computed from

$$L(\delta) = \frac{1}{\sqrt[4]{(\epsilon' - 1)^2 + (60\sigma\lambda)^2}\left(\frac{2\pi kr}{\lambda}\right)^{1/3}} \tag{3.62}$$

The function F_L *versus*

$$L = \sqrt{g(\delta)}\,[2\chi_0 - 2\chi_r] \tag{3.63}$$

is plotted in Figure 3.15, and the parameter $\sqrt{g(\delta)}$ is in Figure 3.16.

When the antenna heights are such that the line of sight is not interrupted by the earth's hump, a situation similar to that in Figure 3.17 arises. There are two propagation paths between the antennas. The length of the direct path shall be designated by R, and the path of the one that reaches the other antenna after a reflection from the ground will be called R_1.

This indirect path consists of two branches, X and Y, given by

$$X = \sqrt{h^2 + krA\alpha^2} \tag{3.64}$$

$$Y = \sqrt{H^2 + krB\beta^2} \tag{3.65}$$

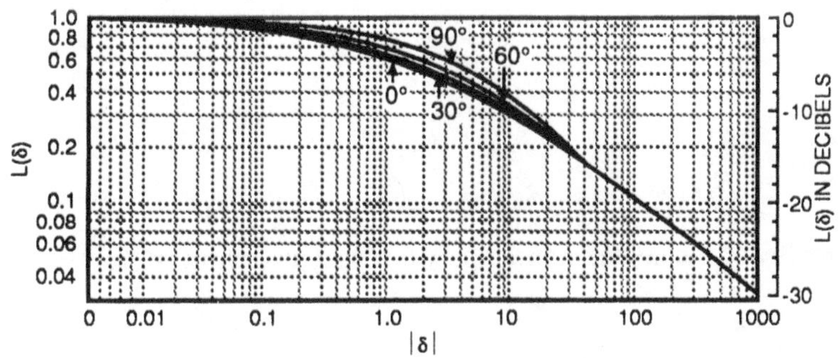

Figure 3.14 The ground parameter L(δ) for vertical polarization as used in expression (3.59) (after Burrows and Gray, 1941, © 1941 IRE).

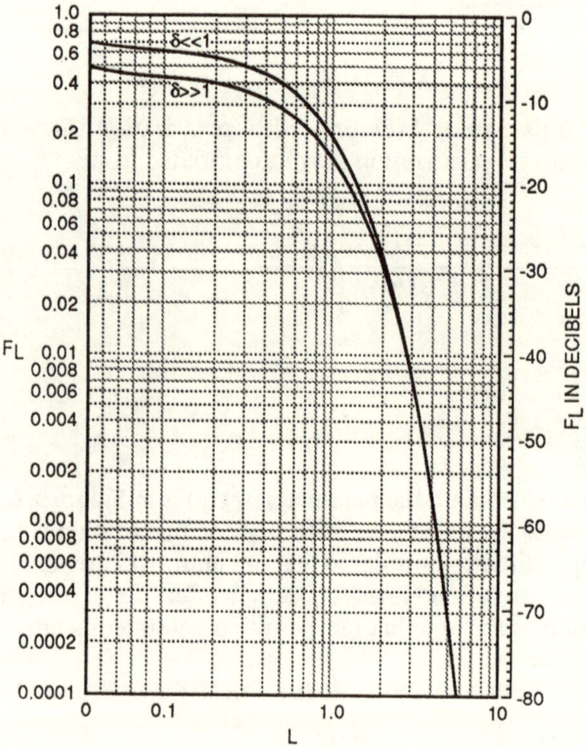

Figure 3.15 The height-distance factor F_L of expression (3.59) as a function of L given by (3.63) (after Burrows and Gray, 1941, © 1941 IRE).

where

$$
\begin{aligned}
A &= kr + h \\
B &= kr + H
\end{aligned} \tag{3.66}
$$

and the smaller of the angles α and β may be computed from [Kerr, 1951]

$$
\begin{aligned}
\beta &= \frac{\xi}{2} + \frac{p}{kr} \cos \frac{\theta + \pi}{3} \\
\xi &= \sqrt{\frac{R^2 - (H - h)^2}{AB}}
\end{aligned} \tag{3.67}
$$

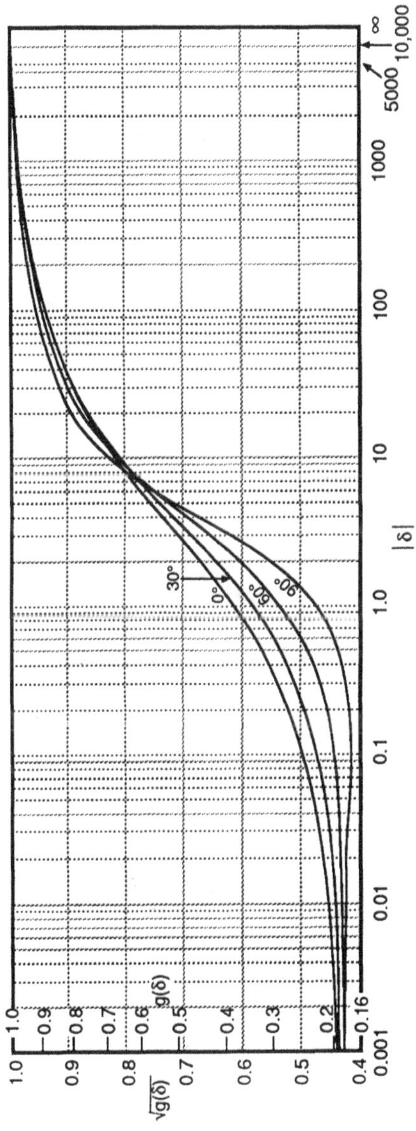

Figure 3.16 The parameter $g(\delta)$ of expression (3.63) as a function of δ given by (3.52) (after Burrows and Gray, 1941, © 1941 IRE).

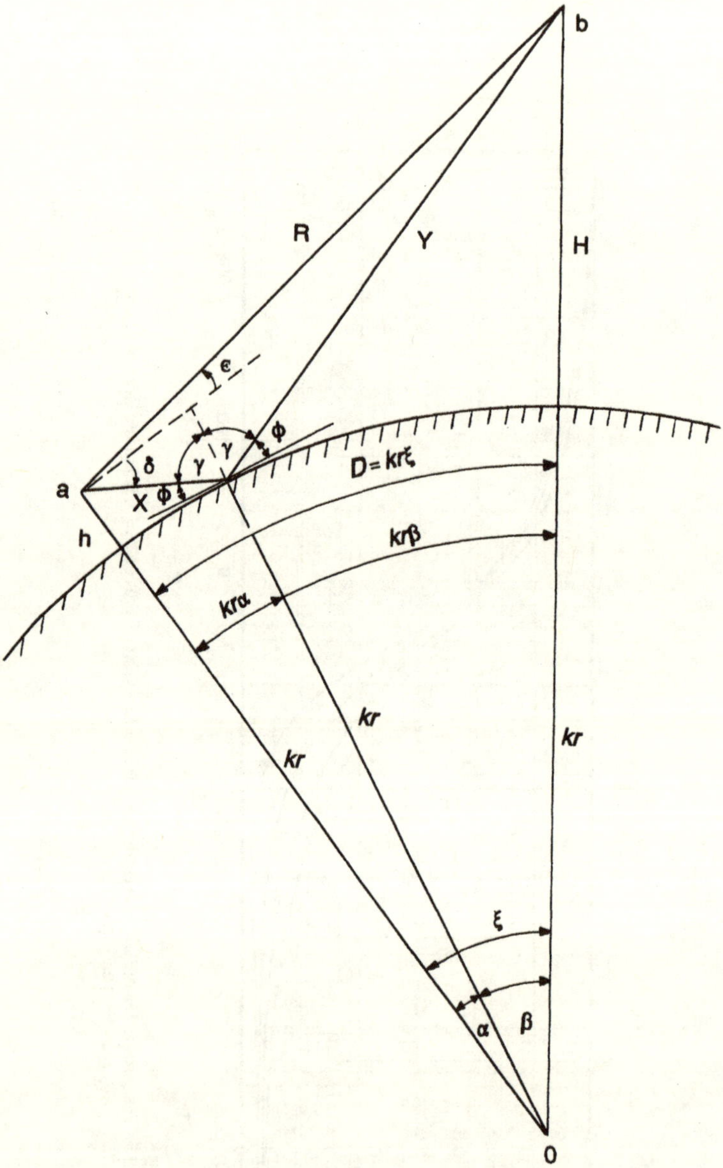

Figure 3.17 The considered geometry.

$$p = \frac{2}{\sqrt{3}} \sqrt{kr(h + H) + (kr\xi/2)^2}$$

$$\theta = \cos^{-1} \frac{2(kr)^2\xi(H - h)}{p^3}$$

The other angle then becomes

$$\alpha = \xi - \beta \tag{3.68}$$

The angle at which the indirect ray leaves the antenna is

$$\delta = \cos^{-1} \frac{kr\alpha}{X} \tag{3.69}$$

so that the grazing angle, the complement of the angle of incidence, becomes

$$\phi = |\delta - \alpha| \tag{3.70}$$

The resultant field E is the vector sum of the field E_1 propagating directly between the antennas and the field E_2 propagating over the other path. On reflection from the ground, this field is affected by the permeability, permittivity, and convexity of the ground as described in Chapter 2. Then,

$$E_1 = \frac{E_0}{R} e^{-jR2\pi/\lambda} \tag{3.71}$$

$$E_2 = \frac{E_0}{R_1} \rho_s \rho_{v,h} e^{-j\mu_{v,h}} D \, e^{-jR_1 2\pi/\lambda} \tag{3.72}$$

and the resultant field then becomes

$$E = \sqrt{E_1^2 + E_2^2} \tag{3.73}$$

3.7 FIELD-STRENGTH–DISTANCE CURVES

To ease the tedium of field-strength computations, the CCIR (Comité Consultatif des Radiocommunications, Geneva) published curves showing the variation of the field of a 1 kW transmitter with distance for propagation over ground of various constants. These curves are based on earth radius corrections appropriate to the fictitious "standard atmosphere" having a linear atmospheric refractive index variation with height.

Bean and Dutton [1966] found, however, that an exponential atmospheric profile in which the refractive index variation with height is described by

$$n(h) = 1 + 315 \times 10^{-6} \exp(-1.361 \times 10^{-4} h) \tag{3.74}$$

when h, the height above ground, is measured in meters, yields a closer representation of average atmospheric conditions than does the "standard atmosphere." The greatest deviation of the results from the true conditions obtained for the linear atmospheric profile occurs at heights above 1 km at all frequencies and below this height at frequencies below about 10 MHz.

The curves in Figure 3.18 computed for a spherical earth and an exponential atmosphere of the preceding refractive index variation by Rotheram for a 1-kW radiated power by a Hertzian vertical electric dipole of $5\lambda/2\pi$ A m dipole moment, which yields a 0.15 V/m field intensity at a distance of 1 km in free space or 0.3 V/m field intensity at the same distance on the surface of a perfectly conducting plane. The dashed straight line is the inverse distance curve showing the value $0.3/R$ V/m, representing the decay of the field strength of the above source placed on a perfectly conducting flat earth, with the distance R.

Rotheram expressed the electromagnetic field of a vertical electric dipole in terms of a scalar Hertz potential U as

$$E = -\frac{\lambda}{2\pi n^2} \nabla \times \nabla \times U\mathbf{r}$$

where n is the refractive index of the troposphere, and $\mathbf{r}$ the radius vector from the center of the earth. The Hertz potential, derived from Maxwell's equations in an integral form, led to practical field-strength calculations over the curved earth's surface.

Three different methods of computation were used for obtaining the curves. Up to nine terms of an appropriate residue series were summed at the largest shown distances. For terminals at or near the ground, this series converges to a distance of about $10\lambda^{1/3}$, with the wavelength in meters, whereas for elevated antennas the convergence extends to just within the radio horizon, close to the vicinity of the last interference maximum of the geometrical optics field. The extended form of Sommerfeld's flat earth theory, which converges out to distances of about $15\lambda^{1/3}$, was used at short distances. Methods of geometrical optics were used for elevated antennas within the radio horizon; that is, at regions covered neither by the residue series nor the extended flat earth theory. These methods were found to be inadequate for regions near the radio horizon or where one antenna is high and the other is close to the ground.

Using the derived mathematical expressions, Rotheram wrote a computer program for plotting the shown curves. Although the program is capable of considering

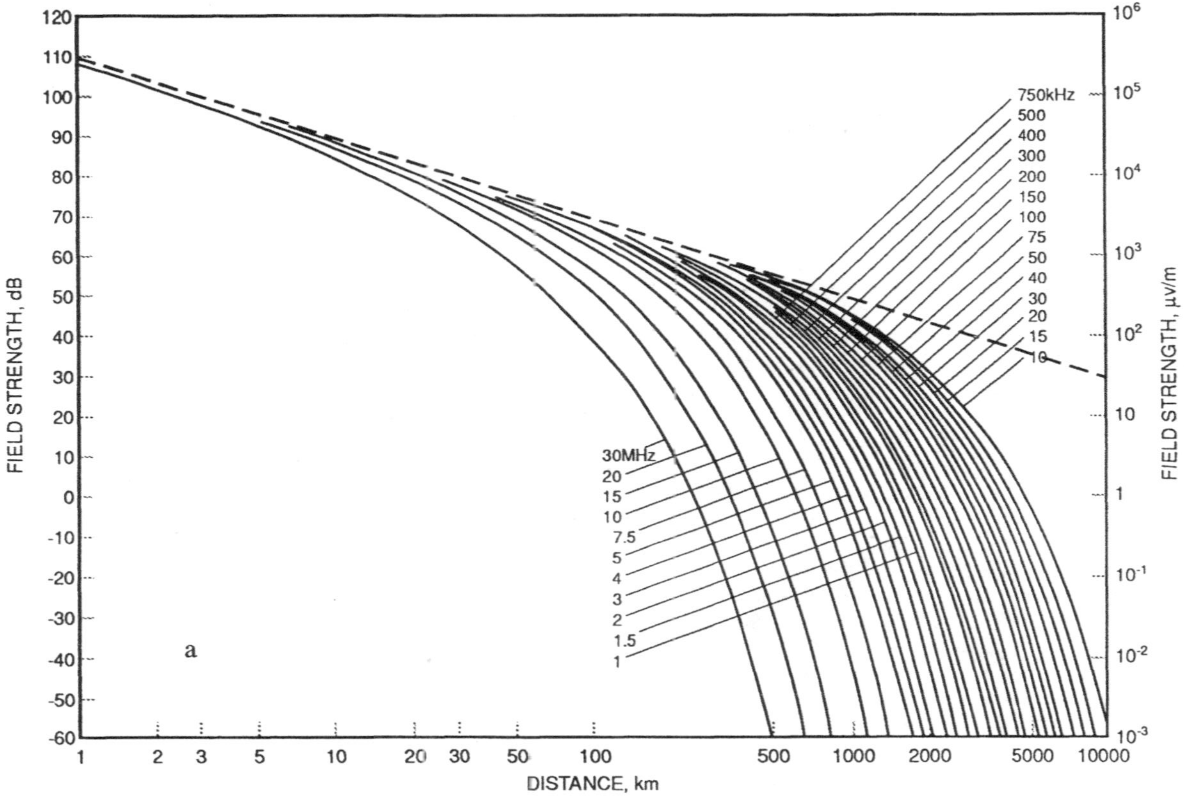

Figure 3.18 Field-strength–distance curves: (a) $\epsilon = 80$, $c = 5$ S/m; (b) $\epsilon = 4$, $\sigma = 3 \times 10^{-2}$ S/m; (c) $\epsilon = 4$, $\sigma = 10^{-2}$ S/m; (d) $\epsilon = 4$, $\sigma = 3 \times 10^{-3}$ S/m; (e) $\epsilon = 4$, $\sigma = 10^{-3}$ S/m; (f) $\epsilon = 4$, $\sigma = 3 \times 10^{-4}$ S/m; (g) $\epsilon = 4$, $\sigma = 10^{-4}$ S/m; (h) $\epsilon = 4$, $\sigma = 3 \times 10^{-5}$ S/m; (i) $\epsilon = 4$, $\sigma = 10^{-5}$ S/m (after Rotheram, 1981, © IEE).

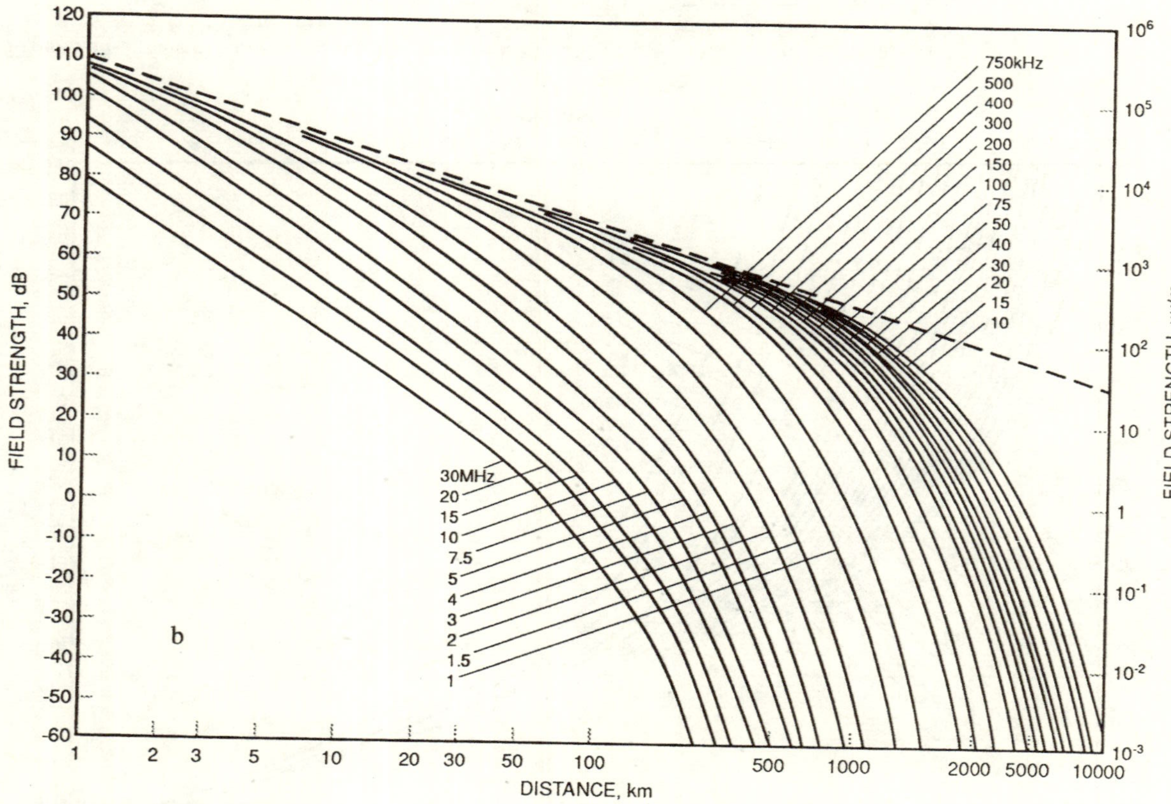

Figure 3.18 continued

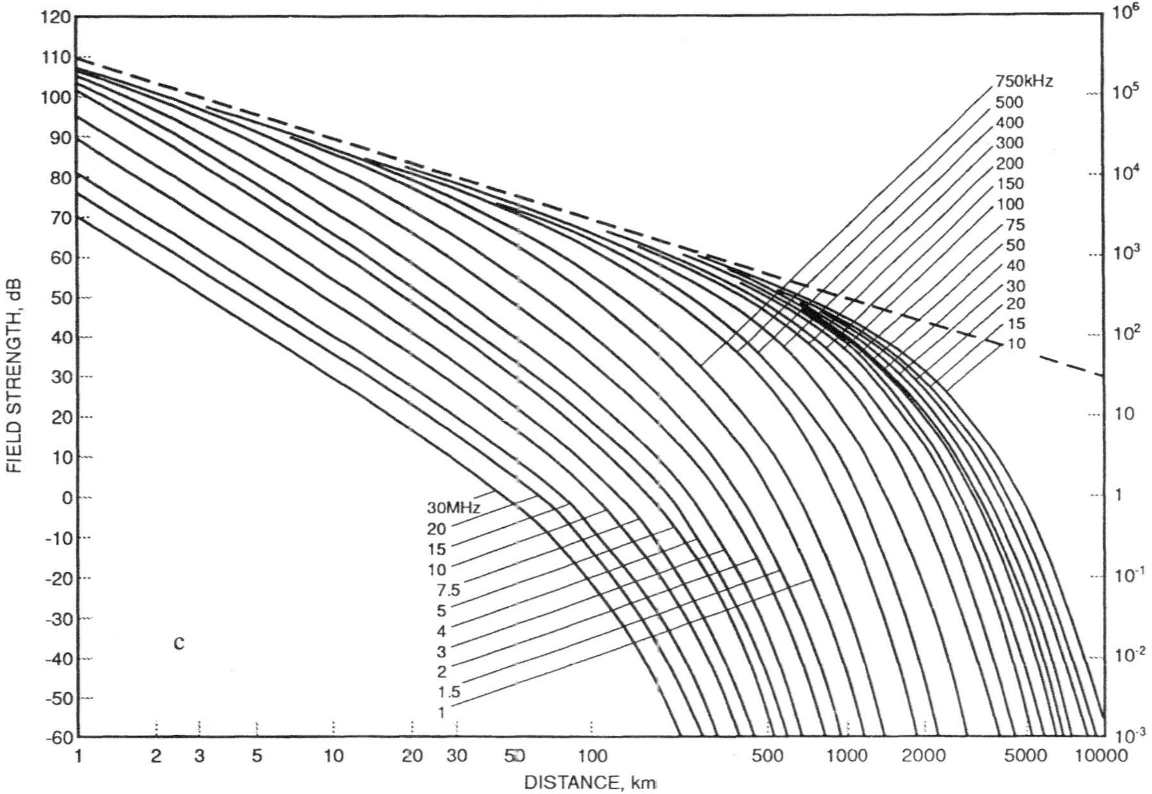

Figure 3.18 continued

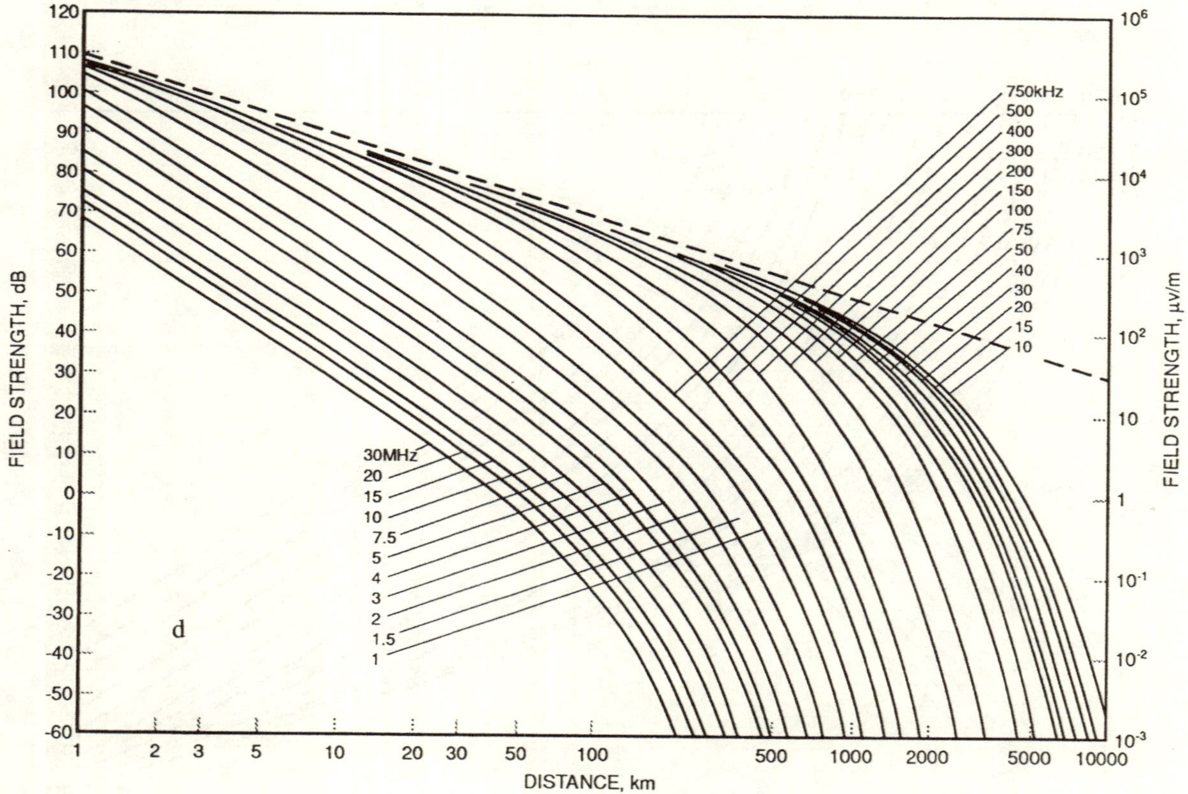

Figure 3.18 continued

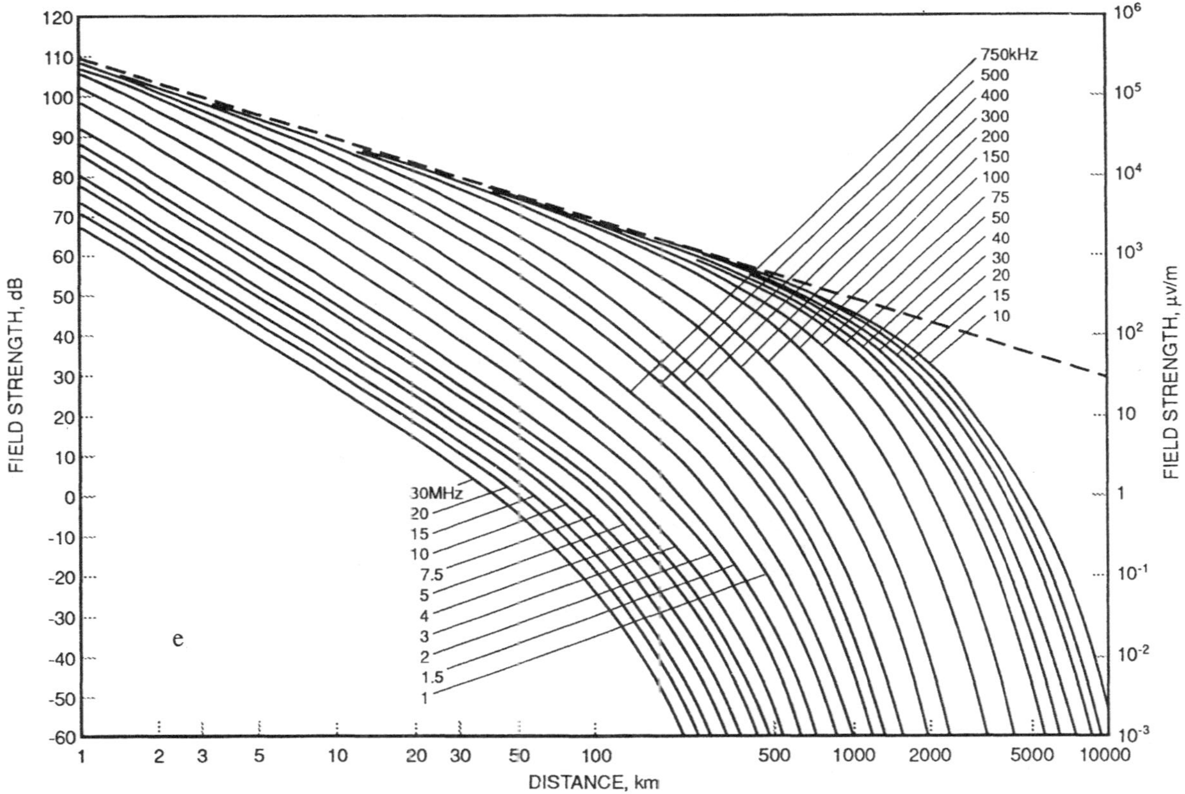

Figure 3.18 continued

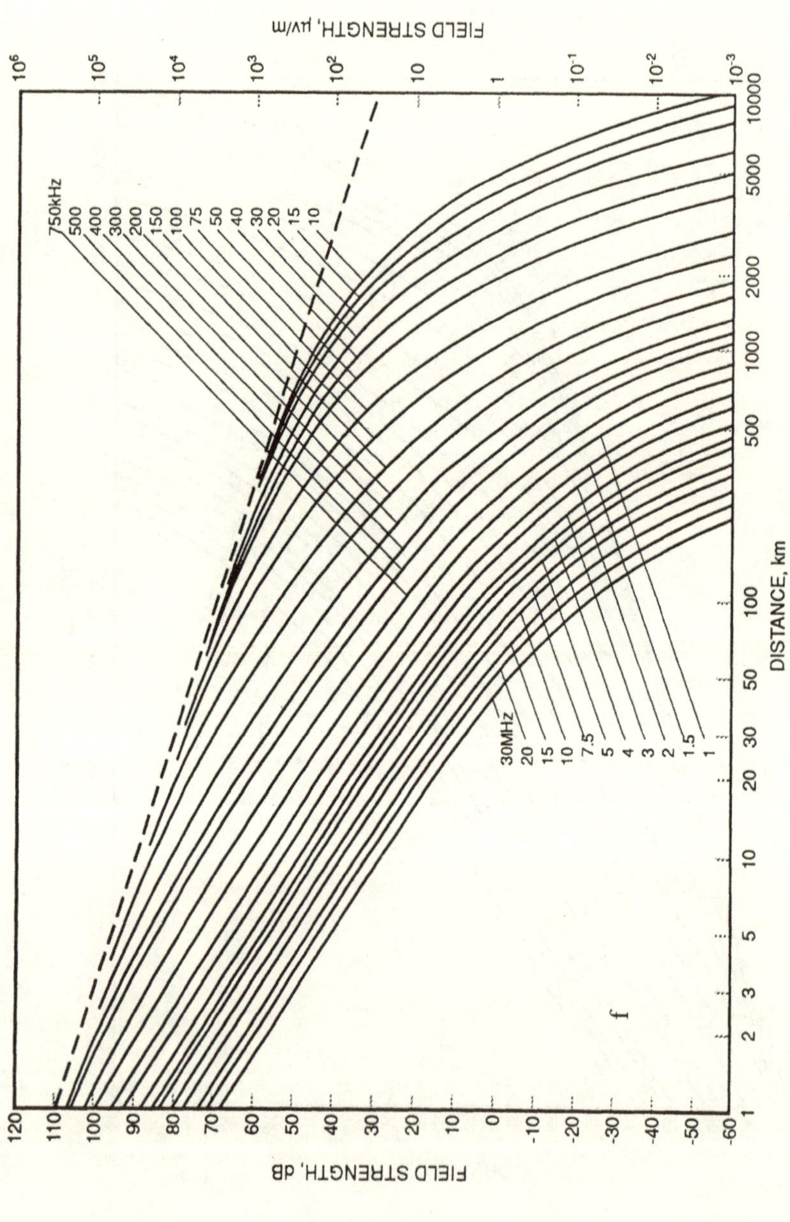

Figure 3.18 continued

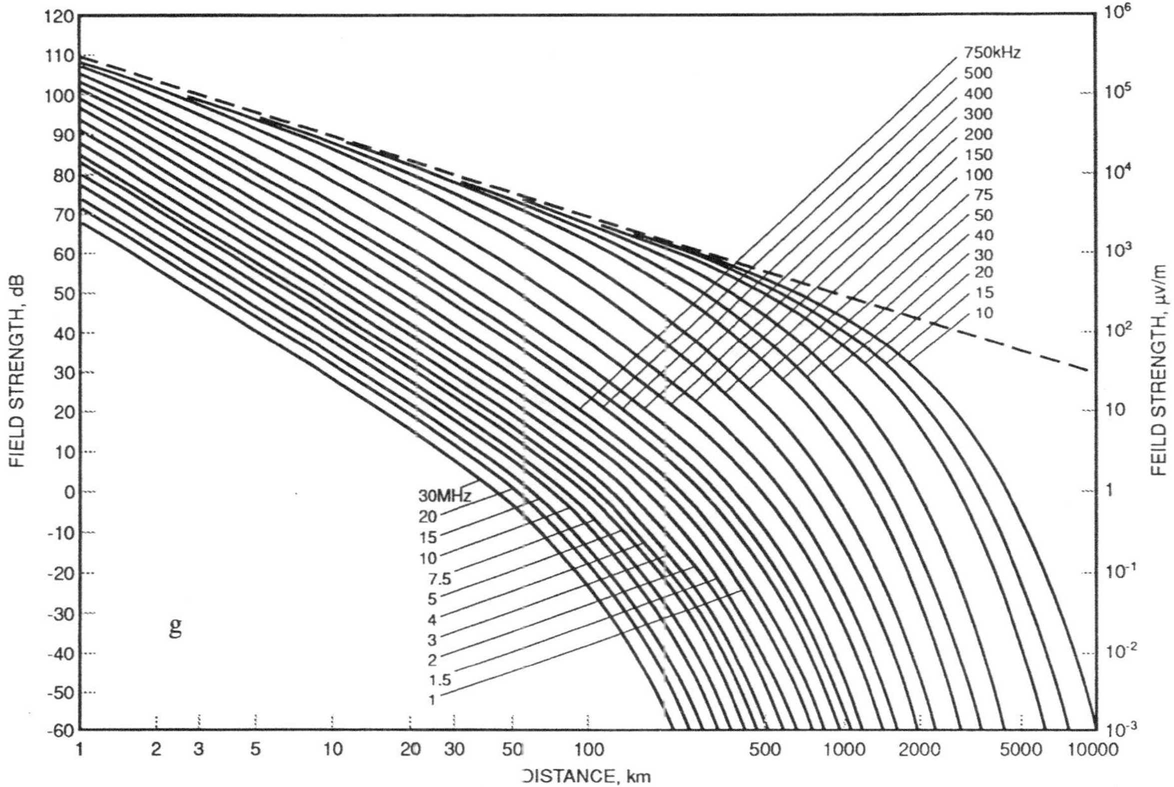

Figure 3.18 continued

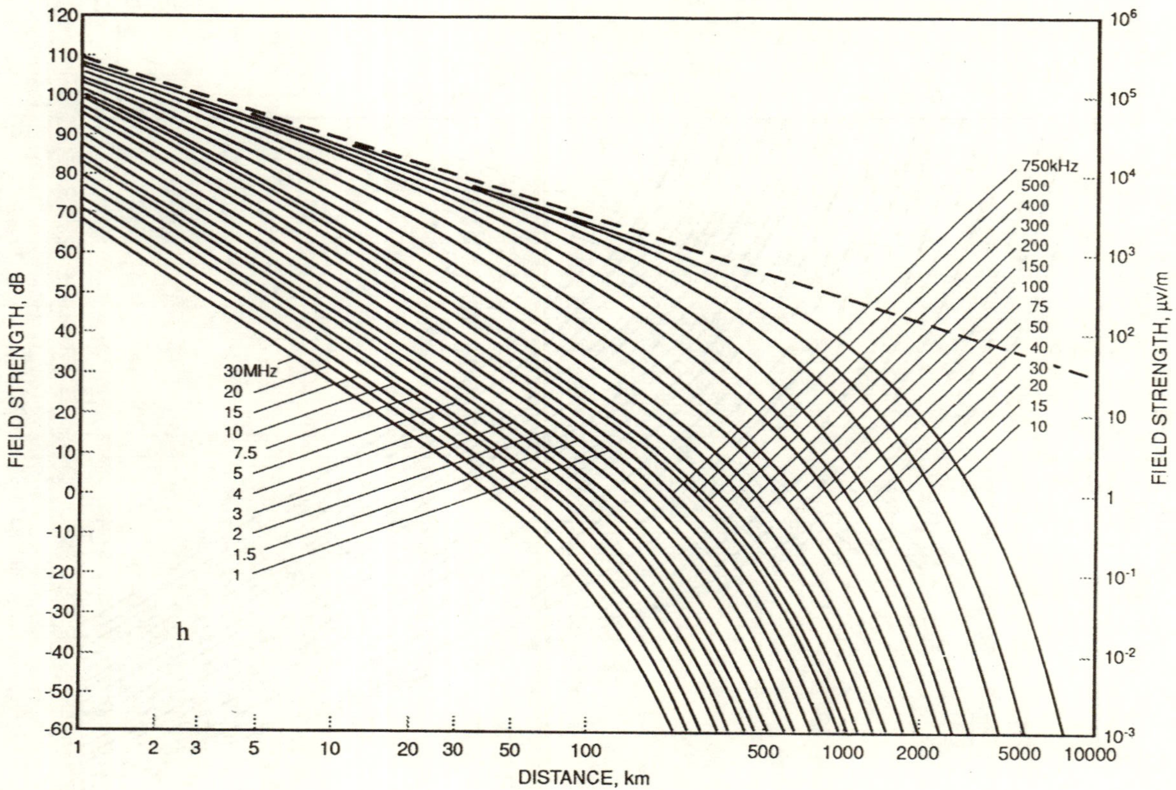

Figure 3.18 continued

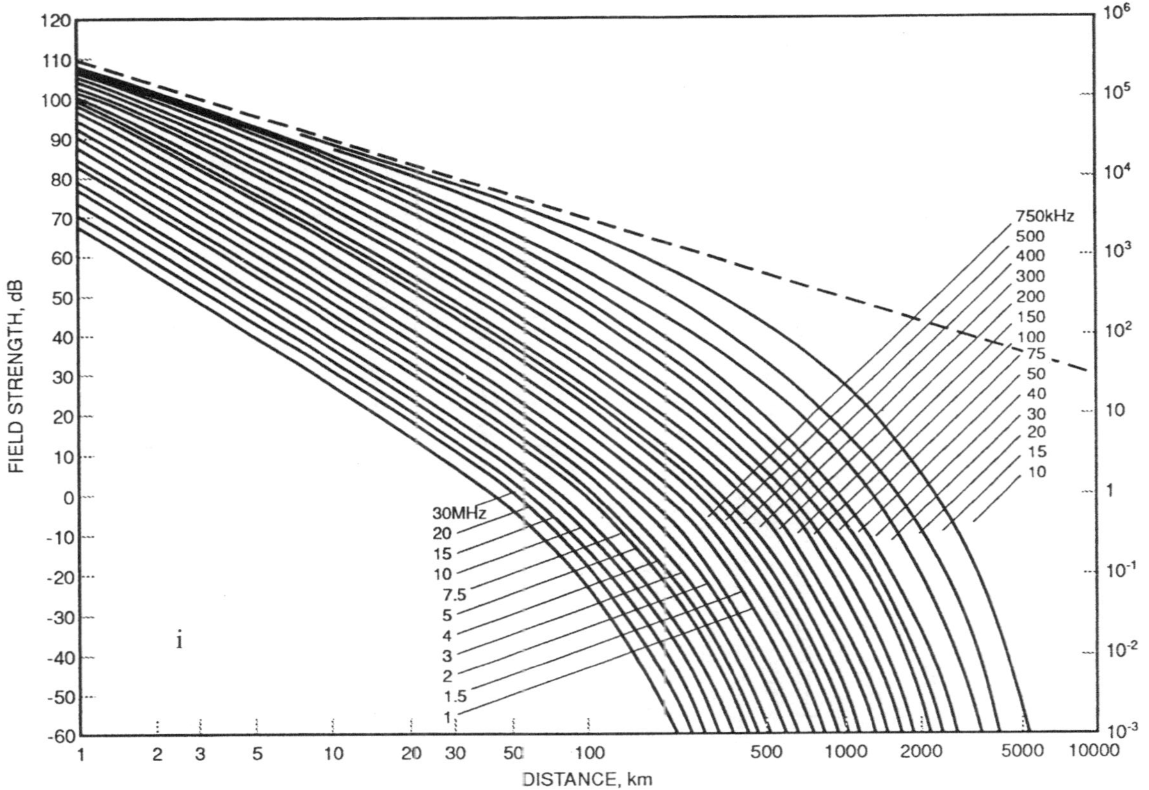

Figure 3.18 continued

the effect of raised antennas, the values of field intensity read from the curves apply only to terminals on the surface of the earth and should be corrected by the height factors given earlier when the antenna heights exceed a wavelength. This is practical only at higher frequencies.

Beyond the distances to which the curvature of the earth may be neglected, the decay of the field strength becomes more pronounced. The curvature of the earth affects higher frequency fields at shorter distances from the transmitter than it does those at lower ones. The predicted field strength must be multiplied by the square root of the used transmitted power in kW, if it differs from the 1 kW power used for the computation of the curves.

These curves, as mentioned earlier, are for vertical polarization. Horizontal polarization that yields too low a field strength for practical applications, especially over grounds of high conductivity, is used seldom below the high frequency range.

The curves show the lowest field intensity values due to ground-wave propagation where the ground wave is considered to be that portion of the wave received on the earth's surface without having been affected by the conducting regions of the upper atmosphere. Sky waves begin to be of importance during daytime at ranges at which diffraction starts to influence the ground wave. At greater distances from the transmitter, the sky wave may be much stronger than the predicted ground wave, during both the day and night.

Example 3.5

Using the appropriate field-strength–distance curve, determine the field-strength of a 300 km distant 100 kW, 1 MHz transmitter when propagation is over the sea.

Solution: From Figure 3.18 (a), a 1 kW transmitter at 300 km produces a field strength $E = 398\ \mu V/m$. A 100 kW transmitter at the same distance yields a field strength

$$\begin{aligned} E &= 398\sqrt{100} \\ &= 3.98 \text{ mV/m}. \end{aligned}$$

PROBLEMS

3.1 Plot the frequency dependence of the wavefront's tilt angle for the frequency range of 3 to 300 kHz for propagation over dry soil.

3.2 Using expression (3.18) determine the frequency dependence of the field strength at a distance of 50 km for propagation over moist soil and a 100 km ionospheric height.

3.3 Determine the angle of tilt and the eccentricity of the ellipse representing the wavefront of a 1-km long wave propagating over rocky soil at a distance where the vertical component of the wave is 5 mV.

3.4 Using expressions (3.12) and (3.20) determine the plane-earth field strength of a 50 kW transmitter operating at a frequency of 800 kHz at a distance of 100 km when the antenna has a 10 dB power gain, propagation takes place over ground with $\epsilon' = 4$, $\sigma = 3 \times 10^{-3}$ mho/m, and the ionospheric height is 100 km.

3.5 Compare the result of problem 3.4 with that obtained from expression (3.21) for vertical polarization.

3.6 Using the result of problem 3.5, compute the attenuation factor from expression (3.29) and compare the obtained field strength with that computed in the preceding problem.

3.7 Compute the phase velocity of a horizontally polarized wave of 400 m length traveling over dry ground at a distance of 1 km from the transmitter.

3.8 Determine the total phase difference between the direct and reflected rays, after reflection, radiated by a horizontally polarized 10 m high antenna at a frequency of 10 GHz at an angle of 10°. Calculate the number of lobes of the vertical coverage diagram.

3.9 What frequency change will shift the twelfth lobe of a horizontally polarized radiation at 10 GHz to ten times its original elevation when the antenna height is 10 m?

3.10 The angle of the lobes of a vertical coverage diagram is affected by two parameters. Identify them and determine their necessary change for doubling the elevation angle of the first lobe when the 10 GHz radiation is from a 10-m high antenna.

3.11 Two 20-m high antennas are 30 km apart at sea. Find the grazing angle at which the reflected ray intercepts the earth's surface. Determine the magnitude and phase angle of the reflection coefficient at vertical polarization when $f = 3$ GHz.

3.12 What power radiated by an antenna with a power gain of 1.76 dB at 500 kHz will produce a 40 mV/m field strength at a distance of 100 km when propagation is over seawater on a spherical earth? The antenna heights are 150 and 10 m, respectively.

3.13 What power will raise the field strength at a point in space by 6 dB?

3.14 Using the appropriate curve of Figure 3.18 (a) determine the power required to produce a field strength of 10 μV/m at 1 MHz at a distance of 150 km.

REFERENCES AND RECOMMENDED READING

Al'pert, Y.L. (1973) *Radio Wave Propagation in the Ionosphere,* Vol. 2, *Propagation of Electromagnetic Waves near the Earth,* Consultants Bureau, New York and London.

Altshuler, E.E., V.J. Falcone, and K.N. Wulfsberg (1968) "Atmospheric Effects on Propagation at Millimeter Wavelengths," *IEEE Spectrum,* Vol. 5, No. 7, July, pp. 83–90.

Bean, B.R., and E.J. Dutton (1966) *Radio Meteorology,* N.B.S. Monograph No. 92, Washington, DC.

Bonnet, G. (1983) "Au-delà d'une vitesse de groupe: vitesse d'onde et vitesse de signal. Première partie: l'opérateur vitesse de groupe en l'absence d'affaiblissement," *Ann. Télécom.,* Vol. 38, No. 9–10, September–October, pp. 345–366.

Bonnet, G. (1983) "Au-delà d'une vitesse groupe: vitesse d'onde et vitesse de signal. Deuxième partie: déformation de l'amplitude et influence de l'affaiblissement." *Ann. Télécom.,* Vol. 38, No. 11–12, November–December, pp. 471–487.

Bremmer, H. (1949) *Terrestrial Radio Waves,* Elsevier Publishing Co., Amsterdam.

Bullington, K. (1957) "Radio Propagation Fundamentals," *BSTJ,* Vol. 36, No. 3, May.

Burrows, C.R. (1937a) "Radio Propagation over Plane Earth-Field Strength Curves," *BSTJ,* Vol. 16, January and October.

Burrows, C.R. (1937b) "The Surface Wave in Radio Propagation over Plane Earth," *Proc. IRE,* Vol. 25, No. 2, February.

Burrows, C.R., and M.C. Gray (1941) "The Effect of the Earth's Curvature on Ground-Wave Propagation," *Proc. IRE,* Vol. 29, No. 1, January.

C.C.I.R. (1963) "Ground wave propagation curves for frequencies below 10 MHz," *Recommendation* 368.

Clemmow, P.C. (1953) "Radio Propagation over a Flat Earth across a Boundary Separating Two Different Media," *Phil. Trans. Royal Scoiety,* London, Vol. 246A, pp. 1–53.

David, P., and J. Voge (1969) *Propagation of Waves,* Pergamon Press, London.

Doluchanow, H.P. (1956) *Die Ausbreitung von Funkwellen,* VEB Verlag Technik, Berlin.

Dobrott, D.R. (1966) "Propagation of vlf Waves Past a Coastline," *Radio Science,* Vol. 1, No. 12, December, pp. 1411–1424.

Eckersley, P.P. (1930) "The Calculation of the Service Area of Broadcast Stations," *Proc. IRE,* Vol. 18, No. 7, July, pp. 1160–1193.

Franke, O. (1937) *Einfuehrung in die physikalischen Grundlagen der Rundfunktechnik,* Springer Verlag, Vienna.

Furutsu, K. (1963) "On the Theory of Radio Wave Propagation over Inhomogeneous Earth," *J. of Research N.B.S.,* Vol. 67D, No. 1, January–February.

Gerks, I.H., and R.M. Anderson (1964) "Propagation over Plane Earth through an Exponential Atmosphere," *J. Research N.B.S.,* Vol. 68D, No. 11, November, pp. 1193–1199.

Hansen, P. (1977) "Measurements of Basic Transmission Loss for h.f. Ground Wave Propagation over Seawater, *Radio Science,* Vol. 12, No. 3, May–June, pp. 397–404.

Kerr, D.E. (1951) *Propagation of Short Radio Waves,* M.I.T. Radiation Laboratory Series, Vol. 13, McGraw-Hill, New York.

Knight, P., and J.A.W. Robson (1984) "Empirical Formula for Groundwave Field-Strength Calculation," *Electronics Letters,* Vol. 20, No. 18, 30 August, pp. 740–742.

Jordan, E.C. (1953) *Electromagnetic Waves and Radiating Systems,* Constable & Co., London.

Millington, G. (1939) "The Diffraction of Wireless Waves Round the Earth," *Phil. Magazine,* Vol. 27.

Millington, G. (1949) "Ground Wave Propagation over an Inhomogeneous Smooth Earth," *Proc. I.E.E.,* Vol. 96, Pt. 3, January, pp. 53–64.

Millington, G. (1971) "Short-Distance Groundwave Propagation," *Electronics Letters,* Vol. 7, No. 3, 11 February, pp. 80–82.

Norton, K.A. (1935) "Propagation of Radio Waves over a Plane Earth," *Nature,* No. 135, June 8.

Norton, K.A. (1936) "The Propagation of Radio Waves over the Surface of the Earth and in the Upper Atmosphere," *Proc. IRE,* Vol. 24, No. 10, October.

Norton, K.A. (1937) "The Physical Reality of Space and Surface Waves in the Radiation Field of Radio Antennas," *Proc. IRE,* Vol. 25, No. 9, September.

Norton, K.A. (1941) "The Calculation of Ground-Wave Field Intensity over a Finitely Conducting Spherical Earth," *Proc. IRE,* Vol. 29, No. 12, December.

Ott, R.H., L.E. Vogler, and G.A. Hufford (1979) "Ground-Wave Propagation over Irregular Inhomogeneous Terrain: Comparison of Calculations and Measurements," *IEEE Trans. on Antennas and Propagation,* Vol. AP-27, No. 2, March, pp. 284–286.

Pressey, B.G., and G.E. Ashwell (1956) "The Deviation of Low-Frequency Ground Waves at a Coastline," *Proc. IEE,* Vol. 103B, July, pp. 535–541.

Pressey, B.G., G.E. Ashwell, and C.S. Fowler (1956) "Change of Phase with Distance of a Low-Frequency Ground Wave Propagating across a Coastline," *Proc. IEE,* Vol. 103B, July, pp. 527–534.

Rohan, P. (1962) "Radio Wave Propagation from the Frequency and Time Standardization Point of View." W.R.E. Tech. Memo. E.I.D. 127, August.

Rohan, P. (1983) *Surveillance Radar Performance Prediction,* Peter Peregrinus, London.

Rotheram, S. (1970) "Ground Wave Propagation at Medium and Low Frequencies," *Electronics Letters,* Vol. 6, No. 25, 10 December, pp. 794–795.

Rotheram, S. (1981a) "Ground-Wave Propagation. I. Theory for Short Distances," *Proc. IEE,* Part F, Vol. 128, No. 5, October, pp. 275–284.

Rotheram, S. (1981b) "Ground-Wave Propagation. II. Theory of Medium and Long Distances and Reference Propagation Curves," *Proceedings IEE,* Part F, Vol. 128, No. 5, October, pp. 285–295.

Sommerfeld, A. (1909) "Ueber die Ausbreitung der Wellen in der drahtlosen Telegraphie," *Annalen der Physik,* Vol. 28, No. 4, March, pp. 665–736.

Stratton, J.A. (1941) *Electromagnetic Theory,* McGraw-Hill, New York.

Terman, F.E. (1943) *Radio Engineers' Handbook,* McGraw-Hill, New York.

Van der Pol, B. (1931) "Ueber die Ausbreitung elektromagnetischer Wellen," Jahrbuch der drahtlosen Telegraphie und Telephonie, *Zeitschrift fuer Hochfrequenztechnik,* Vol. 37, No. 4.

Van der Pol, B., and H. Bremmer (1937) "The Diffraction of Electromagnetic Waves from an Electrical Point Source Round a Finitely Conducting Sphere, with Application to Radiotelegraphy and the Theory of Rainbow," *Phil. Magazine,* Vol. 24, Pt. 1, pp. 141–176, and Pt. 2, pp. 825–864.

Van der Pol, B., and H. Bremmer (1938a) "Ergaebnisse einer Theorie ueber die Fortpflanzung elektromagnetischer Wellen ueber eine Kugel endlicher Leitfaehigkeit," *Hochfrequenztechnik und Elektroakustik,* Vol. 51.

Van der Pol, B., and H. Bremmer (1938b) "The Propagation of Radio Waves over a Finitely Conducting Spherical Earth," *Phil. Magazine,* Vol. 25.

Van der Pol, B., and H. Bremmer (1939) "Further Note on the Propagation of Radio Waves over a Finitely Conducting Spherical Earth," *Phil. Magazine,* Vol. 27.

Wait, J.R. (1960) "Terrestrial Propagation of Very-Low Frequency Radio Waves," *J. of Research N.B.S.,* Vol. 64D, No. 2, March/April.

Wait, J.R. (1961) "On the Theory of Mixed-Path Ground-Wave Propagation on a Spherical Earth," *J. Research N.B.S.,* Vol. 65D, No. 4, July–August, pp. 401–410.

Wait, J.R. (1963) "Oblique Propagation of Ground Waves across a Coastline I," *J. Research N.B.S.,* Vol. 67D, No. 6, November–December, pp. 617–624.

Wait, J.R. (1974) "Recent Analytical Investigations of Electromagnetic Ground Wave Propagation over Inhomogeneous Earth Models," *Proc. IEEE,* Vol. 62, No. 8, August, pp. 1061–1072.

Wait, J.R., and C.M. Jackson (1963) "Oblique Propagation of Ground Waves across a Coastline II," *J. Research N.B.S.,* Vol. 67D, No. 6, November–December, pp. 625–630.

Watson, G.N. (1918) "The Diffraction of Electric Waves by the Earth," *Proc. Royal Society,* Vol. 95A.

Zenneck, J. (1907) "Ueber die Fortpflanzung ebener elektromagnetischer Wellen laengs einer ebenen Leiterflaeche und ihre Beziehung zur drahtlosen Telegraphie," *Annalen der Physik,* Vol. 23, September, pp. 846–866.

Chapter 4
SKY-WAVE PROPAGATION

An antenna, as was shown earlier, radiates in directions other than those causing ground-wave propagation. If the elevation angle at launch is such as to make the wave travel away from the earth, a sky wave ensues. At the early stages of radio communication it was expected that these waves would leave the earth never to return to it; they were considered unsuitable for communication purposes. Therefore, the scientific community of the early years of this century was amazed when Marconi, in 1901, demonstrated successful radio communication across the Atlantic Ocean.

Calculations had shown that diffraction effects around the earth would not yield fields of adequate levels to permit communication over great distance and it was concluded that some other phenomenon must account for the newly observed possibilities. As early as 1902, two scientists independent of each other suggested that there must be a reflecting layer above the earth that returns the impinging electromagnetic radiation. Later experiments confirmed this assumption and found further layers. One of the reflecting layers, the *E*-layer, was named, after those two scientists, the *Kennelly-Heaviside layer,* and the *F*-layer was named after its discoverer the *Appleton layer. Ionosphere* is a cumulative name for all ionized atmospheric layers.

Although reflection from a layer of the upper atmosphere was accepted to be the mechanism that made long-distance communication possible, it was only in 1924 that its height was determined by Appleton and Barnett by measuring the difference of the length of a direct and reflected path between two radio stations. About a year later Breit and Tuve proposed the hitherto used method of ionospheric height measurement based on measuring the time elapsed between the vertical transmissions of a short burst of radio frequency energy and the reception of its echo returned after reflection from an ionospheric layer.

The study of the ionosphere is based on the study of its effects on electromagnetic waves, which either may be reflected by it or not, depending on the frequency of radiation. Many experimenters in this study found that at broadcast frequencies the daytime reflected waves are very weak, inadequate to overcome receiver noise; at night, however, the intensity of the reflected wave increases to levels acceptable for communication. At higher frequencies, between about 5 and 30 MHz, even daytime signals reach levels allowing communication over several thousand kilometers. The ionosphere does not come into regular play at very high frequencies and above. Waves at these frequencies pass through the ionosphere and, though not suitable for long-distance terrestrial communication, are usable for satellite communication.

Generally, sky waves are less stable than ground waves and exhibit various phenomena not encountered with ground waves. They are affected by the state of the ionosphere, which may vary from season to season, day to day, and even from hour to hour. They are affected by such phenomena as sunspots, magnetic storms, or even sudden rainstorms. Many of the variations, however, are fairly regular, and the behavior of the ionosphere may be predicted with good accuracy. Ionospheric observatories operate at various locations of the earth and publish data usable for predicting frequencies suitable for communication over selected paths. Predictions of maximum usable frequencies, to be defined later, are published in various periodicals and used by such bodies as short-wave broadcasting authorities and governments. Figure 4.1 shows examples of such forecasts. A short study of the ionosphere will clarify problems associated with ionospheric propagation.

Example 4.1

On the basis of propagation forecasts for Prague, Czechoslovakia, shown in Figure 4.1, determine the frequencies required for maintaining 24-hour communication between Prague and Bogota, Colombia.

Solution: Using the *optimum usable frequency* (OUF) curve of the appropriate sketch, we can determine that between midnight and about 11 A.M., we can communicate at a frequency of about 10 MHz. We have to shift then to about 22 MHz

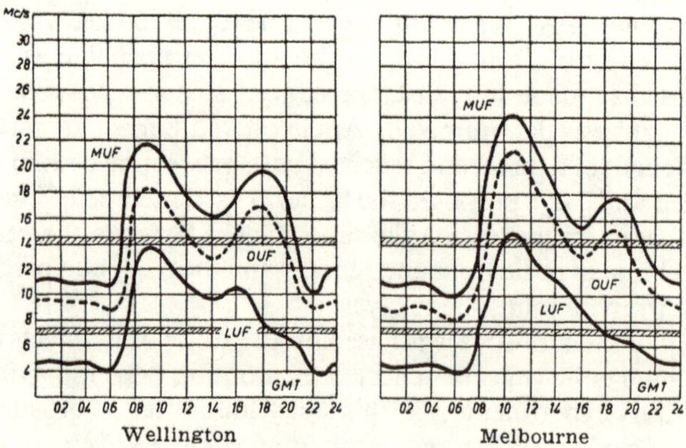

Figure 4.1 Ionospheric propagation forecasts for January 1949 (after Rohan, 1949. © Krátké vlny).

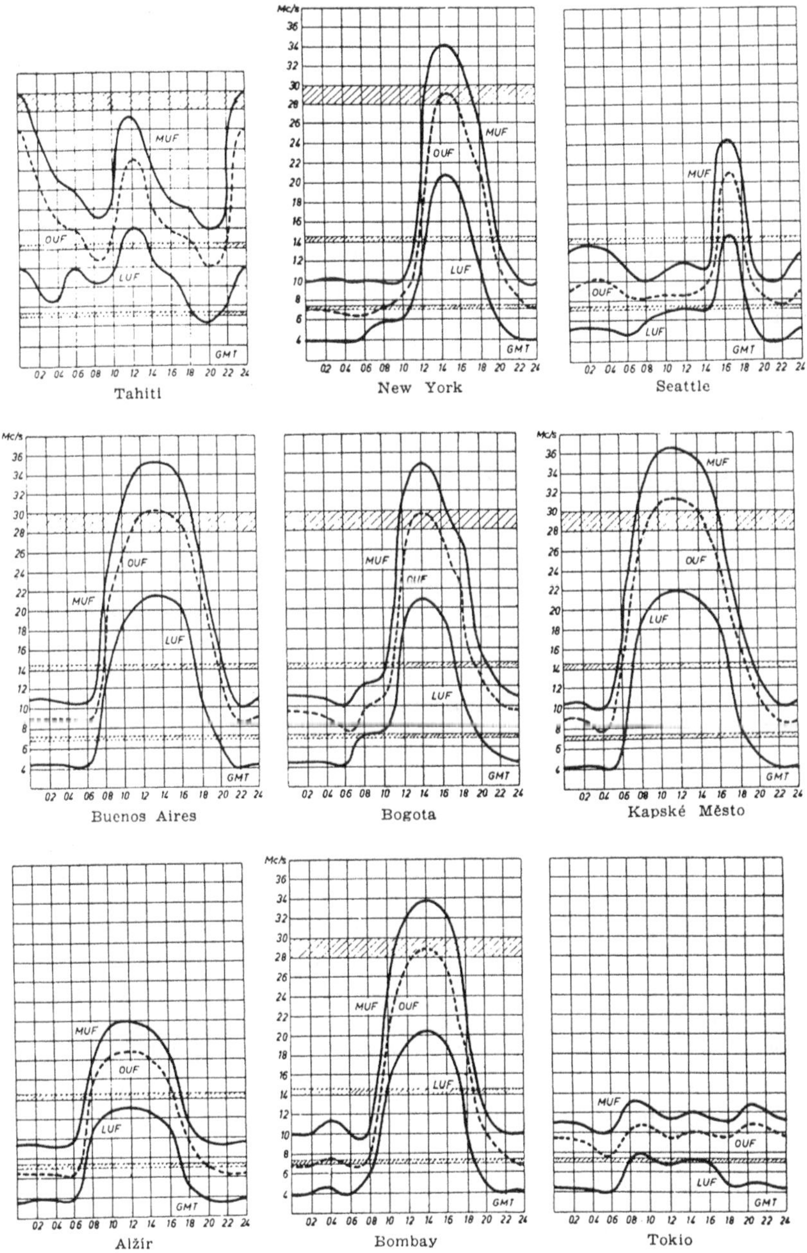

Figure 4.1 continued

until 6 P.M. and revert to 10 MHz until midnight. Some uncertainty about the frequency choice occurs at those times when the shown curves rise and fall steeply.

4.1 THE IONOSPHERE

The *ionosphere* is a cumulative name applied to parts of the earth's upper atmosphere containing free ions in adequate quantity to affect the propagation of electromagnetic waves. It extends in height from about 60 km to several earth radii. Studies of the properties of the ionosphere are intimately connected to those of the upper atmosphere, which aim at determining the composition of the atmosphere, the height dependence of pressure, temperature, molecular density, and so forth.

4.1.1 The Atmosphere

The composition of the atmosphere at heights below about 100 km, where turbulence causes good mixing, is almost uniform. In addition to nitrogen and oxygen, it consists of other ingredients such as water vapor, carbon dioxide, and hydrogen. The existence of oxygen atoms above the height of 100 km was confirmed by aurora studies [Rawer, 1953]. Above this height oxygen atoms result from the dissociation of oxygen molecules by sunlight. A transition region, in which the intensity of the sunlight diminishes and the density increases to such an extent that only some molecules may dissociate, develops below the above height. Recombination of the atoms into molecules during the night is very slow.

Nitrogen dissociates less readily, nevertheless, aurora observations confirm the presence of nitrogen molecules in the upper atmosphere. Cosmic dust is probably responsible for the presence of sodium.

Atmospheric density is given by the number of molecules in a unit volume and is proportional to the weight of the air column above it. Density and pressure at any height are intimately related. The difficulty of determining the height dependence of the atmospheric pressure is increased by temperature variations with height.

A density of 2.7×10^{25} molecules/m^3 and a pressure of 760 mm Hg at the surface of the earth at 0°C falls to 1×10^{20} molecules/m^3 and 2.8×10^{-3} mm Hg at a height of 100 km and 1.4×10^{9} molecules/m^3 and 3.9×10^{-14} mm Hg at a height of 300 km.

The atmospheric temperature falls in the first 10 km of height from 290 K at the surface of the earth to about 220 K to rise slowly toward 235 K at a height of 35 km, then it rises rapidly in the following 15 km to about 350 K. A decrease by about 120 K in the following 30 km is followed by a rise to 370 K at a height of 120 km.

The given temperatures, determined by space probes, become doubtful at greater heights [Rawer, 1953].

4.1.2 Ionization

Ionization, or the removal of an electron from a molecule or atom of atmospheric gas, is caused mainly by natural phenomena, such as the ultraviolet radiation of the sun, cosmic rays, meteors, corpuscular radiation, or by human-made effects like nuclear explosions. All these must reach the upper atmosphere with adequate energy to dislodge electrons from atoms or molecules.

The theoretical study of ionization by ultraviolet light is based on the work of Chapman, who simplified the process by assuming only one type of gas in the planely stratified isothermal atmosphere and a monochromatic radiation of parallel sunbeams.

The radiation enters the atmosphere at a zenith angle χ and its density Q, of dimension quanta/m^{2}s diminishes due to absorption along the ray from its maximum value Q_∞ at the upper atmospheric boundary. Considering the radiation to be composed of light quanta, the problem of absorption becomes that of collision where the probability of a single molecule being hit by an ionized light quantum depends on its effective cross section σ_i. The density of ionized molecules N_i (m^{-3}) can be much smaller than the molecular density.

Because the probability of absorption of a single quantum in a m^3 is $N_i\sigma_i$, there are

$$P = N_i\sigma_i Q \tag{4.1}$$

possible ionizations by Q quanta in the same volume when σ_i is constant [Rawer, 1953].

The molecular density N_i increases along the ray, while the radiation density Q decreases due to ionization. Taking into account other causes of decay, we ascribe a larger cross section σ_a to the quanta. The number of lost quanta/m^{2}s along the length ds of the path is then $N\sigma_a Qds$ and, because the density is height dependent and, according to Figure 4.2 the path length in terms of the height is

$$ds = -\sec\chi \, dh$$

we obtain the differential equation

$$\frac{dQ}{dh} = N\sigma_a \sec\chi Q$$

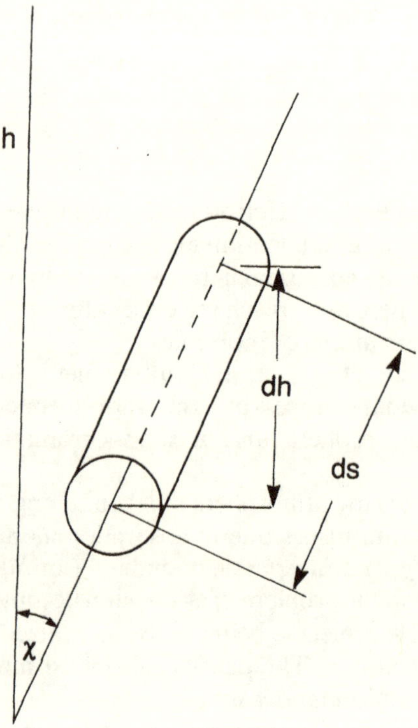

Figure 4.2 Absorption of radiation in a column of gas.

which yields

$$Q = Q_\infty \exp\left(-\sigma_a \sec\chi \int_h^\infty N(h)\,\mathrm{d}h\right) \tag{4.2}$$

for the dependence of the radiation density on height.

Inserting this into expression (4.1) leads to

$$P = \sigma_i N_i Q_\infty \exp\left[-\sigma_a \sec\chi \int_h^\infty N(h)\,\mathrm{d}h\right] \tag{4.3}$$

for the dependence of electron production on height.

The height dependence of the density decay must be known for the evaluation of (4.3). Using Chapman's simplifying assumption of a barometric decay,

$$N = N_0 \, e^{-h/H}$$

acceptable strictly for an isothermal atmosphere only, we obtain

$$P = \sigma N_0 Q_\infty \exp\left[-\frac{h}{H} - \sigma N_0 H \sec\chi \exp(-h/H)\right] \tag{4.4}$$

assuming that absorption occurs only by ionization. The quantity H is a scale factor obtained by considering equilibrium conditions of elementary gas volumes consisting of N molecules of mean mass m subjected to gravitational acceleration g and differential pressure p on their lower and upper faces in an isothermal atmosphere, when the equation of state for a unit volume is $p = NkT$ with k being Boltzmann's constant (1.372×10^{-16} erg/degree). Then, if g is independent of the height h,

$$Nmg = -\mathrm{d}p/\mathrm{d}h = -kT\,\mathrm{d}N/\mathrm{d}h$$

and

$$-\frac{1}{N}\frac{\mathrm{d}N}{\mathrm{d}h} = -\frac{mg}{kT}$$

having the preceding barometric decay with

$$H = -\frac{kT}{mg}$$

for its solution [Ratcliffe, 1972].

Figure 4.3 shows $\sqrt{P}$ for various zenith angles; the variation of maximum-ionization height obviously depends on the zenith angle χ. At smaller angles of incidence of the sun's rays into the atmosphere, the rays penetrate deeper and ionization occurs closer to the earth. The heights of maximum ionization occur at maxima of the exponent in Chapman's equation, given by

$$h_{\max} = H \log(\sigma N_0 H \sec\chi) \tag{4.5}$$

In reality the ionization process is much more complicated, because the simplifying assumptions are never satisfied. Nevertheless, Chapman's approach provides a good approximation of the available observations.

The ordinate of Figure 4.3 is calibrated in terms of the scale height. The heights are measured from the level of the $\chi = 0$ curve's peak, and the ion production rates

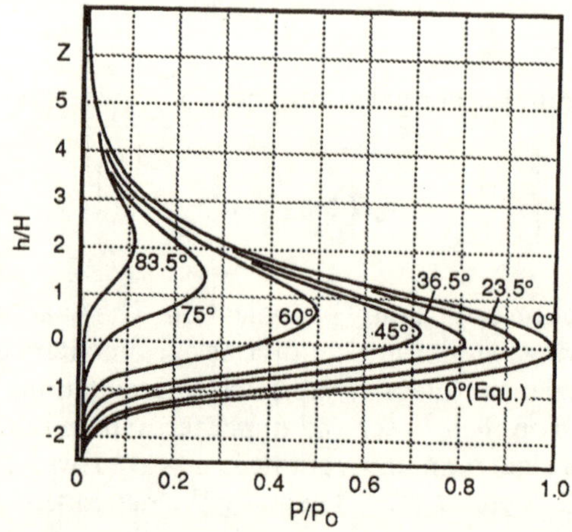

Figure 4.3 The height distribution of the rate of ion production in a plane-stratified exponential atmosphere for different zenith angles of the sun (after Chapman and Bartels, 1940, © Clarendon Press, Oxford).

are expressed in terms of the peak rate denoted by P_0. A decreasing solar altitude causes a decrease in ion production and a rise in its maximum level.

Ionization by corpuscular radiation differs from that by light. Whereas light quanta are destroyed on collision, particles of corpuscular radiation only lose energy but remain in space. Not all collisions cause ionization; the probability of collision is higher for slower particles and restricted to a narrower height interval so that a particle produced layer will be thinner than one due to ionization by light [Rawer, 1953].

Direct ionization of the upper atmosphere is caused only by shortwave ultraviolet radiation; longwave ultraviolet radiation could possess adequate ionizing energy only if the energy of at least two quanta were combined. Metastable atomic states, if prevailing long enough, make such processes possible. The combined energy of two such atoms on collision is adequate for ionization. This process has been used for the explanation of the formation of some of the ionospheric layers, to be discussed later.

A similar energy accumulation occurs at the dissociation of oxygen. Because considerable energy is released on the frequent collisions of two oxygen atoms at heights of approximately 100 to 120 km, they are probably responsible for the appearance of sporadic E-layers, to be described later.

The occurrence of ionization by meteors has been proved. Meteors of various sizes orbit the sun. On entering into the earth's atmosphere, they are slowed down, and intense heat, which induces their burning, is produced. The products of such burning form ionized trails at heights between 80 and 100 km [Picquenard, 1974]. Expansion of the trail is accompanied by a decrease in its ionization to such an extent that it becomes transparent to higher radio frequencies. With large meteors, energy reflected by the meteor itself may interface with the energy reflected by the trail to produce a whistling noise of first decreasing then increasing pitch, a phenomenon referred to as *whistlers.*

Cosmic dust entering the atmosphere also may be responsible for sporadic E-layer formation. Depending on the size of the dust particles, ionization may occur by direct collision or by raised temperature and ultraviolet radiation.

Corpuscular radiation originates in sunspots. It consists of a multitude of charged particles that, under the influence of the earth's magnetic field, travel toward the auroral zones that enclose the magnetic poles. The strong ionization produced by the moving particles is responsible for the auroras and magnetic storms that affect the ionosphere.

The ionizing contribution of the entire star-studded sky is estimated to be about one-thousandth of that of the sun and the contribution of cosmic rays is even smaller. According to Doluchanow, the sun's ultraviolet radiation produces 250 electrons/cm^3/s at a height of 100 km above the earth and 800 electrons/cm^3/s at a height of 250 km, whereas it is estimated that the stars produce only a single electron/cm^3/s at a height of 250 km and cosmic rays produce 10^{-5} electrons/cm^3/s at a height of 100 km. Figures for the sun's corpuscular radiation are not available.

4.1.3 Recombination and Attachment

It has been oberved that, for example, broadcast transmitters have a larger coverage area at night than during daytime. This is due to the disappearance of the shielding effect of the lowest ionospheric layer and hence the layer itself at night. This disappearance of a layer is due to either the recombination of free electrons with positive ions or to their attachment to a neutral molecule to form negative ions.

On collision of a free electron with a positive ion, energy in the form of light is released; however, theory and experiment shows that the probability of the occurrence of such a process is rather low [Rawer, 1953]. It depends only on the ion density and therefore is less probable in higher layers than in lower ones.

The reduction of the electron density per unit of time may be expressed by $\alpha N N^+$, where α is the recombination coefficient having a m^3/s dimension, N is the electron, and N^+ is the ion density. This product becomes equal to αN^2 when no attachment occurs.

No ionization takes place at night, hence the decrease of electron density is given by the differential equation:

$$\frac{dN}{dt} = -\alpha N^2 \tag{4.6}$$

from which results the electron density:

$$N = \frac{N_0}{1 + \alpha t N_0} \tag{4.7}$$

The variation of electron density due to daytime ionization is

$$\frac{dN}{dt} = \sigma_i Q N_i - \alpha N^2 \tag{4.8}$$

where the first term on the right-hand side is the rate of electron production with σ_i the effective cross section of an ionizing light quantum, Q is the radiation density (quanta/m^{2}s), and N_i is the molecular ion density per m^3. At equilibrium this equation becomes zero, so the electron density is given by

$$N = \sqrt{\frac{\sigma_i Q N_i}{\alpha}} \tag{4.9}$$

The electron density for a radiation density dependent on the cosine of the sun's zenith angle is

$$N \approx \sqrt{\cos\chi} \tag{4.10}$$

Attachments on collision between an electron and a neutral molecule or atom produce negative ions. Negative nitrogen ions are not stable; however, those of oxygen are. Recombination of positive and negative ions causes the disappearance of the negative charges. The attachment process depends on the density of oxygen atoms. The ionization decay through attachment is faster the greater is this density.

The nighttime decay of the electron density is

$$\frac{dN}{dt} = -\eta N^0 N,$$

from where the electron density

$$N = N_0 \exp(-\eta N^0 t) \tag{4.11}$$

may be calculated. The term η is the attachment coefficient. It is interesting to note that the electron density given by (4.11) is independent of the zenith angle χ.

The daytime equilibrium electron density is

$$N = \frac{\sigma_i Q N_i}{\eta N^0} \tag{4.12}$$

which is proportional to the ionizing radiation.

Bates et al. [1989], on the basis of extensive theoretical studies, came to the conclusion that direct recombination cannot account for the observed nighttime decay of the ionization and considered attachment with successive ionization to be the most important process. Not even this approach, however, can explain the rapid recombination in the ionosphere and the authors suggested recombination together with the dissociation of molecules to be the more probable process.

4.1.4 Formation of Ionized Regions

Chapman's idealized atmospheric conditions do not actually exist. Although the atmosphere may be taken to be well mixed to a height of about 100 km, the elements constituting the atmosphere become dissociated above this height. The temperature, as indicated earlier, also changes with height so that the height dependence of ionization in the real atmosphere differs from that considered by Chapman. The various atmospheric gasses require different ionizing energies and different ionizing agents may supply them.

The height of distribution of electron density depends on, among other factors, the mechanism of electron disappearance. According to Rawer [1953] the disappearance of electrons in recombination is independent of the gas density, and the electron density is proportional to the square root of the ionizing radiation intensity. The maximum of the region coincides then with the ionization maximum, and the region assumes the shape indicated by Figure 4.3.

The rate of electron disappearance when attachment predominates is very much slower at greater heights than at lower ones. The electron density at equilibrium is given by (4.12), which with expression (4.4) for Chapman's electron production during an exponential decay of density leads, for $N \ll N^0$, to

$$N = \frac{\sigma Q_\infty}{\eta}\frac{N_0}{N^0} \exp[-\sigma N_0 H \sec\chi \exp(-h/H)] \tag{4.13}$$

In an idealized case, assuming a constant attachment coefficient, the exponential term of expression (4.13), the Chapman function, shown by Figure 4.4, depicts regions lacking distinct maxima. Although the mechanism of the depletion of ion-

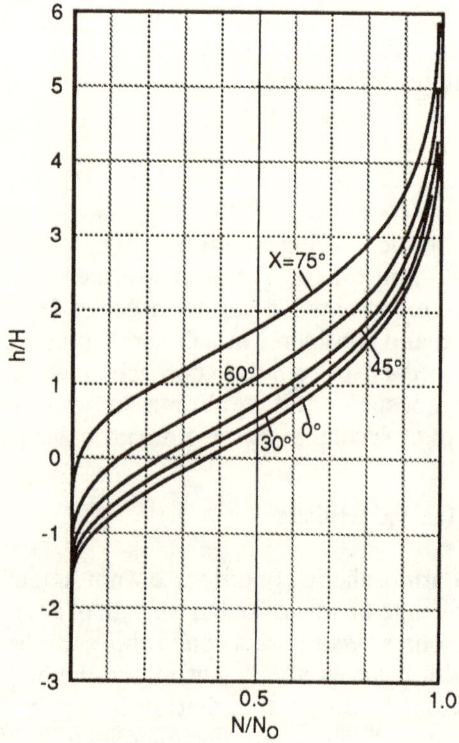

Figure 4.4 Idealized electron density when attachment predominates (electron production according to Chapman; after Rawer, 1953, © P. Noordhoff N.V.).

ization above about a height of 150 km is similar to the attachment process, the molecular density and hence the value of the attachment coefficient diminishes exponentially with height, thus leading, according to Bates et al. [1939], to a two-stage process referred to as *dissociative recombination,* at which a molecule and an electron dissociate and simultaneously recombine. The rate of ionization depletion is determined by the slower of these processes. The first process, being slower at greater heights, causes the formation of layers exhibiting maxima, contrary to the idealized case of Figure 4.4.

The result of the simultaneous action of various ionizing agents is the development of an ionized zone of the atmosphere where several ionization maxima may occur. At least four such maxima have been detected experimentally and, as they are not separated by un-ionized atmospheric regions but only by regions of reduced ionization, they are referred to as *regions* instead of the original designation, *layer.*

There are then the D, E, F_1, and F_2 regions at increasing altitudes. These regions will be described later.

Seddon [1962] studied data obtained from rocket and satellite observations that show that the ionosphere does not consist of discrete regions but is a continuum with an almost monotonic increase in electron density with altitude. This is borne out by Figure 4.5, which gives the average electron densities at various heights as obtained at four different observatories during the International Geophysical Year in 1957. The magnetic inclinations at the observatories are in parentheses.

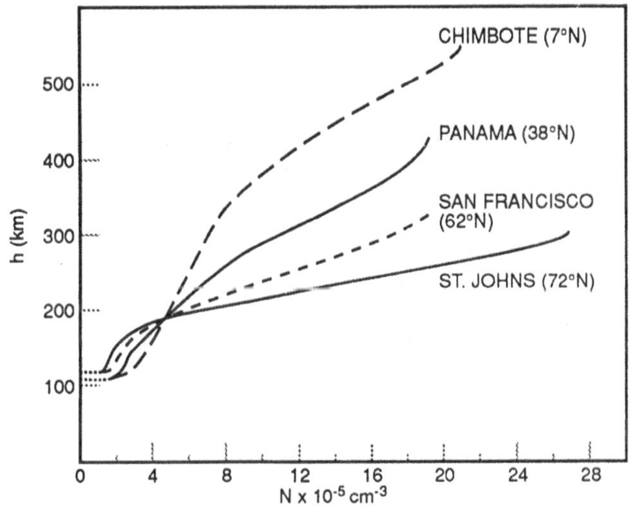

Figure 4.5 Average electron densities at noon during December 1957 (after Thomas, 1962, © Elsevier).

These rocket measurements have shown thin high-electron density gradients, sometimes up to 10^6 electrons/cm^3/km, between 65 and 200 km of altitude. Their thickness variation was found to be from 10 to 100 m in the D-region at altitudes of about 80 to 90 km, to 100–2000m in the E-region at heights of 100 to 120 km. Although electron density increases of about 20% of the ambient are common, doubling of E-region electron density also has occurred. There is a strong tendency for an approximately 6 km separation between regions of high-electron density gradients intimating a relation to atmospheric wind gradients that tend to show the same separation.

Although this short description of the formation of regions with increased ionization is based on highly simplified and idealized assumptions, it shows the possibility of the occurrence of several such regions, as experienced in practice. A more detailed description of the regions themselves will be given later.

Figure 4.6 shows the height distribution and range of variation of electron densities obtained from various experiments. The sharp drop in electron concentration at the upper ionospheric boundary is noteworthy.

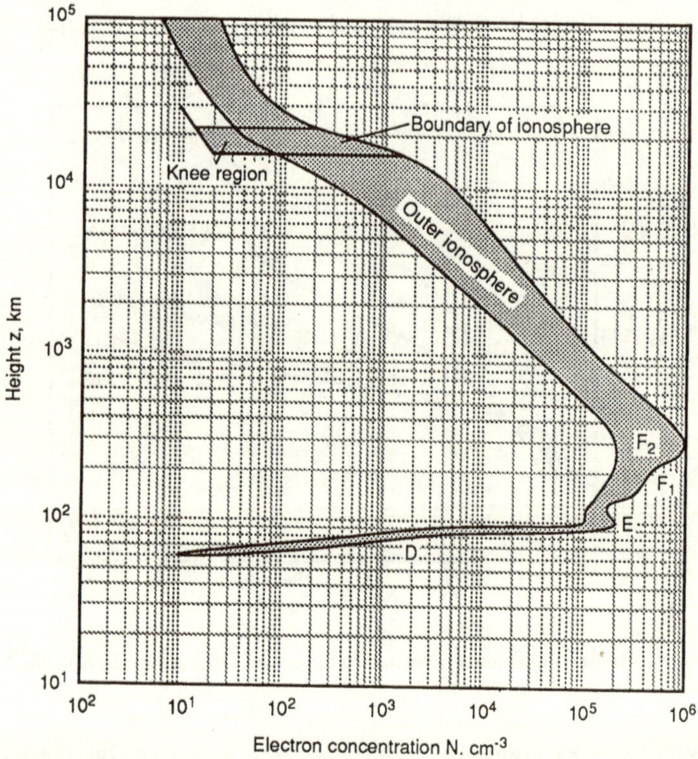

Figure 4.6 Height distribution of electron densities (after Al'pert, 1973, © Vsesojuznoje Agentstvo).

Chapman's formula fails when applied to the *F*-region where Martyn [1959] assumes that ionized plasmas are shifted from one part of the atmosphere to another by winds that shift ionized particles in paths almost parallel with the earth's magnetic field, by electric fields of the charges that together with the earth's magnetic field combine to exert electromagnetic forces on the particles, by pressure of the various atmospheric components, and by gravity.

The variation of ionization density according to Martyn does not occur as given by expression (4.8), but according to

$$\frac{\partial N}{\partial t} = \sigma_i Q N_i - \beta N - \mathrm{div}(v \cdot N) \tag{4.14}$$

where β at the F-region altitude of 300 km is about 10^{-4}/s and varies according to the expression

$$\beta = 10^{-4} \exp \frac{300 - h_{\mathrm{km}}}{50} \tag{4.15}$$

and v is the transport velocity of electrons and ions due to these effects in addition to a vertical ionization drift due to diffusion at the magnetic equator. F-region ionization states exhibit seasonal variations that complicate conditions and make the understanding of F-region phenomena more difficult.

4.2 IONOSPHERIC PERMITTIVITY AND CONDUCTIVITY

The ionosphere, due to its electron and ion content, has a dielectric constant and conductivity that differ from those of free space. A passing electromagnetic wave forces the charge carriers to oscillate. A moving electron represents a current. Each oscillating electron acts as a parasitic antenna that absorbs energy from the passing wave and reradiates it with a different phase. In investigating the ionosphere we can consider it to be a charge-free, lossy dielectric having an effective dielectric constant ϵ' and an effective conductivity σ. These effective values may be obtained, for a given frequency, from the ion density and the collision frequency in the ionized region.

Because the electrons are surrounded by a large number of neutral gas molecules in a continuous thermal motion, the electrons and molecules collide and, assuming that the energy imparted by the electrons onto the molecules is mv, where v is the velocity of the electrons and m their mass, the total energy passed on in a second is $mv\nu$ with ν the average collision frequency per second.

The force acting on an electron in free space is the product of the electric field strength and the electronic charge. Here, however, account must be taken of the energy lost on collision, so that, in accordance with Newton's second law of motion, the complete force equation reads

$$eE - mv\nu = m \frac{\mathrm{d}v}{\mathrm{d}t} \tag{4.16}$$

Assuming that the oscillating field

$$E = E_0 \, \mathrm{e}^{j\omega t}$$

causes the velocity

$$v = v_0 \, e^{j\omega t}$$

the force equation after canceling the exponential terms becomes

$$eE_0 = (j\omega m + m\nu)v_0 \tag{4.17}$$

which may be determined from the velocity:

$$v_0 = \frac{eE_0}{j\omega m + m\nu} \tag{4.18}$$

The current density for N free electrons or ions per m^3 is

$$i = Nev_0 \quad \text{A/m}^2 \tag{4.19}$$

which with (4.18) yields

$$i = \frac{NE_0e^2}{j\omega m + m\nu} \tag{4.20}$$

or when separating the real and imaginary terms,

$$i = \frac{NE_0e^2\nu}{m(\nu^2 + \omega^2)} - j\frac{NE_0e^2\omega}{m(\nu^2 + \omega^2)} \tag{4.21}$$

Inserting this into Maxwell's electromagnetic force equation in an ionized medium [Jordan, 1953]

$$\text{curl}\mathbf{H} = \mathbf{1} + j\omega\epsilon_0\mathbf{E}$$

leads to

$$\begin{aligned} \text{curl}\mathbf{H} &= j\omega\epsilon_0\left[1 - \frac{Ne^2}{\epsilon_0 m(\nu^2 + \omega^2)}\right]\mathbf{E} + \frac{Ne^2\nu}{m(\nu^2 + \omega^2)}\mathbf{E} \\ &= (j\omega\epsilon_0\epsilon' + \sigma)\mathbf{E} \end{aligned} \tag{4.22}$$

from which the relative dielectric constant and the conductivity of the ionized region become

$$\epsilon' = 1 - \frac{Ne^2}{\epsilon_0 m(\nu^2 + \omega^2)}$$
$$\sigma = \frac{Ne^2\nu}{m(\nu^2 + \omega^2)} \quad \text{mho/m} \tag{4.23}$$

respectively.

Expressions (4.23) show that both the relative dielectric constant and the conductivity of the ionized region are affected by the presence of electrons and ions. Both depend on the charge carrier density and collision frequency. The dielectric constant is reduced below the free-space value of one, whereas the conductivity at the considered location becomes maximum at $\nu = \omega$. The height dependence of these two quantities is implicit in the collision frequency, which is a function of temperature and gas pressure, both depending on the altitude. The collision frequency varies between about 10^{12}/s at the surface of the earth, where the molecular density at a temperature of 0°C is 2.7×10^{19} per cm^3 to 1/s at a height of 800 km with a molecular density of 2.7×10^6/cm^3.

Example 4.2

Prove that the ionospheric conductivity at a given location reaches its maximum when the average collision frequency equals the angular frequency of the propagating wave.

Solution: Differentiating expression (4.23) for the conductivity σ with respect to the collision frequency ν and equating the obtained differential to zero, yields the answer.

4.3 IONOSPHERIC REFLECTION AND REFRACTION

Reflection and refraction of electromagnetic waves in the ionosphere depends on the frequency of the waves. Two extreme cases are to be considered. At very low frequencies, where the change in electron density within a wavelength is very large, the ionosphere appears to be an abrupt discontinuity in the medium of propagation and the reflection resembles that from a dielectric. The other extreme occurs at high frequencies, where the change of ionization within a wavelength is small so that the ionosphere resembles a dielectric with a continuously variable refractive index. In between these two extremes, the ionosphere may be considered to consist of thin layers, each with a constant ionization density different from those of the adjacent layers. The incident wave then is partly reflected and partly refracted, with this part entering the neighboring layer where the process repeats itself [Jordan, 1953].

In the case of long waves where the variation of ionization density within a wavelength is large, the reflection coefficients are those given earlier by (2.65) and

(2.69) in which, however, the effective values of the dielectric constant and conductivity, given by expressions (4.23), are to be used.

In accordance with Section 2.5, it becomes obvious that the magnitude and phase of the reflection coefficient will depend on the incident wave's polarization. Where the conductivity is negligible, the ionosphere behaves like a perfect dielectric. For angles of incidence larger than a critical angle, which depends on the refractive index, the reflection coefficient of waves of both linear polarizations will be unity: complete reflection occurs. For angles of incidence below the critical angle, the reflection coefficient will depend on the angle of incidence and be less than unity.

At high frequencies, where the change of electron density within a wavelength and hence that of the phase velocity is small and the refractive index appears to be continuously variable, methods of ray optics are applicable to propagation studies.

The phase velocity of a wave in a medium of negligible losses, as given by (2.15), is $v_p = 1/\sqrt{\mu\epsilon}$. Because the velocity of propagation in vacuo is

$$c = \frac{1}{\sqrt{\mu_0\epsilon_0}}$$

and $\mu = \mu'\mu_0$ and $\epsilon = \epsilon'\epsilon_0$, this expression for the phase velocity becomes

$$v_p = \frac{c}{\sqrt{\mu'\epsilon'}} \tag{4.24}$$

or when assuming the relative permeability to be equal to that of free space, the phase velocity is

$$v_p = \frac{c}{\sqrt{\epsilon'}} \tag{4.25}$$

where ϵ', as shown earlier, is a function of the electron density N.

A high-frequency wave enters then the ionosphere without reflection and bends away from the region of greater electron density where the index of refraction is lower (Figure 4.7).

Snell's law of refraction, according to which

$$n_1 \sin\phi_1 = n_2 \sin\phi_2 \tag{4.26}$$

permits the determination of the angle ϕ between the wave's trajectory and the normal to the ionospheres. The terms n_1 and n_2 are the indices of refraction at heights where the angles ϕ_1 or ϕ_2 prevail.

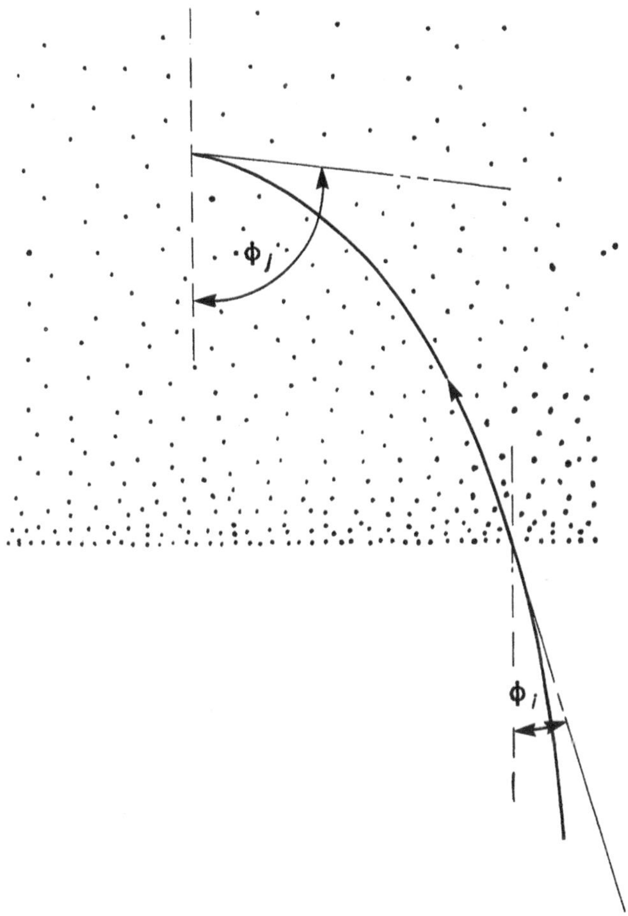

Figure 4.7 Ray path in the ionosphere.

Because the refractive index of any medium is

$$n = \frac{c}{v_p} = \frac{\text{velocity of light in a vacuum}}{\text{phase velocity in the medium}} \tag{4.27}$$

the refractive index of the lossless ionosphere for which the phase velocity is given by (4.25), is

$$n = \sqrt{\epsilon'} \tag{4.28}$$

At F-region heights, where high frequency waves are reflected, the collision frequency and hence the conductivity given by (4.23) is very low so that it may be neglected in initial studies, and the use of (4.23) is justifiable.

When the frequency of the wave greatly exceeds the collision frequency, the relative dielectric constant given by (4.23) becomes

$$\epsilon' = 1 - \frac{Ne^2}{\epsilon_0 m\omega^2} \tag{4.29}$$

which, with $e = 1.59 \times 10^{-19}$ C, $m = 9 \times 10^{-31}$ kg, and $\epsilon_0 = 1/(36\pi 10^9)$, reduces to

$$\epsilon' = 1 - \frac{80.5N}{f^2} \tag{4.30}$$

with N again the number of electrons per m^3 and f the frequency of the radiation in Hz.

Using this expression, the refractive index of the ionosphere in accordance with (4.28) is approximately

$$n = \sqrt{1 - \frac{80.5N}{f^2}} \tag{4.31}$$

This expression is known as the *Appleton formula.* It shows that, if propagation is to take place, the refractive index must be real; that is, it must have a value between zero and one. It is proportional to the frequency and inversely proportional to the electron density. It is independent of the wave's amplitude.

At the bottom of the ionosphere, where the electron density is zero, the refractive index has a unity value. As the incident wave progresses into the ionosphere, the electron density rises and the value of the refractive index decreases to zero. Inserting then $n_1 = 1$ into expression (4.26) shows that at reflection where the wave travels horizontally, or $\phi_2 = \pi/2$ rad, the refractive index must be $n_2 = \sin\phi_1$. The wave reaches the highest point of its trajectory where, on the basis of these considerations, the electron density satisfies the condition

$$\sqrt{1 - \frac{80.5N}{f^2}} = \sin\phi_1$$

or

$$N = \frac{f^2 \cos^2\phi_1}{80.5} \tag{4.32}$$

This means that if the maximum electron density at a given level exceeds that given by (4.32), the wave will be bent enough to return to the earth; if it is less, the wave will propagate beyond this level and might be returned from a greater height, where N is greater.

Reflection at normal incidence requires the greatest electron density. Setting in (4.32) $\phi_1 = 0°$ permits the computation of the so-called critical frequency, that is, the highest frequency that at normal incidence will be reflected from a height where the ionization density in electrons/m³ is N_{max}. This frequency is given by

$$f_c = \sqrt{80.5 N_{max}} \tag{4.33}$$

The refractive index of the layers is close to unity so that expression (4.26) becomes

$$n \sin\phi \approx \sin\phi_0 = \text{const.} \tag{4.34}$$

At the highest point of the trajectory, the ray is horizontal or $\sin\phi = 1$. Using expressions (4.31) and (4.33), we can write

$$n = \sqrt{1 - \left(\frac{f_c}{f}\right)^2} = \sin\phi_0 \tag{4.35}$$

from which we obtain Martyn's secant law [Picquenard, 1974] in the form

$$\left(\frac{f_c}{f}\right)^2 = 1 - \sin^2\phi_0 = \cos^2\phi_0$$

or

$$f = \frac{f_c}{\cos\phi_0} = f_c \sec\phi_0 \tag{4.36}$$

indicating that, at oblique incidence at a point of the ionosphere where the critical frequency is f_c, an ionized layer may reflect waves of higher frequencies than the critical one. In this case it is customary to write f_v instead of f_c and call it the *equivalent vertical-incidence frequency,* corresponding to frequency f [Davies, 1966].

Example 4.3

Given the collision frequency $\nu = 10^3$/s at a height of 300 km, determine the appropriate index of refraction at the highest point of the 10 MHz wave's trajectory for an electron density obtained from Figure 4.6.

Solution: The average electron density at the specified height of 300 km, as read from Figure 4.6, is

$$N = 0.55 \times 10^{11} \text{ electrons/m}^3$$

The critical frequency, from expression (4.32) and (4.33), is

$$f_c = \sqrt{80.5N} = 2.1042 \text{ MHz}$$

Then, from expression (4.35),

$$n = \sqrt{1 - \left(\frac{2.1042}{10}\right)^2} = 0.98$$

Assuming that, due to the ionizing radiation of the sun at zero zenith angle (i.e., at the normal incidence), the number of produced electrons/m^3/s, q, equals the number of electrons disappearing due to recombination at the same height and in the same time, we can write

$$q = \alpha N_{max}^2$$

where α is the recombination coefficient that is almost constant within an ionospheric region.

The relation between the maximum number of produced electrons and those produced at oblique incidence of the ionizing radiation is

$$q_{max} = q \cos\chi$$

from which, due to

$$\sqrt{\frac{q_{max}}{\alpha}} = N_{max}\sqrt{\cos\chi}$$

at oblique incidence, the charge carrier density is

$$N = N_{max}\sqrt{\cos\chi}$$

and the critical frequency, on the basis of (4.33),

$$f_c = \sqrt{80.5 N_{max}\sqrt{\cos\chi}} \qquad \text{(a)}$$

which usually is expressed as

$$f_c = K\cos^n\chi \qquad \text{(b)}$$

with $n \approx 0.3$ instead of the theoretical value of 0.25. Empirical expressions for K and n are available in the relevant literature.

Expression (b) shows good agreement with experimental results for values of $\chi < 1.36$ rad.

In planning communication over a given distance, *maximum usable frequency* (MUF) is of interest. This is the maximum frequency that may be reflected back to earth to reach a corresponding station at a specified distance. The relation of the maximum usable frequency and the critical frequency may be obtained by inserting (4.33) into (4.32) as

$$f_{MUF} = f_c \sec\phi_1 \qquad (4.37)$$

The maximum usable frequency is higher than the critical one. The angle ϕ_1 may be determined from the known ionospheric height and great circle distance of the ends of a communication link. The maximum usable frequency may exhibit small daily variations of up to 15%; it is customary therefore to use slightly lower frequencies than those predicted to secure uninterrupted communication. This frequency, known as *optimum usable frequency* (OUF), is chosen to be between 50 and 85% of the predicted MUF.

We showed earlier that in the ionosphere the wave travels over a curved trajectory. The determination of its highest point is difficult, and we may resort therefore to adopting an equivalent virtual height of a triangular trajectory.

Neglecting the variation of ionization with height and the earth's curvature, the time taken by the ray to travel between the transmitter T and the receiver R (Figure 4.8) along the "optical" trajectory composed of elements of length dl, is

$$t = \int_T^R \frac{dl}{v_p} = \frac{1}{c}\int_T^R n\, dl$$

where the phase velocity, as given by expression (4.27), is used.

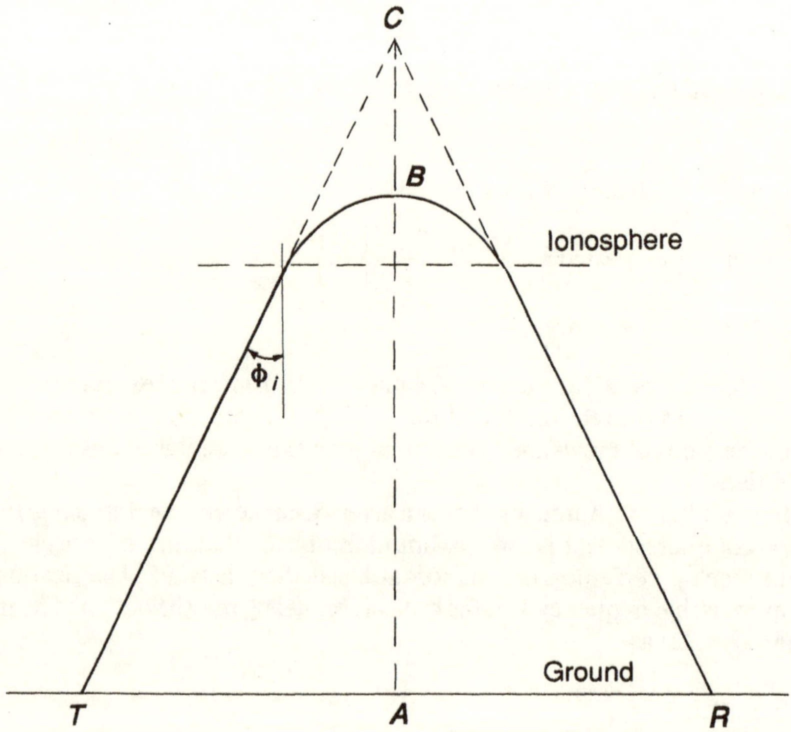

Figure 4.8 Virtual reflection height.

The time taken by the ray traveling over the triangular path consisting of elements of length ds, is

$$t' = \frac{1}{c}\int_T^R n\,\mathrm{d}s$$

Expressing these times in terms of elements of length dx along the surface of the earth, given by dx = dl sinϕ, we obtain

$$t = \frac{1}{c}\int_T^R n\,\frac{\mathrm{d}x}{\sin\phi}$$

for the first trajectory, and similarly

$$t' = \frac{1}{c} \int_T^R n \frac{\mathrm{d}x}{\sin\phi}$$

for the second one. Hence, because these times are equal, the equivalence of the virtual height AC and the true height AB is proven.

Breit and Tuve's theorem is as follows. The true height always is less than the virtual one. The theorem is valid only when both the transmitter and the receiver are outside the ionosphere. If, according to Davies [1966], both are within the ionosphere at a level where the refractive index is n, the right-hand side of the path length given by

$$P = TA + AR$$

is to be divided by this refractive index if P still represents the equivalent virtual path length.

The height of the reflection is obtained from ionospheric soundings that rely on the transmission of a pulse of radiofrequency energy vertically toward the ionosphere and measurement of the time interval between the transmission of the pulse and the reception of its echo. Ionospheric sounding, introduced by Breit and Tuve in 1926, is the first-known application of pulse transmission.

The virtual height of the ionosphere is

$$h' = ct/2 \tag{4.38}$$

with c the velocity of propagation of electromagnetic waves in free space and t the measured elapsed time between the transmission of a pulse and reception of its echo. The factor 2 accounts for propagation to and from the virtual height of reflection.

The curvature of the ionosphere modifies Snell's law, expression (4.26), to read [Davies, 1966]

$$nr \sin\phi = n_0 r_0 \sin\phi_0 \tag{4.39}$$

where r is the radius vector from the center of the earth (Figure 4.9) to the level where n is the refractive index, and ϕ is the angle between the trajectory of the ray and the radius vector; n_0, r_0, and ϕ_0 are the corresponding quantities at a reference level.

Taking the earth's surface to be the reference level makes $n_0 = 1$, $\phi_0 = \pi/2 - \epsilon$, with ϵ the initial elevation angle of the ray, and $r_0 = a$, the modified earth radius, as given in Section 3.6. Expression (4.39) then becomes

$$nr \sin\phi = a \cos\epsilon \tag{4.40}$$

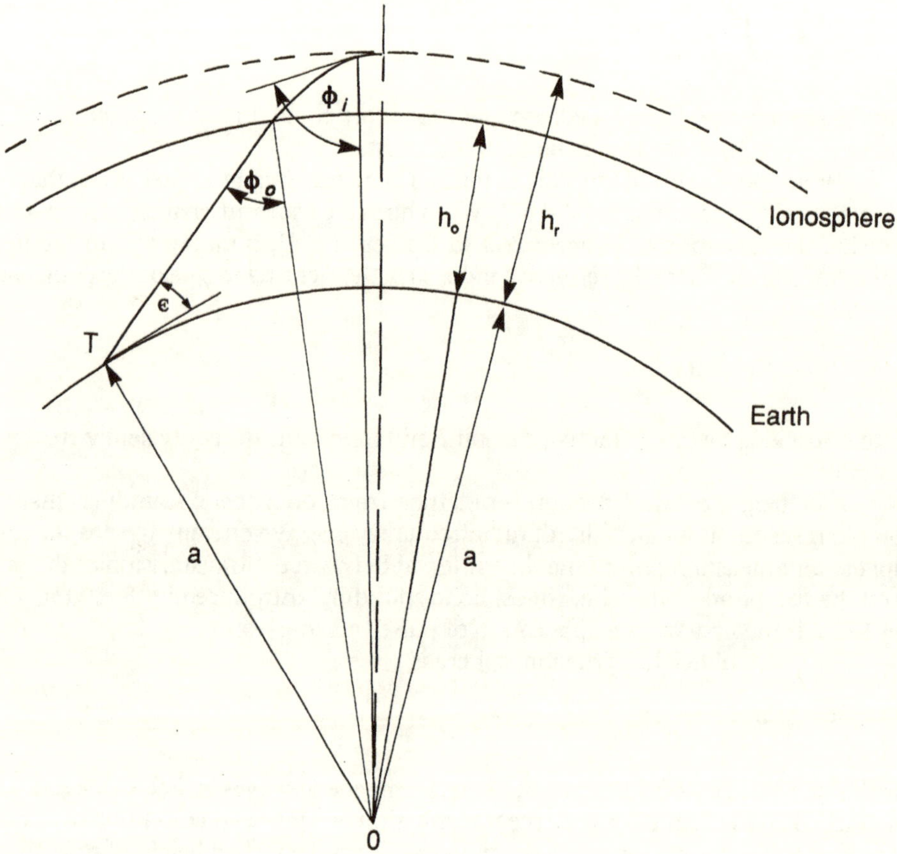

Figure 4.9 Ray path at oblique incidence.

Inserting the refractive index given by (4.35) and replacing f_c by the equivalent vertical-incidence frequency f_v, leads to

$$\left(\frac{f_v}{f}\right)^2 = 1 - \left(\frac{a \cos\epsilon}{a + h_r}\right)^2 \tag{4.41}$$

because $\sin\phi = 1$ when $r = a + h_r$. This expression shows the relation between a frequency f of an obliquely incident wave and the equivalent vertical-incidence frequency f_v reflected from the same height h_r.

Then, from the geometry of Figure 4.9,

$$(a + h_r)\sin\phi = a\cos\epsilon$$

so that

$$f_v = f\cos\phi \tag{4.42}$$

Expression (4.41) shows that the equivalent vertical-incidence frequency depends not only on the initial elevation angle ϵ of the ray trajectory but also on the height of reflection h_r.

4.4 PROPAGATION IN A MEDIUM WITH A SLOWLY VARYING REFRACTIVE INDEX

The previous section considered propagation in an ionosphere consisting of layers of constant refractive indices differing from those of the neighboring layers. In reality, the variation of the refractive index with height is continuous, and at some wavelength it may be assumed to vary only slowly over distances comparable to a wavelength.

We showed in Chapter 2 that an exact solution of Maxwell's equations for propagation in a homogeneous isotropic dielectric leads to a wave equation of the form

$$\frac{\delta^2 E}{\delta t^2} = c^2 \frac{\delta^2 E}{\delta x^2} \tag{4.43}$$

The solution of Maxwell's equations in a medium having properties changing gradually over distances on the order of a wavelength or greater is more involved. In the ionosphere the refractive index, a function of altitude, may be assumed to change over distances very much larger than a wavelength. Restricting ourselves, as before, to propagation in one dimension and expressing the height dependency of the refractive index by $n = n(z)$, expression (4.43) for a wave traveling upward in the ionosphere becomes

$$\frac{\delta^2 E}{\delta t^2} = \frac{c^2}{n^2(z)} \frac{\delta^2 E}{\delta z^2} \tag{4.44}$$

where the phase velocity, according to expression (4.27), is used.

The solution of this equation may be assumed to be of the form

$$E_z = A(z) \exp\left[j\omega\left(t - \int_0^z \frac{n(z)}{c}\,\mathrm{d}z\right)\right] \tag{4.45}$$

where the phase takes the gradual change of the refractive index in the direction of propagation into consideration [Lipson and Lipson, 1969].

Performing the required differentiations and inserting them into (4.44) yields

$$-\omega^2 A = \frac{c^2}{n^2}\left[\frac{\partial^2 A}{\partial z^2} + 2j\omega\frac{n}{c}\frac{\partial A}{\partial z} - \frac{\omega^2 n^2}{c^2}A + j\frac{\omega}{c}\frac{\partial n}{\partial z}A\right]$$

or

$$0 = \frac{c^2}{n^2}\frac{\partial^2 A}{\partial z^2} + 2j\frac{\omega c}{n}\frac{\partial A}{\partial z} + j\frac{\omega c}{n^2}\frac{\partial n}{\partial z}A \tag{4.46}$$

The progress so far follows the rules of exact mathematics. The solution of expression (4.46), however, is greatly simplified by applying the so-called WKB (after its initiators, Wentzel, Kramers, and Brillouin) approximation.

Considering that the variation of the ionospheric conditions is very gradual, we can assume that $\partial n/\partial z$ and $\partial A/\partial z$ are small and therefore $\partial^2 A/\partial z^2$ are even smaller. Expression (4.46) becomes

$$0 = \frac{2}{A}\frac{\partial A}{\partial z} + \frac{1}{n}\frac{\partial n}{\partial z} \tag{4.47}$$

which has a simple solution of the type:

$$A = B/\sqrt{n} \tag{4.48}$$

where B is an arbitrary constant. The wave then is given by

$$E_z = \frac{B}{\sqrt{n}} \exp\left[j\omega\left(t - \int_0^z \frac{n(z)}{c}\,\mathrm{d}z\right)\right] \tag{4.49}$$

provided that n and A vary slowly with the distance along z. Because the refractive index generally is complex, the amplitude of the wave will be complex.

Assuming the frequency to be constant, the up-going wave becomes

$$E_z = \frac{B}{\sqrt{n(z)}} \exp\left[-j \int_0^z \beta n(z)\,dz\right] \tag{4.50}$$

The field due to the wave reflected from the ionosphere, the down-going wave, obtained by the WKB approximation of the wave equation is

$$E_z \frac{B\rho}{\sqrt{n(z)}} \exp\left[j\beta \int_0^{z_0} n(z)\,dz\right] \tag{4.51}$$

with ρ the coefficient of reflection, $\beta = \omega/c$, and z_0 is the height coordinate at ground level [Budden, 1961a].

Total reflection occurs at a level where $n = 0$ and, if the up-going and down-coming waves tend to equality, the coefficient of reflection is

$$\rho = \exp\left[-2j\beta \int_0^{z'} n(z)\,dz\right] \tag{4.52}$$

where z' is the value of z at the level where $n(z) = 0$. This analysis neglects the effect of particle collisions in the ionosphere; otherwise, the assumption of $n(z) = 0$ could not have been used.

4.5 IONOSPHERIC ATTENUATION FACTOR

Maxwell's electromagnetic force equation in a conducting medium when the time variation of currents and fields is $e^{j\omega t}$ was given by (1.11) as

$$\text{curl}\mathbf{H} = (\sigma + j\omega\epsilon)\mathbf{E}$$

which, when differentiated with respect to time, becomes

$$\text{curl}\frac{\partial \mathbf{H}}{\partial t} = j\omega(\sigma + j\omega\epsilon)\mathbf{E}$$

Taking the curl of the Maxwell-Faraday equation (1.3) and substituting it into the obtained expression yields

$$\text{curl curl } \mathbf{E} = -\mu \text{ curl } \frac{\partial \mathbf{H}}{\partial t}$$
$$= -j\omega\mu(\sigma + j\omega\epsilon)\mathbf{E}$$

which, due to the previously used vector relation

$$\text{curl curl} = \text{grad div} - \nabla^2$$

reads

$$\nabla^2\mathbf{E} - j\omega\mu(\sigma + j\omega\epsilon)\mathbf{E} = \text{grad div } \mathbf{E}$$

In a homogeneous medium, where ϵ = const. and $\mathbf{E} = D/\epsilon$, Gauss's theorem (1.9) shows that the divergence of the displacement density at every point equals the charge density, so that the right-hand side of the derived expression becomes

$$\frac{1}{\epsilon} \text{ grad } \rho = 0$$

because there is no net charge within a conductor. Then,

$$\nabla^2\mathbf{E} - j\omega\mu(\sigma + j\omega\epsilon)\mathbf{E} = 0$$

is the wave equation of the electric component.

The wave equation for the magnetic component obtained on the basis of a similar deliberation reads

$$\nabla^2\mathbf{H} - j\omega\mu(\sigma + j\omega\epsilon)\mathbf{H} = 0$$

The term

$$\gamma = \sqrt{j\omega\mu(\sigma + j\omega\epsilon)} = \alpha + j\beta$$

usually is referred to as the *propagation constant.* Its real part indicates the rate of the wave's decay with distance from the source and is called the *attenuation constant.* The imaginary part yields the phase shift per unit length of the wave.

Performing the multiplication in the squared propagation constant and equating real and imaginary terms leads to

$$\alpha = \omega \sqrt{\frac{\mu\epsilon}{2}\left(\sqrt{1 + \frac{\sigma^2}{\omega^2\epsilon^2}} - 1\right)} \tag{4.53a}$$

$$\beta = \omega \sqrt{\frac{\mu\epsilon}{2}\left(\sqrt{1 + \frac{\sigma^2}{\omega^2\epsilon^2}} + 1\right)} \tag{4.53b}$$

Using $\epsilon = \epsilon'\epsilon_0$ with ϵ' and σ as given by (4.23), we arrive at the attenuation factor:

$$\alpha = \frac{\omega}{c} \sqrt{\frac{\epsilon'}{2}\left(\sqrt{1 + \frac{\sigma^2}{\omega^2\epsilon^2}} - 1\right)}$$

which with $\sigma = \nu\epsilon_0(1 - \epsilon')$, obtained from (4.23), yields

$$\alpha = \frac{\omega}{c} \sqrt{\frac{1}{2}\sqrt{\epsilon'^2 + \left[(1 - \epsilon')\frac{\nu}{\omega}\right]^2} - \frac{\epsilon'}{2}} \tag{4.54}$$

Expanding the inner square root of (4.54) by assuming that for certain parts of the ionosphere $\sigma/\omega\epsilon \ll 1$, we arrive at a simplified expression for the attenuation factor:

$$\alpha = \frac{\sigma}{2}\sqrt{\frac{\mu}{\epsilon}}$$

which, with

$$\sqrt{\frac{\mu_0}{\epsilon_0}} = 120\pi$$

given in Section 2.1 and with σ as given by (4.23) becomes

$$\alpha = \frac{60\pi\sigma}{\sqrt{\epsilon'}} = \frac{60\pi Ne^2\nu}{\sqrt{\epsilon'}\,(\nu^2 + \omega^2)m} \tag{4.55}$$

Thus, at operating frequencies much higher than the collision frequency, the attenuation is approximately inversely proportional to the square of the operating frequency, intimating that, to reduce absorption, we should choose as high an operating frequency as possible, far enough from the maximum usable frequency.

When $\omega^2 \ll \nu^2$ (i.e., the wave's frequency is very much lower than the collision frequency), expression (4.55) becomes

$$\alpha = \frac{60\pi Ne^2}{\sqrt{\epsilon'}\nu m}$$

showing that under these conditions absorption may decrease with an increase in collision frequency. This probably is caused by electronic motions so confined that little energy is absorbed from the wave. This type of absorption is thought to occur during certain solar flare effects [Davies, 1966]. Differentiating (4.55) with respect to the collision frequency shows that maximum absorption occurs when the wave's frequency and the collision frequency are equal.

Martyn has shown [Davies, 1966] that the relation between the absorption of an obliquely incident wave and that of a vertically incident one is

$$L_{\phi 0} = L_v \cos\phi_0$$

where $L_{\phi 0}$ is the loss suffered by the wave incident at an oblique angle ϕ_0, and L_v is that suffered by the vertically incident one. This formula, often referred to as *Martyn's absorption theorem,* applies only to a plane ionosphere in the absence of the earth's magnetic field.

Ionospheric absorption depends on the collision frequency and occurs mainly in the lower layers at operating frequencies at which the waves pass through these ionospheric layers in the possibility of being reflected by higher ones. Because waves in these layers are not affected by refraction and emerge to travel in a direction parallel with the incident ones, their absorption is termed *nondeviative,* in distinction from deviative absorption occurring during curving of the trajectory due to refraction. The total attenuation computable from the integral of the absorption along the wave's trajectory depends on the length of the path through the layer.

As the *F*-region collision frequency is low, the deviative absorption of waves reflected there is small, except at frequencies near the maximum usable one. Nondeviative absorption predominates when the refractive index is near unity and the product of the electron density and collision frequency is large. Maximum nondeviative absorption occurs in a narrow region of the lower ionosphere, in the *D*-region, and extends in height from about 60 to about 100 km. Waves suffer from deviative absorption when the wave's trajectory bends and near the level of reflection or when the refractive index tends toward zero.

Example 4.4

Determine the ionospheric attenuation factor for the parameters obtained in Example 4.3.

Solution: The required quantities to be used with expression (4.55) are σ, ϵ', ν, and ω. The collision frequency of Example 4.3 is 10^3/s and the angular frequency $\omega = 2\pi 10^7$ rad/s. The conductivity and permittivity may be computed from expressions (4.23), as

$$\epsilon' = 1 - \frac{N\omega^2}{\epsilon_0 m(\nu^2 + \omega^2)}$$

$$\sigma = \frac{Ne^2}{m(\nu^2 + \omega^2)} \text{ S/m}$$

which, for $N = 0.55 \times 10^{11}/\text{m}^3$, $m - 9.11 \times 10^{-31}$ kg, $e = 1.59 \times 10^{-19}$, and $\epsilon_0 = 10^{-9}/36\pi$ F/m, yields $\epsilon' = 0.96$; $\sigma = 3.87 \times 10^{-10}$ S/m.

The attenuation factor then becomes

$$\alpha = \frac{60\pi Ne^2\nu}{\sqrt{\epsilon'}\,(\nu^2 + \omega^2)m} = 7.443 \times 10^{-5}/\text{km}$$

or -82.52 dB/km.

The solution of the wave equation leads to an expression for the phase velocity of the propagating wave and hence for the refractive index of the medium [Jordan, 1953]. A possible solution of the wave equation is

$$\mathbf{E} = \mathbf{V}\,\mathrm{e}^{\pm\gamma x}$$

where

$$\mathbf{V} = E_0\,\mathrm{e}^{j\omega t}$$

Then

$$\begin{aligned}\mathbf{E} &= \mathbf{E}_0\, e^{j\omega t}\, e^{\pm\gamma x}\\ &= \mathbf{E}_0\, e^{j\omega t}\, e^{\pm\alpha x} e^{\pm j\beta x}\\ &= \mathbf{E}_0\, e^{\pm\alpha x}\, e^{j(\omega t \pm \beta x)}\end{aligned}$$

Because the amplitude of the traveling wave must diminish with distance, the positive sign in the exponent may be disregarded and then

$$\mathbf{E} = \mathbf{E}_0\, e^{-\alpha x}\, e^{j(\omega t - \beta x)}$$

describes a wave traveling in the positive x-direction. The first exponential term yields the distance dependency of the wave's amplitude; the second one shows its time variation and phase.

Writing $\omega t = \beta x$, or

$$\frac{\omega}{\beta} = \frac{x}{t}$$

which has the dimension of velocity, when using expression (b), we obtain the phase velocity as

$$v_p = \frac{\omega}{\beta} = \frac{c}{\sqrt{\epsilon'\mu'}\sqrt{\frac{1}{2}\left(1 + \sqrt{1 + \frac{\sigma^2}{\omega^2\epsilon^2}}\right)}}$$

Further, according to expression (4.27), the refractive index $n = c/v_p$, so that the refractive index of the medium in which the wave progagates, is

$$n = \sqrt{\epsilon'\mu'}\sqrt{\frac{1}{2}\left(1 + \sqrt{1 + \frac{\sigma^2}{\omega^2\epsilon^2}}\right)}$$

Its value depends on the electrical properties of the medium and the frequency of the wave.

Most of the considered propagation occurs in the troposphere, where the relative permeability $\mu' = 1$ and the conductivity $\sigma \approx 0$, so that the expression for the refractive index reduces to

$$n = \sqrt{\epsilon'}$$

a value identical with that obtained earlier in expression (4.28).

Example 4.5

Determine the phase velocity and refractive index for the parameters of Example 4.4.

Solution: Because the conductivity obtained in Example 4.4 is very small, the expression for the phase velocity reduces to

$$v_p = \frac{c}{\sqrt{\epsilon'}} = \frac{3 \times 10^8}{0.98}$$

$$v_p = 3.06122 \times 10^8 \text{ m/s}$$

For the considered conditions, the phase velocity in the medium appears to be greater than the velocity of light. Actually, it means only that the wavelength in the medium is greater than it is in free space and not that energy is traveling faster than light.

4.6 PROPAGATION IN THE EARTH'S MAGNETIC FIELD

The presence of the earth's magnetic field complicates ionospheric propagation phenomena. Due to the influence of the magnetic field, the atmospheric plasmas become doubly refracting with the two refractive indices depending on the direction of the wave normal. The ionized gas in the upper regions of the atmosphere assumes the properties of an anisotropic medium. Due to the influence of the electric and magnetic fields the free electrons of the plasmas move in complicated paths, the character of which depends on the direction of propagation with respect to the direction of the external fields.

A free particle of mass m carrying a charge e in the ionosphere is acted on by two forces. The electric field of intensity $\mathbf{E}$ set up by a passing electromagnetic wave exerts a force $\mathbf{F}_e = e\mathbf{E}$, and the earth's magnetic field having an induction $\mathbf{B}_0$ exerts a force $\mathbf{F}_m = e(\mathbf{v} \times \mathbf{B}_0)$.

The total force $\mathbf{F}$, when neglecting frictional forces due to collision between electrons and gas molecules, is given by Lorentz's force equation:

$$\mathbf{F} = \mathbf{F}_e + \mathbf{F}_m = e(\mathbf{E} + \mathbf{v} \times \mathbf{B}_0) = m\frac{\partial \mathbf{v}}{\partial t}$$

The influence of the magnetic field on a particle moving with velocity **v** perpendicular to the magnetic field $\mathbf{B}_0$ is of interest for this study.

Assuming that the velocity and the field strength are constant, the radius of curvature r of the path will be constant and the electron will move on a circular path. The centrifugal force acting on the electron is mv^2/r. For equilibrium, the two forces to which the electron is subjected must be equal, or

$$\frac{mv^2}{r} = evB_0$$

from which the radius of the electron's trajectory's curvature is

$$r = \frac{mv}{eB_0}$$

The velocity of the particle, when the time required for one revolution over a circumference $2\pi r$, is

$$t = 2\pi r/v$$

leads to

$$t = 2\pi \frac{m}{eB_0}$$

The reciprocal of this time yields the frequency of revolution:

$$f = \frac{1}{t} = \frac{eB_0}{2\pi m}$$

For an electron, the ratio e/m is $1.59 \times 10^{-19}/9 \times 10^{-31} = 1.77 \times 10^{11}$ C/kg, so that the frequency becomes

$$f_{\mathrm{MHz}} = 2.817 \times 10^4 B_0$$

where B_0 is to be inserted in $\mathrm{Wb/m^2}$($1\,\mathrm{Wb/m^2} = 10^4$ G).

Because the horizontal intensity of the earth's magnetic field H varies considerably among different geographical locations and is inversely proportional to the distance from the center of the earth, the gyrofrequency varies accordingly. The magnetic field intensity near the magnetic poles is about (0.6 to 0.7) $\times 10^{-4}$ $\mathrm{Wb/m^2}$; it drops to 0.25×10^{-4} Wb/m^2 at a place 20° south of the equator and east of South America.

For the range of values of the earth's magnetic field, the gyrofrequency is on the order of 700 to 2000 kHz. Ions having a mass 1800 times greater than that of electrons exhibit a gyrofrequency in the audio frequency range outside the range of frequencies of interest. This indicates that we need to consider only electrons in the ionosphere when investigating the influence of the earth's magnetic field on electromagnetic wave propagation.

The trajectory of particles becomes more complicated when influenced by a variable electric field. Additional longitudinal and transverse components of the electron movements may develop, which may give rise to longitudinal and transverse components of the electric field. These produce a rotation of the plane of polarization and thus cause double refraction, a characteristic of an anisotropic medium.

When the frequency of the electromagnetic radiation is near the gyrofrequency, the electrons move along a spiral trajectory with a velocity higher than that in the absence of the earth's magnetic field. The increase in velocity raises the possible number of collisions of the electrons with neutral molecules and ions. Each charge carrier absorbs a large amount of energy from the wave and dissipates it completely on collision. This causes attenuation over most of the broadcast band to become much higher than without the presence of the magnetic field. The attenuation at frequencies removed from the gyrofrequency is given by expression (4.55).

Because the properties of an anisotropic medium change from point to point, the electric displacement, according to Doluchanow [1956], becomes

$$D_i = \Sigma\, \epsilon_{ik} E_k$$

with $k = 1, 2, 3$ denoting the rectangular coordinates x, y, and z. Only the terms of identical indices are to be summed.

The quantities ϵ_{ik} are tensor permittivities in the direction of the three axes that, in an anisotropic medium, differ from one another.

If $\mathbf{r}$ is the electron displacement from its initial position, the polarization of an ionized gas may be expressed by

$$\mathbf{P} = -eN\mathbf{r} \tag{4.56}$$

and the relation between $\mathbf{D}$ and $\mathbf{P}$ becomes

$$\mathbf{D} = \epsilon_0 \mathbf{E} + \mathbf{P} \tag{4.57}$$

Assuming propagation in the x-direction (Figure 2.1) causes the differentials of the field quantities in the y- and z-directions to be equal to zero; and the longitudinal and transverse magnetic field components, where propagation along an axis is at an angle β with the geomagnetic field H_0 to be

$$H_L = H_0 \cos\beta$$
$$H_T = H_0 \sin\beta$$

The aim of this study is to find expressions for the index of refraction, on which the phase velocity depends and field amplitudes under the assumed conditions. Taking the curl of the Faraday-Maxwell equation:

$$\nabla \times \nabla \times \mathbf{E} = -\mu_0 \left(\nabla \times \frac{\partial \mathbf{H}}{\partial t} \right)$$

and inserting the Ampère-Maxwell equation leads to

$$\nabla \times (\nabla \times \mathbf{E}) = -\mu_0 \frac{\partial^2 \mathbf{D}}{\partial t^2}$$

Hence, using the previously given vector equality:

$$\nabla^2 \mathbf{E} - \nabla(\nabla \cdot \mathbf{E}) = -\nabla \times (\nabla \times \mathbf{E}) \tag{4.58}$$

yields

$$\nabla^2 \mathbf{E} - \nabla(\nabla \cdot \mathbf{E}) = \mu_0 \frac{\partial^2 \mathbf{D}}{\partial t^2} \tag{4.59}$$

thus eliminating the variable **H** from Maxwell's equations.

Because the divergence of the electric displacement for the considered conditions, as given by (4.57), is zero, we obtain

$$\nabla \cdot \mathbf{E} = -\frac{1}{\epsilon_0} (\nabla \cdot \mathbf{P}) \tag{4.60}$$

which, with expression (4.59), yields

$$\nabla^2 \mathbf{E} + \frac{1}{\epsilon_0} \nabla(\nabla \cdot \mathbf{P}) = \mu_0 \frac{\partial^2 \mathbf{D}}{\partial t^2} \tag{4.61}$$

This expression represents the propagation of an electromagnetic wave in an anisotropic medium.

In a uniformly polarized medium, the E_y and E_z components of the electric field in the y- or z-direction are constant; hence, expression (4.60) intimates that

$$\nabla_y(\nabla \cdot \mathbf{P}) = 0$$
$$\nabla_z(\nabla \cdot \mathbf{P}) = 0$$

so that expression (4.61) may be given by the sum of the appropriate projections of two harmonically varying vectors as

$$\begin{aligned} \frac{\partial^2 E_y}{\partial x^2} + \mu_0 \omega^2 D_y &= 0 \\ \frac{\partial^2 E_z}{\partial x^2} + \mu_0 \omega^2 D_z &= 0 \end{aligned} \tag{4.62}$$

Further, because the charges are assumed to be distributed with statistical uniformity so that there is no permanent space charge (however, the contributions of all electrons and ions to the electric polarization **P** are included), the differential form of Gauss's theorem that results from the inverse square law of force in electrostatics reads

$$\nabla \cdot \mathbf{D} = \frac{\partial D_x}{\partial x} = 0$$

and the x-component of the electrical displacement follows from (4.57) as

$$D_x = \epsilon_0 E_x + P_x = 0 \tag{4.63}$$

Expressing the movement of electrons in the field of an electromagnetic wave by the Lorentz force equation when $\omega \gg \nu$ (i.e., the wave's frequency is very much higher than the collision frequency so that collision may be neglected [Budden, 1985]), as

$$m \cdot \frac{\mathrm{d}^2\mathbf{r}}{\mathrm{d}t^2} = -e\mathbf{E} - e\left(\frac{\mathrm{d}\mathbf{r}}{\mathrm{d}t} \times H_0\right)$$

and assuming that ν is the same for all electrons, the force equation may be multiplied by the product of the electron density N and the electronic charge e that, with $\mathbf{r} = \mathbf{R}\, \mathrm{e}^{j\omega t}$ and the previously defined polarization of an ionized gas, expression (4.56), leads to

$$-\omega^2\mathbf{P} = \frac{Ne^2}{m}\mathbf{E} - j\frac{\omega e}{m}\left(\mathbf{P} \times \mathbf{H}_0\right) \tag{4.64}$$

Expressing components of the earth's magnetic field coinciding with the coordinate axes by the previously defined longitudinal and transverse field components as

$$H_{0x} = H_L$$
$$H_{0y} = 0$$
$$H_{0z} = H_T$$

and introducing

$$\frac{Ne^2}{\epsilon_0 m} = \omega_0^2$$
$$\frac{eH_T}{m} = \omega_H \sin\beta = \omega_T$$
$$\frac{eH_L}{m} = \omega_H \cos\beta = \omega_L$$

we obtain, due to $H_0 = \sqrt{H_L^2 + H_T^2}$,

$$\frac{eH_0}{m} = \omega_H$$

an expression similar to the expression for the gyrofrequency.

Because only perpendicular magnetic field components are of interest in the Lorentz force equation and $H_{0y} = 0$, only the z- or x-component of this field appears in the first and third of the following expressions. The x- and z-components are used in the second one.

The components of (4.64) in the direction of the three axes then are

$$-\omega^2 P_x = \omega_0^2\epsilon_0 E_x - j\omega\omega_T P_y$$
$$-\omega^2 P_y = \omega_0^2\epsilon_0 E_y - j\omega\omega_L P_z + j\omega\omega_T P_x$$
$$-\omega^2 P_z = \omega_0^2\epsilon_0 E_z + j\omega\omega_L P_y$$

The first one of these, with $E_x = -P_x/\epsilon_0$ obtained from expression (4.63), results in

$$P_x = j\frac{\omega\omega_T}{\omega^2 - \omega_0^2} P_y \tag{4.65}$$

which, when inserted in the remaining two expressions, provides formulas for the polarization components of the electric field:

$$\begin{aligned} P_x &= -\epsilon_0 E_x \text{ (from [4.63])} \\ P_y &= \epsilon_0 \frac{X}{Y} E_y + j\epsilon_0 \frac{\omega_L}{\omega}\frac{X}{Y} E_z \\ P_z &= -j\epsilon_0 \frac{\omega_L}{\omega}\frac{X}{Y} E_y + \epsilon_0 \frac{Z}{Y} E_z \end{aligned} \tag{4.66}$$

Here,

$$\begin{aligned} X &= \omega_0^2(\omega^2 - \omega_0^2) \\ Y &= (\omega^2 - \omega_0^2)(\omega_L^2 - \omega^2) + \omega^2\omega_T^2 \\ Z &= \omega_0^2(\omega^2 - \omega_0^2 - \omega_T^2) \end{aligned}$$

Expression (4.57) with those of (4.66) leads to

$$\begin{aligned} D_x &= 0 \\ D_y &= \epsilon_0 E_y + P_y = \epsilon_0\left(1 + \frac{X}{Y}\right) E_y + j\epsilon_0 \frac{\omega_L X}{\omega Y} E_z \\ D_z &= \epsilon_0 E_z + P_z = -j\epsilon_0 \frac{\omega_L}{\omega}\frac{X}{Y} E_y + \epsilon_0\left(1 + \frac{Z}{Y}\right) E_z \end{aligned} \tag{4.67}$$

thus describing the components of the displacement in the direction of the axes.

Equation (4.67) inserted into (4.62) yields

$$\begin{aligned} \frac{\partial^2 E_y}{\partial x^2} + \epsilon_0\mu_0\omega^2\left[\left(1 + \frac{X}{Y}\right) E_y + j\frac{\omega_L X}{\omega Y} E_z\right] &= 0 \\ \frac{\partial^2 E_z}{\partial x^2} + \epsilon_0\mu_0\omega^2\left[-j\frac{\omega_L}{\omega}\frac{X}{Y} E_y + \left(1 + \frac{Z}{Y}\right) E_z\right] &= 0 \end{aligned} \tag{4.68}$$

For harmonic waves propagating in a homogeneous, nonabsorbing medium, the coefficients, X, Y, and Z are constant and the solution of expressions (4.68) may be obtained by adopting

$$E_y = E_{0y} \exp\left(-j \frac{\omega}{c} nx\right)$$

$$E_z = E_{0z} \exp\left(-j \frac{\omega}{c} nx\right)$$

where n is the index of refraction of the ionized gas, x is the distance from the source, and c the velocity of propagation of electromagnetic waves in free space. Then, as $\epsilon_0\mu_0 = 1/c^2$ (2.15), we arrive at the expressions:

$$\left[\left(1 + \frac{X}{Y}\right) - n^2\right] E_{0y} + j \frac{\omega_L}{\omega} \frac{X}{Y} E_{0z} = 0 \tag{4.69}$$

$$-j \frac{\omega_L}{\omega} \frac{X}{Y} E_{0y} + \left[\left(1 + \frac{Z}{Y}\right) - n^2\right] E_{0z} = 0 \tag{4.70}$$

This homogeneous system of equations has nontrivial solutions if and only if its coefficient determinant is zero [Kreyszig, 1967]; that is,

$$\begin{vmatrix} 1 + \frac{X}{Y} - n^2 & j \frac{\omega_L}{\omega} \frac{X}{Y} \\ -j \frac{\omega_L}{\omega} \frac{X}{Y} & 1 + \frac{Z}{Y} - n^2 \end{vmatrix} = 0$$

The indices of refraction then are solutions of the appropriate biquadratic equation given by the determinant.

Then,

$$n_{1,2}^2 = 1 - \frac{\omega_0^2}{\omega^2 - \dfrac{\omega^2\omega_T^2}{2(\omega^2 - \omega_0^2)} \mp \sqrt{\left[\dfrac{\omega^2\omega_T^2}{2(\omega^2 - \omega_0^2)}\right]^2 + \omega_L^2\omega^2}} \tag{4.71}$$

The computed indices of refraction apply only when collisions of electrons with neutral molecules and ions are neglected. Consideration of these complicates the preceding formulas to a great extent, as shown by the Appleton formula [Davies, 1966].

The polarization of an electromagnetic wave propagating in an ionized gas is given by the ratio of the amplitudes of the two constituting components of the field that follows from expression (4.69) as

$$\left(\frac{E_{0y}}{E_{0z}}\right)_{1,2} = j\left[\frac{\omega_T^2\omega}{2(\omega^2 - \omega_0^2)\omega_L} \pm \sqrt{\left[\frac{\omega_T^2\omega}{2(\omega^2 - \omega_0^2)\omega_L}\right]^2 + 1}\right] \tag{4.72}$$

The longitudinal component of the electric field may be obtained from expressions (4.63), (4.65), and (4.66):

$$E_x = -\frac{P_x}{\epsilon_0} = \frac{\omega\omega_T\omega_0^2}{Y}\left(-jE_y + \frac{\omega_L}{\omega}E_z\right) \tag{4.73}$$

The two solutions for the index of refraction and amplitude ratio are due to the occurrence of double refraction. The top sign and the index 1 in the expression for the refractive index belong to the extraordinary ray, whereas the lower sign and the index 2 to the ordinary one. In accordance with expression (4.27) we find that, due to double refractivity of the medium, the ordinary and extraordinary waves travel with different phase velocities.

In agreement with definitions used in optics, the linearly polarized ray, having an E_{0z} component unaffected by the magnetic field, is termed an ordinary ray, whereas one having an E_{0y} component is an extraordinary one. Several variants of propagation of electromagnetic waves in a homogeneous ionized gas in the presence of a permanent magnetic field may be studied [Doluchanow, 1956].

4.6.1 Absence of the Magnetic Field

Because the longitudinal and transverse magnetic fields, in this case, equal zero, the corresponding angular frequencies ω_L and ω_T also must disappear, so that the expression for the refractive index (4.71), reduces to

$$n^2 = \epsilon_i' = 1 - \left(\frac{\omega_0}{\omega}\right)^2 = 1 - \frac{Ne^2}{\epsilon_0 m\omega^2} \tag{4.74}$$

and the components of the displacement given earlier by expressions (4.67) become

$$D_x = 0$$

$$D_y = \epsilon_0\left[1 - \left(\frac{\omega_0}{\omega}\right)^2\right]E_y$$

$$D_z = \epsilon_0\left[1 - \left(\frac{\omega_0}{\omega}\right)^2\right]E_z$$

Then, because the tensor permittivities of these two electrical displacement components are equal, we may conclude that the medium is isotropic.

4.6.2 Direction of Propagation Parallel with the Magnetic Field

In this case there is neither a transverse field nor a transverse angular frequency; that is, $H_T = 0$ and $\omega_T = 0$. This causes the refractive index to become

$$n_{1,2}^2 = 1 - \frac{\omega_0^2}{\omega(\omega \mp \omega_L)} \tag{4.75}$$

and the ratio of the amplitudes, as given by (4.72), reads

$$\left(\frac{E_y}{E_z}\right)_{1,2} = \pm j \tag{4.76}$$

indicating that the field is due to two opposite circularly polarized waves. The top sign in these two expressions refers to a clockwise polarization, whereas the bottom one refers to a counterclockwise one.

Inserting $\omega_T = 0$ into expression (4.73) for E_x causes $E_x = 0$, thus showing that under the considered conditions no longitudinal component of the field exists.

A linearly polarized wave may result from the combination of two circularly polarized ones. Whereas the previous two constituting waves at every point in space are in phase and the result is linearly polarized, the plane of polarization changes with the distance traveled. When the direction of propagation coincides with that of the magnetic field, electrons rotate in a plane perpendicular to the direction of propagation. Only one of the two opposite circularly polarized waves will agree in sense with that of the rotating electrons. The propagation conditions of the two waves differ, and this causes a rotation of the plane of polarization.

4.6.3 Direction of Propagation Perpendicular to the Magnetic Field

In this case, the longitudinal field component H_L, as well as the angular frequency ω_L, is zero [Doluchanow, 1956].

Expression (4.71) for the refractive index then yields

$$n_1^2 = 1 - \frac{\omega_0^2}{\omega^2 - \dfrac{\omega^2\omega_T^2}{\omega^2 - \omega_0^2}} \tag{4.77a}$$

$$n_2^2 = 1 - \frac{\omega_0^2}{\omega^2} \tag{4.77b}$$

where the second expression is identical with that obtained for the case without a magnetic field.

Expression (4.72) for the ratio of the amplitudes of the field components results in

$$\left| \left(\frac{E_{0y}}{E_{0z}} \right)_1 \right| = \infty$$

indicating that although $E_{0y} \neq 0$, the z-component E_{0z} is zero. Further,

$$\left| \left(\frac{E_{0y}}{E_{0z}} \right)_2 \right| = 0$$

or $E_{0y} = 0$ and $E_{0z} \neq 0$.

The index n_1 describes the propagation conditions of a linearly polarized wave having its electric field vector E_{0y} and direction of propagation perpendicular to the direction of the magnetic field. Index n_2 belongs to a wave traveling also in a direction perpendicular to the magnetic field, but having a field vector pointing in the direction of the magnetic field.

A linearly polarized ray, the electric field vector of which is at an angle between zero and $\pi/2$ rad to the magnetic field splits into linearly polarized ordinary and extraordinary rays. The ordinary one propagates as if the magnetic field were not present. The extraordinary one, which has a longitudinal electric field component shifted by $\pi/2$ rad with respect to the transverse one, propagates with a velocity depending on the appropriate refractive index.

No splitting occurs when the electric field coincides with either the z- or y-axis. Assuming the electric field and electron movement to be in the direction of the z-axis, we find that the previously defined Lorentz force becomes zero, thus indicating that the magnetic field does not influence the propagation. The ray propagates only as an ordinary one. When the electric field coincides with the y-axis, the electrons rotate in the x,y plane and cause an alteration of the velocity of propagation and the appearance of a longitudinal component of the electric field. The ray propagates only as an extraordinary one.

A linearly polarized wave propagating in any direction with respect to the direction of a permanent magnetic field generally splits into two elliptically polarized ones possessing longitudinal components of the electric field. The rays propagate with velocities determined by their appropriate refractive indices. The major axes of the ellipses are mutually perpendicular and the resulting vectors rotate in opposite senses.

Example 4.6

Determine the refractive indices for a 10 MHz wave propagating in a direction perpendicular to the earth's magnetic field when the electron density is taken to be $N = 0.55 \times 10^{11}$ electrons/m^3.

Solution: The two indices are given by expressions (4.77a) and (4.77b). The parameters ω_0 and ω_T may be determined from

$$\omega_0^2 = \frac{Ne^2}{\epsilon_0 m}, \quad \omega_T = \frac{eH_T}{m}$$

where $H_T = H_0 \sin\beta$ with $H_0 = 0.5 \times 10^{-4}$ Wb/m^2, the average value of the geomagnetic field, and β, the angle between this field and the axis of the propagation path.

In the considered case, $\beta = \pi/2$ rad; hence, $H_T = H_0$ and then

$$\omega_0^2 = 1.7262 \times 10^{14}$$

$$\omega_T = 8726673.99 \text{ rad}$$

and the refractive indices $n_1 = 0.98$ and $n_2 = 0.98$.

4.7 SOLAR ACTIVITY

Field strength and critical frequency observations show a marked correlation with solar activity that exhibits almost regular periodic variations over the years. The observed period is about 11.1 years with a deviation of ± 4 years. Variations of solar activity manifest themselves in the changing number of sunspots (Figure 4.10), intensity and number of solar flares, the frequency of solar eruptions, and various other phenomena. Because the relative Zurich or Wolf number of sunspots R is easily observed, many mathematical expressions concerning propagation phenomena are based on it. It may be expressed in terms of f, the total number of sunspots on the solar hemisphere facing the earth and g, the prevailing number of spot groups, by

$$R = k(f + 10g) \tag{4.78}$$

where k, usually near unity, is a factor depending on the power of the telescope used for the observations. It reduces all solar spot observations to the oldest available sun-

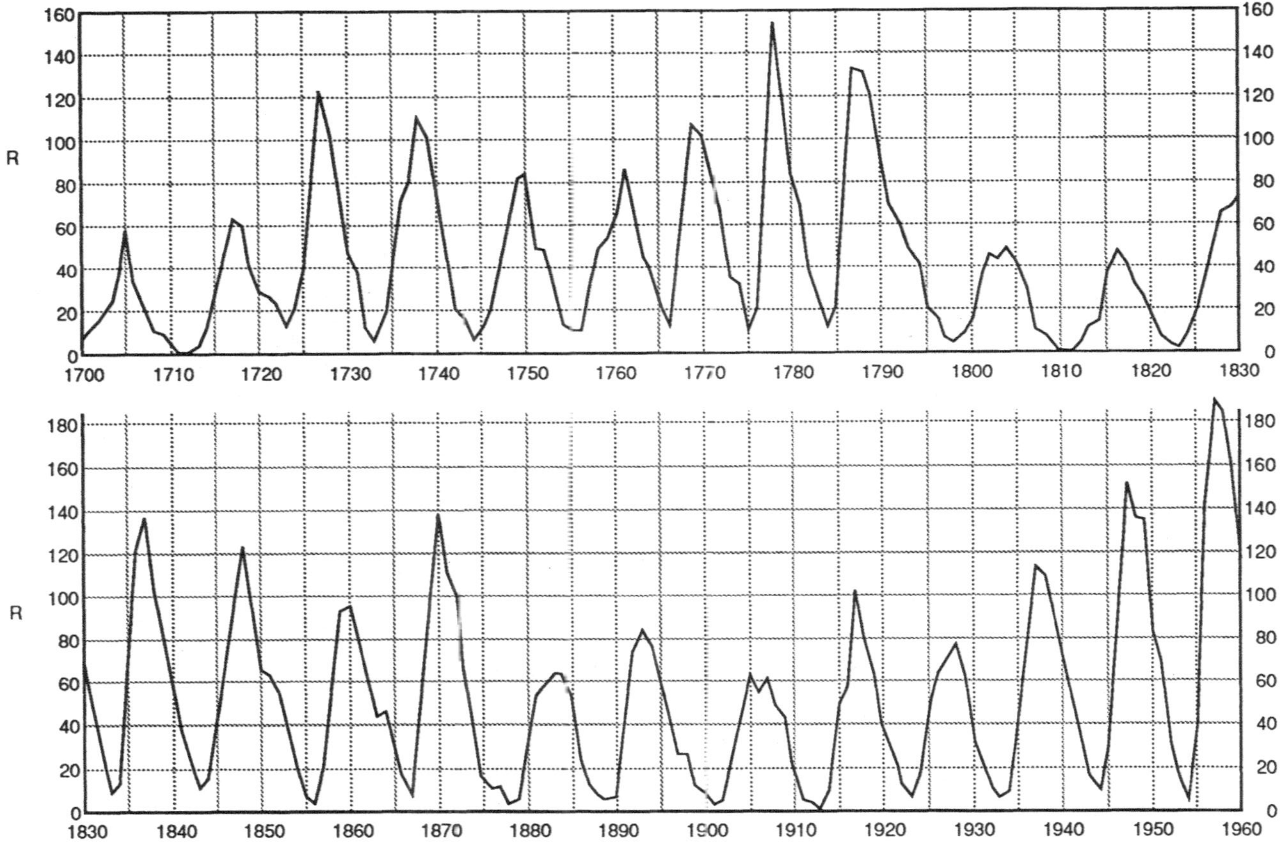

Figure 4.10 Sunspot number variation over the years (after Waldmeier, 1961, © Schulthess Polygraphischer Verlag).

spot data obtained at the Zurich observatory in 1749 with a 3-inch telescope for which $k = 1$. The value of R is not the same at all sunspot maxima but varies widely, for example, it reached a count of 154 in 1778 but only about 50 in 1804; relative sunspot numbers at minima exhibit smaller variations [Doluchanow, 1956].

A general formula for the dependence of the critical frequency of an ionospheric region on the Zurich number, as given by Al'pert, is

$$f_c = f_{c0}(1 + aR) \tag{4.79}$$

where f_{c0} is the critical frequency of the region of interest at the epoch of minimum solar activity ($R = 0$) and a is a coefficient that varies but little with the seasons and usually is given as a constant, so for example for the E-region

$$a_E \approx 0.001 - 0.002$$

for the F_1-region

$$a_{F_1} \approx 0.01$$

and, for the F_2-region, it is

$$a_{F_2} \approx 0.005 - 0.01$$

The influence of the 11-year period of solar activity is most pronounced in the ionization of the F_2-region, as illustrated by Figure 4.11, showing average ionograms for June and December of a year of low solar activity in 1933 through years of increasing solar activity to a maximum in 1937. We can see that over the years the critical frequency of the F_2-region almost doubled corresponding to a fourfold increase of the charge density in the region's maximum. Figure 4.11 shows the dependence of region height on the sunspot cycle; the middle of the region is higher during the sunspot maximum than during the minimum; however, because the height of the bottom of the region does not change to a great extent, it must be assumed that the region's thickness increases during a sunspot maximum.

A similar increase in the critical frequency of the F_1- and E-regions during sunspot maxima also may be observed. In addition, an increase of D-region attenuation has been noted. These phenomena may be ascribed to the extensive variation of the ionizing radiation (i.e., basically that of the solar ultraviolet) during the solar spot period.

Note that the solar ultraviolet and corpuscular radiation is not directly related to the sunspots. This follows from observations showing that the ionospheric variations and sunspot relative numbers correspond only when the analysis extends over

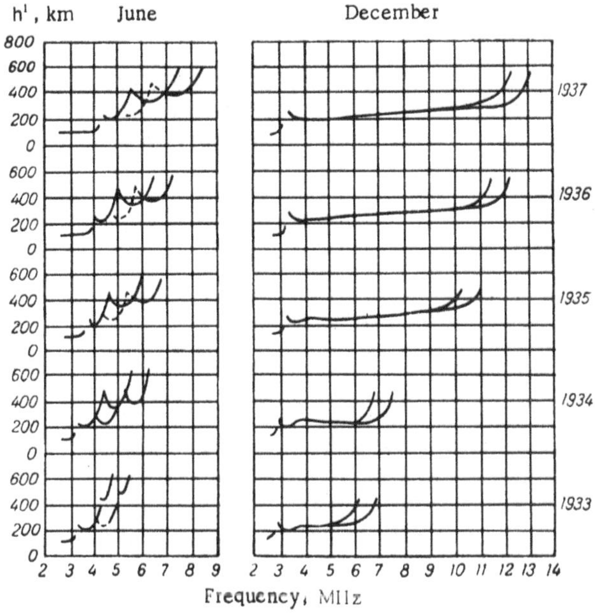

Figure 4.11 Monthly northern mean noon ionograms for June and December for years of increasing solar activity (after Doluchanow, 1956, © VEB Verlag).

periods of at least one year. The correspondence of unique events is less good, and this leads us to believe that the sunspots are not the prime cause but, together with other observable solar phenomena that cause more or less sporadic effects, are part of a common periodically variable agent designated by the cumulative name of solar activity [Doluchanow, 1956]. An interesting 27-day variation of the ionization of the higher atmosphere, corresponding to the sun's rotation around its own axis, may also be observed.

Although these empirical relations of the critical frequency and the sunspot numbers yield results very close to those obtained experimentally, the nature of the relations has not been adequately investigated so far. No exhaustive explanation of the cyclic variation of the ionospheric processes has been found to date.

Increased intensity solar X-ray radiation occurs during chromospheric eruptions. The annual number of eruptions is proportional to the sunspot number [Picquenard, 1974]. Solar X-ray radiation, depending on its wavelength and hence height of absorption, contributes to the formation of ionized regions. Longwave X-rays, absorbed at greater heights, contribute to the formation of the *F*-region, whereas shorter ones, which penetrate deeper into the atmosphere, give rise to the lower ionized regions.

4.8 THE IONOSPHERIC REGIONS

As mentioned earlier, the ionosphere consists of a number of regions that affect the propagation of electromagnetic waves at various frequencies in different ways. The following few words are devoted to the description of region behavior in a quiet and also disturbed ionosphere.

4.8.1 The *D*-Region

Little is known about the ionosphere below heights of 60 km. The *D*-region, the lowest ionospheric region at a height of between 60 and 90 or 100 km, manifests itself practically only in daytime, when it exhibits very strong absorption of electromagnetic radiations at frequencies around 1 MHz. Due to a low charge carrier density of about $1 \times 10^8/m^3$ to about $1 \times 10^{10}/m^3$ and its low critical frequency of between 100 and 700 kHz, it does not reflect high-frequency radiations above about 5 MHz, which generally are used for ionospheric soundings, but it reflects very efficiently very low and low frequency waves and thus accounts partly for their propagation to very great distances.

Long-term observations [Gibbons and Waynick, 1957] show that reflection of very low-frequency radiations occur at heights of about 76 km in daylight and about 90 km at twilight. Very low-frequency waves reaching the region at low grazing angles may be reflected at lower heights. The daytime *D*-region appears as a weakly defined ionization maximum at a height between 75 and 90 km followed by a decrease in ionization above 90 km where the daytime *E*-region begins. During periods of sunspot minima, a subsidiary ionization maximum, at heights of 65 to 70 km, may occur at mid-latitudes during the predawn and morning hours; this has been named the *C*-region in contrast to the earlier assumption of the *D*-region consisting of two part-regions at these heights.

According to Al'pert, measured predawn and early morning *C*-region electron concentrations show a maximum of 8×10^7 electrons/m^3 at a height of 65 km and a minimum of 1×10^7 electrons/m^3 at 80 km. However, as the day progresses, the *C*-region disappears as the electron concentration of the lower atmosphere increases.

The experimental investigation of the *D*-region relies on measuring the daytime attenuation in the short-, medium-, and long-wave ranges and on the observation of the reflection of waves in the very low-frequency range. Weak reflections of shortwaves at heights of between 60 and 80 km occasionally are observed during periods of low disturbances in winter.

The electron concentration in the *D*-region is determined by the sun; substantial rises in electron concentration due to increased solar X-ray emission were observed during periods of increased solar activity. The influence of the sun on nighttime *D*-region ionization has been observed as well; this is due to scattered Ly-α radiation.

The *D*-region ionization was found to vary with the seasons and geographical latitude. In winter, when the atmosphere circulates better and complex molecular particles may be transferred from the lower atmosphere into the *D*-region, a greater variability of electron density occurs at fixed heights. This causes an increased variation in ionospheric absorption. A seasonal asymmetry was observed, too; it is caused by the electron densities at equinoxes being more in agreement with those of the preceding season, thus exhibiting a certain inertia in adjusting to the densities of the new season.

Although there is little difference between the height distribution of electron density profiles for the same zenith angle at different geographical latitudes, at zenith angles below 80° the height distribution of electron densities is less strongly controlled in high latitudes than in lower ones. Some anomalies of this effect exist at the polar regions.

The regular, almost complete, disappearance of the *D*-region at sunset and the rapid recombination of charge carriers can be attributed to the large collision frequency of between $1 \times 10^7/\mathrm{m}^3/\mathrm{s}$ and $1 \times 10^8/\mathrm{m}^3/\mathrm{s}$. The reappearance of the region at sunrise and its daytime behavior's dependence on the sun's position suggest that the source of ionization is the sun's ultraviolet radiation, although it is not known which atmospheric component is being ionized and what is the ratio of electrons and ions. Observations favor the view that electrons are the active charge carriers.

4.8.2 The *E*-Region

The *E*-region, sometimes called the *Kennelly-Heaviside region,* is at a height of about 120 km above the earth and is relatively sharply defined in a thickness of between 20 and 40 km. The maximum daytime charge carrier density at middle geographical latitudes, generally, is on the order of 1×10^{11} electrons/m^3 and, although the layer persists at night, the charge carrier density is greatly reduced. It varies by 50 to 60% during the sunspot cycle [Appleton, 1957]. This charge carrier density corresponds to a critical frequency of about 4 MHz; the nighttime critical frequency drops to about 250 kHz during periods of sunspot minima and 500 kHz at sunspot maxima. These correspond to a charge carrier density on the order of 10^9 electrons/m^3 [Dolchanow, 1956].

The dependence of the *E*-region critical frequency on the Zurich sunspot number is given by the empirical formula

$$f_c = 0.9[(180 + 1.44R)\cos\chi]^{0.25} \tag{4.80}$$

which yields the diurnal and seasonal critical frequencies usually within 0.2 MHz of the observed value [Davies, 1966]. The exponent may vary between 0.1 and 0.4 for diurnal variations, but 0.25 yields a reasonable approximation for seasonal and diurnal variations.

Because the collision frequency is rather large, with values of between 5×10^3 and 2×10^4 m^3/s, the absorption is correspondingly high [Picquenard, 1974].

The diurnal electron density variation exhibits a direct dependence on the cosine of the sun's zenith angle and hence a maximum at local noon. A slight pre-noon depression of the $N = f(t)$ curve has been observed by Appleton, Lyon, and Turnbull [Appleton, 1957]. A similar seasonal regularity of the electron density profile with some variation of the mean noon value over the year also has been reported.

The *E*-region is not a quiescent medium and Chapman's theory does not apply directly to it. Chapman's theory has been modified by Appleton and Lyon to account for the more complicated *E*-region conditions. A vertical drift of electrons from the *E*-region, according to Martyn, perturbs the F_2-region.

E-region ionization is of solar origin. Observations have shown that, although the bulk of the ionizing radiation comes from the sun's visible disk, some radiation from the active spots and from the corona contribute to it. Soft X-rays seem to cause part of the *E*-region ionization. The recombination coefficient that has a value of between 1.5×10^{-2} and 4×10^{14}/m^3/s was determined by these observations. It causes the region's rapid response to variations of illumination.

Appleton found that a value of the constant a_E in expression (4.79) of approximately 0.004, which varies very little with the seasons, offers a better correlation between the critical frequency and sunspot number variations. At times, another, slightly greater maximum of ionization is noticeable at a height of about 140 km.

The *E*-region persists at night, however, with a greatly reduced ion density. So far there is no known nighttime ionizing source. Similarly, winter densities also are lower due to the obliquity of the sun's ionizing effect during this season. Ionization of molecular oxygen accounts most probably for the *E*-region's existence.

4.8.3 The *F*-Region

The *F*-region appearing at heights above about 150 km is known to split in summer into two subregions the behavior of which differ to a great extent. So far there is no known reason for this splitting. During periods of sunspot minima the two subregions are believed to be formed by the same solar ionizing radiation [Martyn, 1959].

The ionization peak of the lower region, usually called the F_1*-region,* is found at heights of about 160 km, whereas that of the higher one, the F_2-region, at about 250 km. Thus, the F_1-region is closer to the *E*-region than to the F_2 one. The F_1-region does not seem to exist at night or in winter.

The F_1-region substantially satisfies Chapman's conditions. A peak noon electron density of 2.5×10^{11} electrons/m^3 was found at the equator at equinoxes during sunspot minima. Decay of ionization follows the recombination law, expression (4.6), with $\alpha \approx 8 \times 10^{-9}$. A delay of about 4 minutes between noon and the occurrence of the ionization maximum indicates that this is proportional to $\sqrt{\cos\chi}$, where χ is the zenith angle of the sun at the chosen location. Long-term observations show some small deviations of the ionization levels from Chapman's theory. These are assumed to be due to local electric currents associated with solar and lunar magnetic variations [Martyn, 1959].

In the first approximation the critical frequency of this subregion, according to Davies, is

$$f_c = (4.3 + 0.01R)\cos^{0.2}\chi \tag{4.81}$$

with the exponent varying not only from place to place but also with the seasons and sunspot numbers.

The higher F_2-region accounts for long-range shortwave radio communication. It may be found at heights beyond about 250 km above the earth, and its structure is more complicated than that of the F_1-region. It was found to depend on the magnetic latitude and therefore appears to be asymmetrical about the geographic equator [Martyn, 1959].

The critical frequencies that can be computed from expression (4.79) exhibit pronounced differences between summer and winter behavior, as shown by Figure 4.12. Whereas the summer critical frequency varies but little during the day, in winter the critical frequency rises sharply at sunrise to its maximum value at about noon and drops more gradually to its nighttime value. The region's height differences can be seen from Figure 4.13. Although the F_1-region stays in summer at the winter heights of the F-region, the F_2-region height rises at sunrise to drop at sunset and reunites with the F_1-region at about the all-year nighttime height of the F-region. A nighttime appearance of the F_2-region has been experienced during times of auroras.

Contrary to the F_1-region critical frequency's dependence on the sun's position, the critical frequency of the F_2-region is independent of sun position; however, it depends on the relative sunspot number, as shown by (4.79), and may exhibit fluctuations of up to 15% of the mean value even during a quiet ionospheric day.

In addition to the described time dependency, the F_2-region is found also to vary with geographical location. The critical frequency drops at or near the magnetic equator during daylight hours at sunspot minima. At sunspot maxima the drop is smaller and persists until after midnight. Curves joining locations of equal critical frequencies show that the critical frequency rises in an inverse relation to the geographical latitude. Bunchings of the curves caused by the earth's magnetic field may be observed on either side of the geographic equator, as shown by Figure 4.14 depicting world maps of isopleths for the monthly mean critical frequency of the F_2-region at midnight and noon (April 1958) during maximum solar activity. Due to the difference in the positions of the magnetic and geographic equators, a longitudinal effect on the critical frequency is present. These phenomena indicate that the earth's

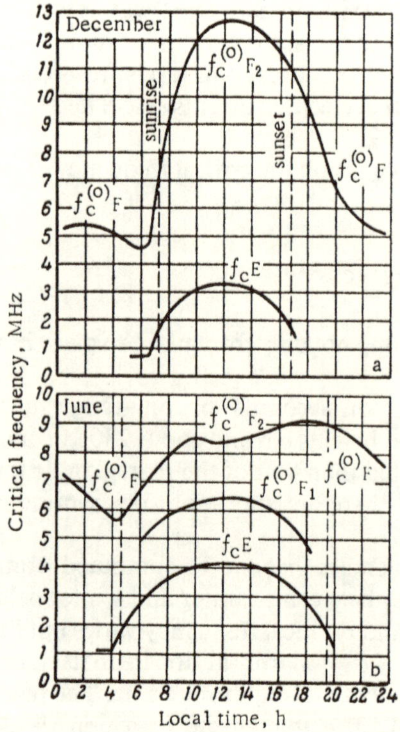

Figure 4.12 Monthly mean diurnal variation of ordinary wave critical frequencies in mid-latitudes during the northern winter and summer (after Doluchanow, 1956, © VEB Verlag).

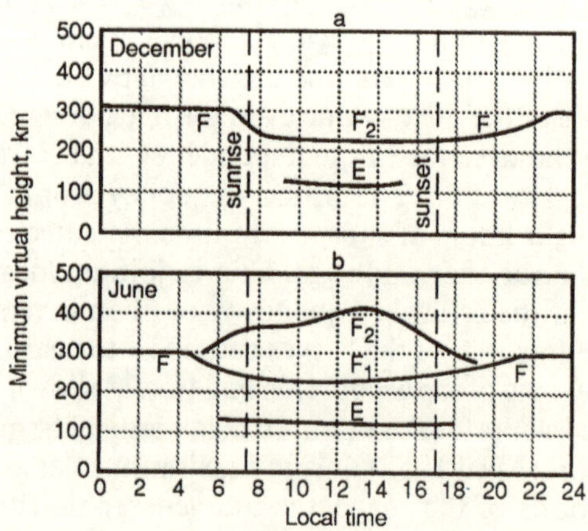

Figure 4.13 Mean diurnal variation of the minimum virtual heights in northern mid-latitudes (after Doluchanow, 1956, © VEB Verlag).

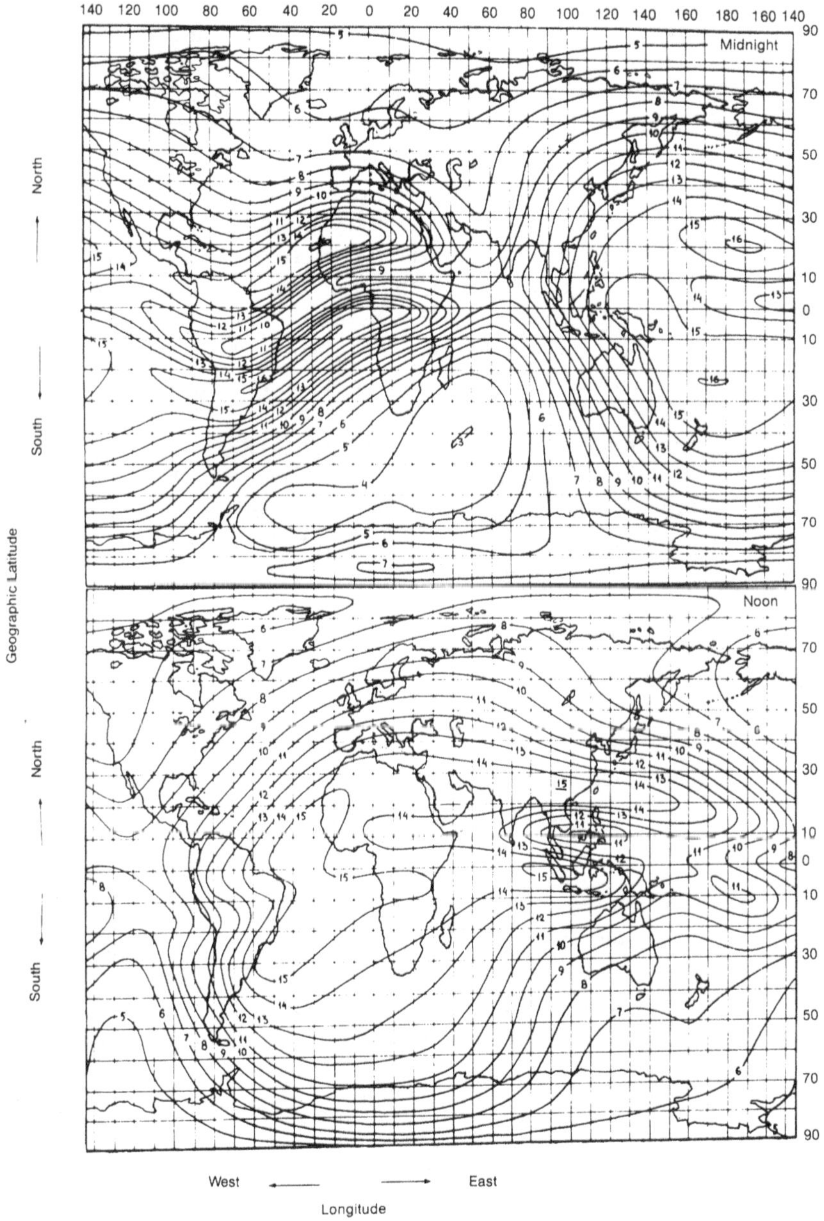

Figure 4.14 Mean midnight and noon F_2-region critical frequency distribution for April 1958 during a period of maximum solar activity (after Al'pert, 1973, © Vsesojuznoje Agentstvo).

magnetic field, by influencing the charge carrier movement at the higher levels of the ionosphere, has an effect on the structure of the F_2-region. Some dependence on the earth's land masses also has been observed [Doluchanow, 1956]. It is interesting to note that the total ionization of the F_2-region (i.e., that of the northern and southern hemispheres), rises during the northern winter. No explanation for this is available so far.

Ionograms of the F_2-region show magneto-ionic splitting that, due to the great difference in the apparent heights of the ordinary and extraordinary rays, becomes particularly prominent at the vicinity of the critical frequency. A third component of the splitting occasionally also is noticeable; this may be ascribed to the more or less pronounced inhomogeneities of the F_2-region. Propagation then is determined uniquely by the electronic component of the F_2-region, the charge carrier distribution of which from the lower edge to the maximum of the region may be described approximately by a parabolic law [Doluchanow, 1956].

The F_2-region temperature is between 1000 and 2000 K depending on the consideration of the atomic or molecular ionization of nitrogen. The influence of the sun's corona's shortwave radiation, at wavelengths of between 500 and 700 Å, was confirmed during solar eclipses. A very weak region, the *G*-region, occasionally is observable above the F_2-region, however, due to its rare appearance, very little is known about it.

4.9 IONOGRAMS

Traces of the type given by Figures 4.11 and 4.13, showing the virtual height of reflection of electromagnetic waves from the various ionospheric regions, are called *ionograms.* They reveal the involved electron density variations with height and the involvement of the various regions in the propagation of waves of different frequencies.

The June traces of Figure 4.11 show, at a frequency of about 3 MHz, a distinct jump from the *E*-region at an amplitude of approximately 120 km to the *F*-region at about 200 km. The *F*-region then, at about 4 MHz, exhibits splitting into two independent traces, intimating the splitting of the electromagnetic wave into an ordinary and an extraordinary wave marked *O* respectively *X* in the ionogram for 1934. Maxima of the virtual height occur at distinct frequencies; the virtual heights rise sharply before the region becomes transparent; and both waves disappear completely when the frequency exceeds the critical frequency f_c.

The traces show that the *F*-region already is birefringent at its bottom, and it has two electron concentration maxima at elevated heights. The propagation time then increases near the maxima, thus accounting for the increase in virtual height, to fall again and proceed without interruption through a minimum to the upper layers of the *F*-region. The region becomes transparent to radiations at frequencies above f_c.

The E-region also may exhibit birefringence at times. The rarity of E-region birefringence is caused by the large electron density gradient below the region's maximum and the small thickness of the region resulting in small time differences between the ordinary and extraordinary waves, on the one hand, and on the strong absorption of the extraordinary wave, on the other [Al'pert, 1973]. The electron density at the critical height of the ordinary wave is given by expression (4.33). That for the extraordinary one, which is usually higher, is more complicated to evaluate.

Both the June and December traces of Figure 4.11 show how the critical frequency of the regions rises in successive years when the solar activity tends to a maximum in 1937. Figure 4.13 shows that the F_1-region does not develop in winter; the virtual height of the F-region remains at an approximately constant altitude during daylight to rise to a greater one at night. The drop from the night altitude at local sunrise is faster than the increase from the daylight altitude to the night altitude at local sunset. This is probably due to the inertia of adjustment of the electron densities to the nighttime conditions.

The June curves show splitting of the nighttime F-region into the lower F_1- and higher F_2-regions at sunrise and their recombination into a single region at sunset. The height of the F_1-region remains approximately at the winter's daylight height of the F-region. The nighttime altitude of this region does not change with the seasons. The winter and summer altitudes of the E-region remain largely constant, however; the region prevails longer in summer than in winter due to the solar origin of its ionization.

Multiple reflections from the different regions may lead to the duplication of the ionogram's traces. Figure 4.15 shows multiple reflections from the F-region. The branching of the ionograms is caused by the magnetoactive anisotropy of the ionosphere responsible for birefringence [Al'pert, 1973]. Due to the diffuse character of the reflections from the irregular structure of the ionosphere, ionograms obtained at high geographic latitudes, particularly in polar regions, may be rather complex and often difficult to interpret.

Radio soundings of the ionosphere are usually at frequencies above 1 to 2 MHz so that the presence of the D-region manifests itself mainly by the absorption of radiation reflected by other regions. Ionosondes operating at frequencies around 50 kHz are used for sounding the D-region [Al'pert, 1973].

4.10 IONOSPHERIC DISTURBANCES

The preceding sections described the diurnal and seasonal variations of the ionosphere and the influence of the 11-year solar cycle on it. Besides these, periodic fluctuations of the virtual height, height of maximum electron concentration, and critical frequency, within a half-lunar-day period, were discovered at various heights. They are subject to tidal fluctuations.

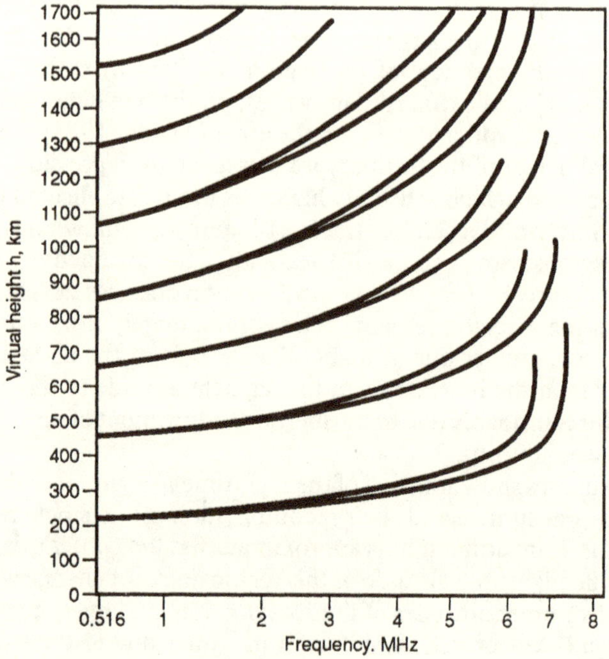

Figure 4.15 Ionogram of multiple reflections (after Al'pert, 1973, © Vsesojuznoje Agentstvo).

Measurements indicate that the lunar influence on ionospheric tides is much larger than it is at the earth's surface [Al'pert, 1973]. It is larger in the higher ionospheric regions than in the lower ones. Various ionospheric parameters vary in sympathy with the moon's phases and exhibit two periods every 27 days of lunar time. Correlation between stratospheric temperature and *D*-region absorption also has been established.

Besides the more or less regular ionospheric phenomena, others manifest themselves irregularly from time to time; these are referred to cumulatively as *ionospheric disturbances.* They are observed during eruptions of the solar chromosphere; that is, the solar layer of reduced temperature surrounding the sun's photosphere. Sunspots appearing from time to time on the photosphere, according to Al'pert, have a temperature of about 2000 K; that is, lower than that of the photosphere. Magnetic fields of the sunspots, which are proportional to the spots' area, may reach intensities of several thousand Gauss. They are rather complex in structure and behavior. Variations in the intensity of the ionizing solar radiations account for disturbances in the behavior of the ionospheric regions. The periodic solar activity is accompanied by variations in the frequency of the occurrence of solar flares.

During a solar flare, the intensity of the sun's X-ray radiation increases and raises the *D*-region's ionization, particularly at altitudes of about 60–70 km above the earth. Whereas the ionization of the *E*-region may be affected by soft X-ray radiation, that of the *F*-region is influenced by increased ultraviolet radiation.

A strong solar flare is followed by cosmic rays, protons, and α-particles, falling mainly onto the earth's polar regions and ionizing the lower portions of the *D*-region. Some 20 to 40 hours after the solar flares, electrons and protons accelerated in the magnetosphere begin to influence the ionosphere to cause an increase in *D*-region absorption. It may take a day or more for the particles ejected by the sun to influence the ionosphere.

Ionospheric disturbances are of interest not only as natural phenomena, but also due to their interference with radio communication. They may influence communication over intervals extending from a few minutes to several days. The changed ionospheric conditions may cause a complete disappearance of signals.

The *D*- and *F*-regions of the ionosphere are affected more than the *E*-region. Solar flares affect the *D*-region instantaneously, whereas their effect on the *F*-region is often delayed by a day or more [Davies, 1966].

The following disturbances are ascribed to solar flares:

(a) ionospheric storms,
(b) *solar flare effects* (sfe),
(c) *sudden ionospheric disturbances* (sid),
(d) *polar cap absorption* (pca).

In addition to these, the appearance of a sporadic *E* region may be classed as an ionospheric disturbance. The description of the disturbances follows.

4.10.1 Ionospheric Storms

According to Doluchanow, ionospheric storms are caused by the penetration of solar corpuscles into the higher regions of the atmosphere. During solar activity large corpuscular clouds fully ionized by the sun's ultraviolet radiation, although externally neutral, envelope the whole earth. In places where they move along the earth's magnetic field's lines of force, they penetrate into the ionosphere, and after moving along rather complicated trajectories, they arrive, predominantly on the nightside of the earth, at regions about 2500 km from the magnetic poles. Strong corpuscular penetrations may cause auroras. Measurements indicate that corpuscles may penetrate to about 80 km above the earth and hence cause ionospheric disruptions. The ionosphere becomes very unstable, radio communication at frequencies below about 20 MHz is disrupted, and complete fadeouts may occur. Polar zones experience the strongest ionospheric storms, but these may extend to mid-latitudes during particularly disturbed periods.

Although the earth's magnetic field, at any point on its surface, varies during periods of smooth and regular ionospheric conditions, diurnally and seasonally, ionospheric disturbances usually are accompanied by geomagnetic storms. Great variations are seen in the horizontal and vertical magnetic intensity and the declination, caused by the additional currents in the auroral zones; these are called *magnetic storms.* They are stronger in higher latitudes than in lower ones.

The geographic distribution of ionospheric and magnetic storms is not uniform; they tend to occur most frequently in regions between about 40° and 90° of latitude of both hemispheres. Local disturbances affecting only a limited area have been observed, too.

A drop in the F_2-region critical frequency, sometimes accompanied by a similar drop in the F_1-region and an apparent height increase, is the main indication of an apparent ionospheric storm. These are accompanied by strong absorption in the *D*-region and often, for several hours, by a complete absence of ionospheric reflections over a wide frequency range.

Maximum F_2-region disturbances in high latitudes occur most often during morning and evening, whereas at mid-latitudes at midnight. These disturbances are closely related to solar activity, the strongest and most frequent ionospheric disturbances appear shortly after the sunspot maximum. The duration of disturbances varies widely and may last from a few hours to several days. Several periods of disturbances may follow one another, thus disrupting communication for several weeks. The ionospheric storms exhibit yearly variations similar to those of the earth's magnetic activity; that is, an increase in frequency about the time of the equinoxes. The reason for this is assumed to be that corpuscular eruptions originate in a band of 5–20° of the sun's latitude and the corpuscular streams are emitted as beams vertically from the sun's surface. Due to the inclination of the earth's axis with respect to the ecliptic, the most favorable condition for intercepting the corpuscular radiations occur at about the equinoxes. Weak ionospheric storms tend to a 27-day periodic appearance corresponding to the sun's rotation around its own axis.

4.10.2 Solar Flare Effects

The main contributors to disturbances in the ionosphere are intensified X-ray, ultraviolet, and corpuscular radiations of the sun. These are caused by solar flares. The eruptionlike brightenings in the hydrogen spectral line at times of solar flares may increase the normal ultraviolet solar radiation up to 10^6 times and thus cause, mainly in the polar regions, a sudden rise of the *D*- and *E*-region ionization, often resulting in a complete absorption of shortwave radiation and thus the cessation of shortwave communication over the whole illuminated hemisphere of the earth for a few minutes and occasionally longer than an hour. An enhancement of very low-frequency propagation due to improved reflection from the *D*-layer occurs simultaneously [Doluchanow, 1956].

Flares strengthen the auroras and magnetic storms by producing streams of electrons accelerated in the earth's magnetic field. The mechanics of interaction between the corpuscular streams and the earth's magnetic field is not fully understood as yet.

A day or more after the occurrence of solar flares the particles ejected by the sun begin to influence the ionosphere. Solar flares then provide a good forewarning of ionospheric disturbances.

4.10.3 Sudden Ionospheric Disturbances

Sudden ionospheric disturbances, or the Dellinger effect, cause an instantaneous cessation of ionospheric reflection and hence shortwave radio communication. These disturbances occur only over the illuminated hemisphere of the earth and may last from a few minutes to a few hours. They are caused by abnormally high absorption in the *D*-region with usually, though not always, a very sudden onset but relatively slow recovery. Lower frequencies are affected first. Signals reflected from the *F*-region then reappear gradually, followed by weak reflections from the *E*-region until normal conditions of all ionospheric regions are restored.

Ionograms taken during sudden ionospheric disturbances show that, although reflections from the *E*-region and *F*-region disappear, the minimum usable frequency rises sharply to values exceeding the F_2-region's critical frequency, thus indicating a very large rise of the *D*-region's charge carrier density. Absorption of radio waves over the whole frequency range and cessation of reflection from all the regions ensue. Normal conditions are restored first at higher frequencies because an increased absorption prevails longer at lower ones.

Increased *D*-region ionization lowers the effective height of reflection by up to 15 km, thus advancing the phase of long and very long waves reflected from there. This is aided also by an increased electron density in the region below the reflection level, resulting in a slight change of the wave's frequency.

4.10.4 Polar Cap Absorption

Increased *D*-region ionization due primarily to the ionizing effect of high-energy solar protons may cause polar cap absorption in auroral zones [Al'pert, 1973]. It is observable mostly around the equinoxes and in summer over a region around the pole extending sometimes to latitudes of 70° and seldom as far as 60°. The effect may persist continuously for three or four days and is not accompanied by geomagnetic or auroral activity. The daytime absorption is 5 to 10 times higher than the nighttime one. Polar cap absorption occurs more often at periods of high solar activity.

Bailey and Harrington [1963] gathered information over the years 1952 to 1960 on the occurrence of polar cap absorption by monitoring the signal strength of very high-frequency ionospheric forward scatter links between Labrador, Greenland, and Iceland. They observed a marked effect of a polar cap absorption on the signal intensities. Ionospheric scatter, to be dealt with later, relies on the scattering of incident electromagnetic radiation by irregularities in the electron distribution in the ionosphere.

The researchers found that the scattering heights are situated at altitudes as low as 60 km. Further, if the proton flux causing polar cap absorption starts during darkness, the very high-frequency signal intensity increases abruptly at the start to decrease greatly during the following daylight. Because no direct relation between the intensity of the polar cap absorption and the magnitude of the solar cosmic ray effect was observed at ground level, we can assume a great event-to-event variability of the form of the energy spectrum of the arriving solar protons. In high latitudes protons have been observed at balloon altitudes.

The onset of a magnetic storm and often a strong decrease in the intensity of cosmic rays may be followed, many hours after the onset of polar cap absorption in regions closer to the poles, by absorption outside the polar cap region. Polar cap absorption usually is heralded by a major solar flare on the visible hemisphere of the sun [Davies, 1966] a fraction of an hour to several hours beforehand. It reaches its maximum within a few hours and exhibits diurnal variations in sympathy with the level of solar activity during its decay.

4.10.5 Sporadic *E*-Region

Ionsopheric soundings intimate that in addition to the regular ones, another region is appearing sporadically at about the *E*-region heights, usually referred to as the *sporadic E* or the *E_s-region.* Ionogram analyses show a range of different types of reflections from this region, indicating the presence of different structural peculiarities, probably caused by various mechanisms of region formation. Vertical soundings prove that critical frequencies in the range of 10 to 12 MHz are not unusual, although they were found to be influenced not only by the region's properties but, as Rawer [1953] shows, also by those of the used equipment. Diurnal limiting frequency variations exhibit a regular pattern depending on the latitude.

The E_s-region appears to be rather thin, some 500 to 2000 m. At mid-latitudes the region may have a radius of 200 to 600 km. The region consists of ionized clouds, about 50 m thick, having a horizontal diameter of about 200 m. However, according to Al'pert [1973] these dimensions may vary over a wide range. Due to the region's irregular occurrence, statistical methods must be used for the analysis of observations. These show that the E_s-region is more likely to appear at lower geographical latitudes and during daytime, although no solar control of the region's behavior has been noticed. The high stability of the region is comparable with that of the regular *E*-region.

The E_s-region's lifetime at a particular location varies widely and usually is on the order of one hour. There is no evidence, so far, on whether the region arises locally or drifts there from some adjacent location. Although drift of the region has been established, it is not known whether it moves as a whole or individual clouds move independently. The velocity of the drift varies from about 1000 m/s in the polar regions to about 150 m/s and 30–60 m/s at mid-latitudes and equatorial zones, respectively. Vertical motion has a lunar tide character.

E_s-region formation probably is due to changing rates of ion formation caused by meteor ionization, by the injection of fast particles, by the formation of negative ions or by the transfer of rapidly ionized particles from lower atmospheric layers. Although the probability of the region's occurrence as well as the hour-to-hour and day-to-day variability of its parameters depends on the season and geographic location, some regularity in pattern has been detected. Al'pert [1973] quotes that the average scatter in the individual values of the critical frequency about the monthly median may be as high as ±50% exceeding that of the regular *E*-region about 10 times and 2–3 times that of the *F*-region. Although the critical frequency's variation with local time follows a definite pattern at various seasons, this depends to a great extent on the geographic latitude. Some evidence suggests that the E_s-region's occurrence probably is higher during periods of minimum solar activity.

The cloudlike consistency of the region permits the passage of electromagnetic waves through gaps between the clouds to the *F*-region. Waves reflected from there may strike the upper side of the E_s-region and be reflected back to the *F*-region. The process may be repeated several times until the *F*-region reflected wave passes through another gap in the E_s-region to the ground. The wave moves then in an *M*-shaped trajectory.

4.11 ARTIFICIAL DISRUPTIONS OF THE IONOSPHERE

Today, when humankind experiments with nuclear devices, one may ascertain the effect atmospheric nuclear explosions have on the ionosphere and hence on radio communication. On July 9, 1962, the United States performed some high-altitude nuclear tests at Johnston Island, which is near the great circle arc between Hawaii and Adelaide. A nuclear bomb was exploded at a height of about 360 km, or right in the *F*-region of the ionosphere. This offered a rare opportunity for observing the effect of the ensuing ionization on the propagation of electromagnetic waves.

The signal strength of the 10 MHz transmitter WWVH (Hawaii) was observed and recorded over three days (Figure 4.16), with the second day being the day of the test. The instant of the explosion at 090009 U.T.2 is marked by the letter *H*. We can see that the explosion caused a sudden drop in the signal strength. It recovered after about 15 minutes to drop again a half-hour later. After that it rose to a more or less

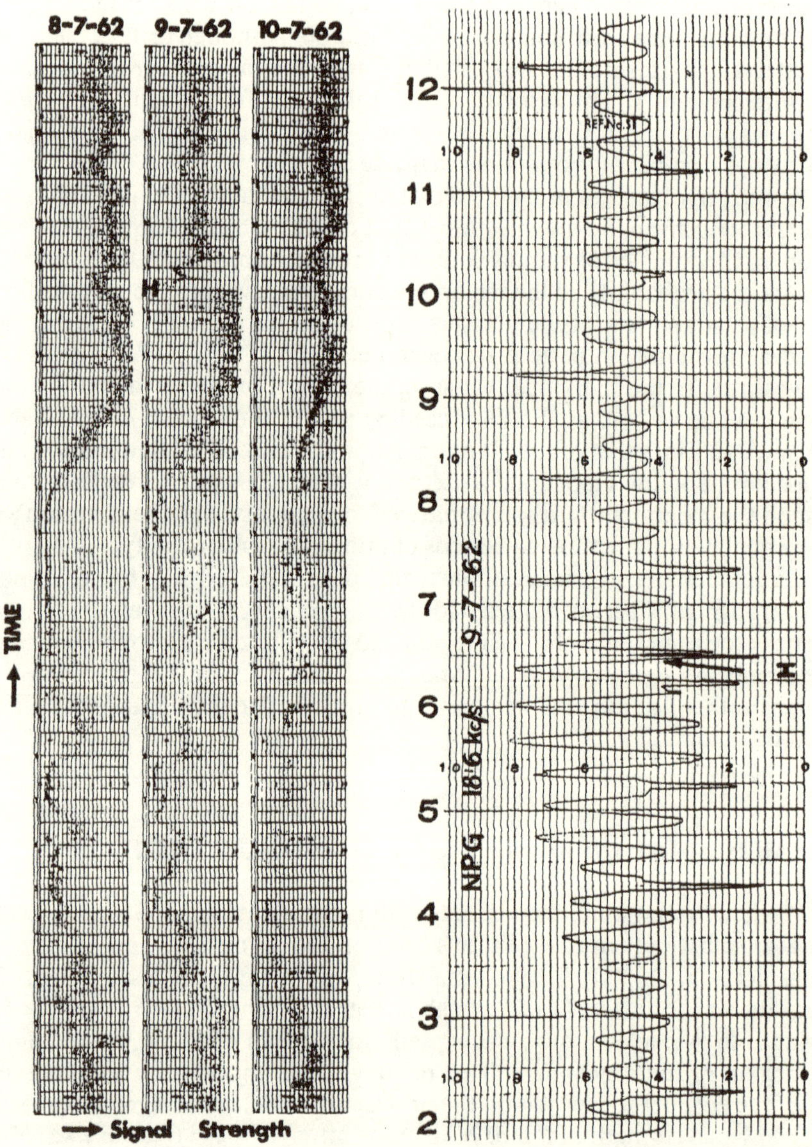

Figure 4.16 The effect of a high-altitude nuclear explosion on electromagnetic wave propagation (after Rohan, Anderson, and Cooke; reprinted by permission from *Nature,* Vol. 197, No. 4870, pp. 887–888. © 1963 Macmillan Magazines Ltd.).

steady value, which, however, was lower than that on the previous and following days and which could be expected here at that time of the year and day.

The same figure also shows the beat signal record of the very low frequency transmitter NPG, at Seattle, operating on 18.6 kHz, for July 9. It was obtained by beating the received signal with that of a highly stable local oscillator used for frequency standardization. This method of phase monitoring does not follow changes of a complete 360°, and it cannot be ascertained therefore whether a phase shift of such magnitude had occurred.

No effect of the explosion was observed on signals of the very low-frequency transmitter GBR in Rugby, England. However, those of NBA in Balboa, Panama Canal Zone, with a great circle far removed from the circle between Hawaii and Adelaide, Australia, have been affected. Note that, although the WWVH signals come from the F_2-region, those of NPG come from the *D*- or *E*-regions, which are below the height of the explosion.

4.12 PROPAGATION FORECASTS

Many ionospheric observatories scattered over the world issue data on the basis of which predictions may be computed of propagation conditions between chosen locations. Computations usually are made three months in advance. Predictions permit the planning not only of equipment requirements but also the frequency choice for different times of the day. These predictions usually are in the form of graphs showing the variation of the maximum usable frequency with the time of the day. They may include graphs of the optimum and minimum usable frequency.

A sample of such predictions of propagation conditions for January 1949 between Prague (Czechoslovakia) and various places in the world is shown in Figure 4.1, which shows the frequency variations with Greenwich mean time (one hour behind Prague local time) and exhibit the characteristics mentioned in the preceding sections of this chapter. The optimum frequency to be used for communication between Prague and, for example, New York from noon to about 19 h GMT is about 14 MHz. It must be reduced for reliable communication before noon and at nighttime.

It is interesting to note that the curves predicting the recommended frequency choice for communication between Prague and places west of it peak shortly after GMT noon, whereas the Melbourne and Wellington curves peak before noon. Curves for places near the Prague meridian peak at noon. This is due to ionospheric reflections occurring somewhere above places midway between the communication link's ends, in the propagation by a single ionospheric reflection, and at several intermediate points, in multihop propagation. Description of the procedures used for propagation predictions is beyond the scope of this book. Those interested in this topic are referred to Davies [1966].

4.13 OTHER IONOSPHERIC PHENOMENA

Two additional ionospheric phenomena that affect electromagnetic waves of different wavelengths deserve to be mentioned briefly: ionospheric scattering and focusing. Ionospheric scattering manifests itself at very high frequencies; focusing occurs at longer wavelengths.

4.13.1 Ionospheric Scattering

The maximum electron densities of the described ionospheric regions are seldom such as to reflect electromagnetic radiations at frequencies above about 30 MHz; however, waves at these frequencies may be affected by the irregularities in electron density distribution in the ionosphere [Davies, 1966]. These may be caused by the following:

- Turbulent mixing at *D*-region heights leading to regular but weak scattering of waves at frequencies between 30 and 60 MHz, thus enabling communication over ranges of 1000–2000 km.
- Ionized meteor tracks reflecting intermittently radiations at frequencies between 40 and 80 MHz and at times making communication over ranges of up to 2000 km possible.
- Occasional scattering at frequencies up to 50 MHz from region irregularities at times of high sunspot numbers.
- Reflections of radiations at frequencies up to 80 MHz from the E_s-region.

Very high-frequency waves propagating via the ionosphere suffer from scattering, resulting in low signal strengths. They are subject to rapid fading and possible interference from other channels.

Turbulent mixing means the redistribution of the electrons. Irregularities in electron density cause corresponding irregularities in the refractive index (expression [4.31]). Stratification in one place means condensation in another. Sudden changes in electron density cause abrupt refraction index changes leading to the deviation of the direction of propagation of the energy from the direction of the incident one. If the direction of propagation does not differ much from its original direction, we speak of "forward scatter"; if, on the other hand, it deviates so much as to permit part of the energy to travel toward its source, the scatter is called *backscatter.*

The mentioned *D*-region irregularities occur at altitudes of 70 to 90 km and are greatly influenced by the sun's action. Although daytime scattering of very high-frequency waves, probably due to turbulence, seems to occur at an altitude of about 70 km, nighttime scattering, attributed to meteor activity, appears to take place at altitudes of 85 to 90 km.

Diurnal and seasonal signal strength variations depend on the geographic latitude. At mid-latitudes maxima occur at about noon, whereas at high latitudes they occur earlier in the day. Summer maxima are stronger than those in winter. Fading rates lie between 0.2 and 3 Hz and the rate increases with frequency and beamwidth. Negligible variation in signal strength with sunspot number has been observed.

Reflections at frequencies between 20 and 100 MHz from ionized meteor trails at altitudes of 80 to 120 km cause sudden bursts of signal strength. According to Davies billions of such trails are produced daily; however, their lifetime is only a few seconds. Trail ionization results from the conversion of the meteor's kinetic energy into potential energy of ionization on the meteor's deceleration by its collisions with neutral molecules.

The height distribution of trails depends on particle mass and velocity; their length depends on the particle mass and on the sun's zenith angle. The energy reflected by a meteor trail also depends on such factors as the ionization density of the trail, the frequency of the radiation and the polarization of the incident wave, and the transmitter-trail-receiver geometry.

Trails may be underdense (i.e., having an electron density so low that the incident wave passes through the trail, which appears to consist of independent scatterers) or overdense, with such electron density that the incident wave does not penetrate completely but reflects back like ionospheric reflection.

Communication by reflection from meteor trails is restricted to times when reflections offer signals of adequate strength to ensure acceptable error rates. For a more comprehensive treatise of meteor trail reflections, which is beyond the scope of this book, the reader is referred to the book on ionospheric propagation by Davies [1966].

F-region scatter, according to Bailey [1962], usually occurs during hours of darkness from sunset to midnight. It occurs mostly around the autumn equinox and may raise the signal strength by 30 to 40 dB above the level due to forward scatter. The rise depends on the wavepath orientation relative to the earth's magnetic field.

The signals are subject to fading with a 10 Hz or more fading rate, and they exhibit doppler shifts appropriate to a 100 m/s magnetic east-west motion of the scattering surface. Flutter fading, which makes speech communication impossible, occurs shortly after sunset and lasts for two to three hours.

The occurrence of equatorial spread-*F* was found to be closely related to *F*-region scattering. It consists of a thin sheet of irregularities of about 50 km thickness. Broadening of echo pulses by about 10 times when compared with the transmitted pulse width also was observed. Propagation by reflection from the E_s-region was mentioned earlier.

Radar observations confirmed the existence of coherent ionospheric scattering by free electrons, as described by Bowles [1962]. Its cause was assumed to be by free electrons contributing echo power in random phase according to their individual scattering cross section. However, discrepancies between the theoretical and measured pulse spectrum intimates that scattering more likely is to be due to irregularities in the electron density and that the greater mass of the ions accounts for their dominating the statistics of the irregularities. Excellent agreement between theory and experiment lead to the adoption of incoherent scattering techniques for the determination of vertical electron density profiles.

4.13.2 Focusing

As mentioned earlier, this phenomenon occurs at frequencies below the very high-frequency range. It manifests itself by an enhanced or diminished power flux on the earth caused by the curvature of the ionosphere and possibly by the change of the ground range with the ray elevation angle. Focusing, or the convergence of originally neighboring rays, causes enhancement, whereas reduction of power flux is due to defocusing. The amount of focusing or defocusing is difficult to determine without a detailed knowledge of the ionosphere along the ray's trajectory.

Budden [1961a] considered reflection from an ionosphere with a parabolic electron density profile and showed that near the minimum distance at which a ray can return from the ionosphere to the ground (i.e., the "skip distance") the rays are focused by a caustic surface. A ray reaching the ground at the skip distance arrives there at a certain angle of incidence. Rays arriving at neighboring angles reach the ground at greater distances. The resulting signal is zero within the skip zone but, due to the proximity of a very large number of rays, high beyond it. This is referred to as *skip distance focusing.* The envelope of the rays forms the caustic similar to the configuration obtained for parallel light rays reflected or refracted by a spherical mirror or lens of imperfect focusing properties.

The signal strength beyond the skip zone exhibits an oscillatory character with an amplitude diminishing with distance, a pattern similar to the one obtained at a fixed distance for frequency variations near the maximum usable frequency causing maximum usable frequency focusing. Focusing factors of 6 to 9 dB have been observed.

The geometrical optics approximation of propagation within the waveguide constituted by a concentric spherical earth and ionosphere would intimate the occurrence of antipodal focusing. However, this is counteracted by the irregularities of both surfaces of the waveguide, which causes this type of focusing to be of no importance.

If, according to Bremmer [1958], the elevation angle of the ray trajectory approaches $\pi/2$ rad, the geometrical optics approximation of the divergence coefficient tends to infinity. This corresponds to receiving points close to the horizon formed by the ray trajectory being tangential to the earth. The resultant focusing partially cancels the reduction of the field caused by the ground attenuation factor.

PROBLEMS

4.1 Plot the height dependence of the atmospheric temperature using the data given in Section 4.1.1.

4.2 Determine the value of the ionospheric relative permittivity, conductivity, and refractive index at a frequency of 10 MHz at a height of 300 km where, according to Ratcliffe, the collision frequency is $\nu = 10^3$/s. Obtain the mean value of the electron density from Figure 4.6.

4.3 The great circle distance between a transmitter and receiver on a spherical earth is 2000 km. Considering a concentric spherical ionosphere at an altitude of 400 km above the earth, determine the maximum usable frequency for communication between these locations and also the equivalent vertical incidence frequency.

4.4 What critical frequency will be reflected at an altitude of 400 km, and what is the maximum reflected frequency reaching this height at an angle of 80°?

4.5 Collision frequency data given by various researchers vary to a great extent. Assume, according to Davies, a collision frequency profile

$$\nu = 6 \times 10^4 \times \exp\left(-\frac{h - 100}{6.7}\right) \text{ Hz}$$

with h the ionospheric height of interest in km applicable to the lowest two ionospheric regions. Compute the attenuation and phase constant, phase velocity, and refractive index in the D-region at a frequency of 600 kHz.

4.6 What is the difference between the velocities of the two circularly polarized waves due to the splitting of a wave of 10 MHz frequency traveling parallel with the earth's magnetic field of 0.5×10^{-4} Wb/m^2 induction when the considered ionospheric height is 400 km?

4.7 To what is the time dependency of the maximum usable frequency ascribable?

4.8 Find an analytic expression for the maximum ground range of a one-hop ionospheric ray path.

4.9 Plot the height dependency of the critical frequency when the sun's zenith angle is $\chi = 45°$. (Use Figure 4.6.)

4.10 Describe and explain propagating at an oblique angle from the ground to the ionosphere and back.

REFERENCES AND RECOMMENDED READING

Al'pert, Y.L. (1973) *Radio Wave Propagation and the Ionosphere*, Vol. 1, *The Ionsophere*, Consultants Bureau, New York and London.

Appleton, Sir Edward V. (1957) "The Normal E Region of the Ionosphere," *Proc. IRE,* Vol. 47, No. 2, February.

Bailey, D.K. "Ionospheric 'forward' scattering," *see* W.J.G. Beynon.

Bailey, D.K., and J.M. Harrington. "The Occurrence of Polar Cap Absorption," *see* G.J. Gassmann.

Bates, D.H., R.A. Buckingham, H.S.W. Massey, and J.J. Unwin (1939) "Dissociation, Recombination and Attachment Processes in the Upper Atmosphere," Part 2, "The Rate of Recombination," *Proc. Royal Society,* Vol. 170, No. A942, 3 April, pp. 322–340.

Beynon, W.J.G., ed. (1962) *Monograph on Ionospheric Radio,* XIIIth General Assembly of URSI, London, [September 1960]; Elsevier Publishing Co., Amsterdam and New York.

Bowles, K.L. "Recent Radar Observations of New Forms of Ionospheric Scatter," *see* W.J.G. Beynon.

Breit, G., and M.A. Tuve (1926) "A Test of Existence of the Conducting Layer," *Physics Rev.,* Vol. 28, September, p. 554.

Bremmer, H. (1958) "Propagation of Electromagnetic Waves," in S. Flügge, ed., *Encyclopedia of Physics,* Vol. 16, *Electric Fields and Waves,* Springer-Verlag, Berlin, Göttingen, and Heidelberg.

Budden, K. (1953) "The Propagation of Radio Waves to Great Distances," *Phil. Magazine,* Vol. 44.

Budden, K. (1961a) *Radio Waves in the Ionosphere,* Cambridge University Press, Cambridge, England.

Budden, K. (1961b) *The Wave-Guide Mode Theory of Wave Propagation,* Logos Press, London.

Budden, K.G. (1985) *The Propagation of Radio Waves,* Cambridge University Press, Cambridge, England.

Budden, K.G., and E. Cooper (1962) "Ionospheric Equivalent Heights of Reflection Calculated by a Full Wave Method and by the Phase Integral Method," *J. Atm. Terr. Physics,* Vol. 24, July, pp. 609–618.

Casaverde, H., A. Giesecke, and R. Cohen (1963) "Effects of the Nuclear Explosion over Johnston Island Observed in Peru on July 9, 1962," *J. Geophysical Research,* Vol. 68, No. 9, May 1, pp. 2603–2611.

Chapman, S. (1931) "The Absorption and Dissociative or Ionizing Effect of Monochromatic Radiation in an Atmosphere on a Rotating Earth," *Proc. Physics Society,* Vol. 43, pp. 26–45, 483–501.

Chapman, S., and J. Bartels, (1940) *Geomagnetism,* Clarendon Press, Oxford, England.

C.C.I.R. (1966) "Estimation of Sky-Wave Field Strength and Transmission Loss Between the Approximate Limits of 1.5 and 40 MHz," Report 252-1.

Cohen, E.A. (1984) "The Study of the Effect of Solar Eclipse on the Ionosphere Based on Satellite Beacon Observations," *Radio Science,* Vol. 19, No. 3, May–June, pp. 769–777.

Davies, K. (1966) *Ionospheric Radio Propagation.*" Dover Publications, New York.

Doluchanow, M.P. (1956) *Die Ausbreitung von Funkwellen,* VEB Verlag Technik, Berlin.

Gassmann, G.J., ed. (1963) *The Effect of Disturbances of Solar Origin on Communications,* Pergamon Press, New York.

Gibbons, J.J., and A.H. Waynick (1957) "The Normal D Region of the Ionosphere," *Proc. IRE,* Vol. 47, No. 2, February.

Golton, E. (1961) "Skip-Distance Ray-Focusing in the Ionosphere," *Nature,* Vol. 189, January 7, pp. 48–49.

Jordan, E.C. (1953) *Electromagnetic Waves and Radiating Systems,* Constable & Co., London.

Kelso, J.M. (1964) *Radio Ray Propagation in the Ionosphere,* McGraw-Hill, New York.

Kreyszig, E. (1967) *Advanced Engineering Mathematics,* 2d Ed., John Wiley and Sons, New York.

Lipson, S.G., and H. Lipson (1969) *Optical Physics,* Cambridge University Press, London.

Martyn, L.F. (1959) "The Normal F Region of the Ionosphere," *Proc. IRE,* Vol. 47, No. 2, February.

Nichols, W., and J.C. Schelleng (1925) "Propagation of Electric Waves over the Earth," *BSTJ,* Vol. 4, April, p. 215.

Picquenard, A. (1974) *Radio Wave Propagation,* Macmillan Press Ltd., London.

Raemer, H.R., and Y.P. Verma (1970) "An Analytical Numerical Approach to Analysis of Ionospheric Wave Propagation," *Radio Science,* Vol. 5, No. 8–9, August, pp. 1175–1183.

Ratcliffe, J.A. (1972) *An Introduction to the Ionosphere and Magnetosphere,* Cambridge University Press, London.

Rawer, K. (1953) *Die Ionosphere,* P. Nordhoff N.V., Groningen, the Netherlands.

Rohan, P. (1949) "Předpověď' dx podmínek na leden," *Krátkê vlny,* Vol. 8, No. 1, January.

Rohan, P., L.L. Anderson, and D.J. Cooke (1963) "The Effect of the U.S. High Altitude Nuclear Explosion on Radiowave Propagation," *Nature,* Vol. 197, No. 4870, 2/3.

Seddon, J.C. (1962) "Summary of Rocket and Satellite Observations Related to the Ionosphere," *see* Beynon, 1962.

Sonayajulu, Y.V., (1963) "Changes in the F Region During Magnetic Storms," *J. Geophysical Research,* Vol. 68, No. 7, April, pp. 1899–1922.

Thomas, J.O. (1962) "The Electron Density Distribution in the Ionosphere: Some Results for the I.G.Y. Deduced from $h'(f)$ Observations," *see* Beynon, 1962.

Thomas, J.O. (1963) "The Electron Density Distribution in the F_2 Layer of the Ionosphere in Winter," *J. Geophysical Research,* Vol. 68, No. 9, May 1, pp. 2707–2718.

Wait, J.R. (1959) "Diurnal Change of Ionospheric Heights Deduced from Phase Velocity Measurements at vlf," *Proc. IRE,* Vol. 47, No. 5, May.

Waldmeier, M. (1961) *The Sunspot Activity in the Years 1610–1960,* Schulthess Polygraphischer Verlag, Zurich.

Wu, D.M., and N. Marcuvitz (1983) "Ionospheric Scintillation," *Radio Science,* Vol. 18, No. 4, July–August, p. 589.

Yuen, P.C., and T.H. Roelofs (1965) "Diurnal Variation of the Ionospheric Total Electron Content," *J. Geophysical Research,* Vol. 71, No. 3, February, pp. 849–854.

Chapter 5
PROPAGATION OF ELECTROMAGNETIC WAVES OF THE VARIOUS FREQUENCY RANGES

The behavior of the ionosphere at the various frequency ranges depends on its properties at the appropriate frequencies. As far as their electrical properties are concerned, there are three types of materials: conductors, semiconductors, and dielectrics. These are distinguished by their conductivity and relative dielectric constants, which determine the magnitudes of the conducting and displacement current in the material subjected to an electromagnetic field.

We mentioned in Section 2.2 that the ratio of these two currents determines the behavior of a medium at a given frequency. Formulas for the determination of the relative dielectric constant and conductivity of the ionosphere are given by (4.23). They show that, in addition to the frequency of the propagating electromagnetic wave, both parameters depend on the charge carrier density and collision frequency. Because both the charge carrier density and the collision frequency depend on their heights above the earth, the ionospheric parameters will be different at different heights and different frequencies.

The ratio of the displacement and conduction currents, as given earlier,

$$\frac{i_{\text{displ}}}{i_{\text{cond}}} = \frac{\epsilon'}{60\sigma\lambda} \tag{5.1}$$

permits then to gain insight into the ionosphere's behavior and hence influence on the propagation of electromagnetic waves at the various frequency ranges.

5.1 VERY LOW-FREQUENCY WAVE PROPAGATION

Experience has shown that disturbances to radio communication caused by lightning discharges, referred to generally as *atmospherics,* even at very long distances, are of such levels at frequencies below about 10 kHz that they make communication impossible at these frequencies.

According to Picquenard [1974], generally, about 100 lightning flashes occur every second of the day. Each flash consists of two discharges at an altitude of 2 to 4 km, with the discharge currents of 10 to 100 kA setting up damped oscillations with a duration of about 10^{-2} ms. There is then a continuous sequence of irregular pulses of the above duration with a maximum power of up to several tens of GW at about 10 kHz. Their frequency spectrum covers a very wide range, and they may cause disturbances to radio communication up to about the very high-frequency range. Then, because below 10 kHz the disturbances make communication impossible, these frequencies are of no practical interest. They permit the location of storm centers by direction finding techniques, however. Statistical data collected over many years were used to compile maps and tables of the distribution of atmospherics (Figures 6.4 and 6.5). They are used to determine the field strength required for acceptable communication. Further, although the very low-frequency designation applies strictly to the 3 to 30 kHz frequency range, the propagation of waves at frequencies up to 100 kHz will be considered in this section because there is but little difference, as far as propagation is concerned, among waves at frequencies below 100 kHz.

The ground wave, as it was established by Pressey, Ashwell, and Hargreaves [1961], does not exhibit the phase variations encountered in the sky wave. The ground wave travels along the surface of the earth and is guided by it. Part of the energy of the wave is absorbed due to losses in the ground; however, the attenuation at these frequencies is much lower than that at higher ones. Ziem [1960] has shown that the ground-wave field exceeds the sky-wave field even at distances of about 1000 km from the transmitter.

The lengths of waves at 10 and 100 kHz (namely, 30,000 or 3,000 m) are comparable with the height of the lowest ionospheric regions. Because at high angles of incidence in the ionosphere only a low electron density, such as is found at the lower boundaries of the *D*-region by day and the *E*-region by night, is sufficient for reflection, these waves propagate in the narrow waveguide created by the surface of the earth on one side and the ionospheric regions on the other, in a waveguide mode. Due to their very low attenuation and hence large ground-wave range and their reflections from stable ionospheric layers, at adequate sender powers, these waves exhibit reliable means of communication and are used by special services for communication over the whole world. These services include navigational aids, the transmission of meteorological data, time signals, communication with ships at sea and submerged submarines, *et cetera.* As a matter of interest, some measured figures of depth penetration of a 16 kHz radiation of a 1 MW transmitter in seawater, as given by Ziem, are shown in Table 5.1.

Table 5.1
Depth of Penetration of Very Low-Frequency Radio Waves in Seawater

Area	*Approximate Total Distance in km*	*Distance over Land in km*	*Depth of Loop Antenna under the Water's Surface, m*
North Sea to Peterhead-Stavanger	1000	300	15–25
Barents Sea	2300	1000–1600	13–18
Mediterranean	2300	700–800	13–18
North Atlantic, including New York, Gulf of St. Lawrence	5200–7000	700–1400	8–26
Caribbean	7600	1000–1200	8–20
South Atlantic to 750 km South of Cape Town	9800	8200	8–12
Indian Ocean, Strait of Malacca	8300	7400	8–15

To determine how the ionosphere behaves at these frequencies, according to Doluchanow [1956], we must obtain the two parameters required for the computation of the ratio of the two currents, as mentioned earlier. The daytime *D*-region electron density is about 10^9 electrons/m^3 with collision frequencies in the range of 10^7–10^8/s, whereas the nighttime *E*-region electron density is about 10^9 electrons/m^3 and the corresponding collision frequency is about 10^4/s. Then by using (4.23), we find that the *D*-region relative permittivity for the whole considered frequency range is $\epsilon' \approx 1$, and the conductivity is about $\sigma = 5.55 \times 10^{-7}$ mho/m. The *D*-region ratio of displacement to conduction current is about 1 at the lower considered frequency and about 10 at the higher one.

The *E*-region relative permittivity is negative. Its conductivity for the considered frequency range is between about 3.5×10^{-5} and 3.5×10^{-7} mho/m.

The *D*-region behavior is similar to that of a semiconductor or dielectric. The negative *E*-region relative permittivity intimates that reflection occurs at a lower altitude than that appropriate to the chosen electron density and collision frequency.

The reflection coefficients given by (2.64) or (2.69) show that complete reflection occurs when $\epsilon' = \sin^2\phi$, where ϕ is the angle of incidence in the ionosphere. Assuming that the nighttime height h of the lower boundary of the *E*-region is about 90 km, the angle of incidence of a horizontally radiated wave is given by

$$\sin\phi = \frac{r}{r + h} \approx 1 - h/r = 1 - 0.014$$

where r is the radius of the earth.

Then, by using the obtained value for σ and (4.23) for the computation of the permittivity and writing

$$\epsilon' = \sin^2\phi \approx 1 - 0.028 = 1 - 3141 \frac{N}{\omega^2 + \nu^2}$$

we could compute, for known values of ν or N, either the electron density or the collision frequency at the height where reflection occurs. General considerations suggest that the collision frequency and not the electron density determines the conditions for very low-frequency reflection from the ionosphere.

Because at the considered frequencies the properties of the medium change considerably over a wavelength, methods of wave optics are applicable to the investigation of reflection phenomena. Investigation of polarization shows that the reflected very low-frequency waves are elliptically polarized. This may be caused only by the earth's magnetic field, which therefore must not be neglected in very low-frequency propagation studies.

The strict investigation of very low-frequency propagation is complicated by the fact that the inner surface of the waveguide (i.e., the surface of the earth) at these frequencies behaves like a semiconductor or almost like a conductor, and the outer one, the lower boundary of the ionosphere, has the properties of a dielectric. The investigation may be made easier by making two simplifying assumptions. First, both the earth's surface and the lower ionosphere must behave like ideal reflectors. Second, the energy flow of the waves is assumed to be an adequate distance from the transmitter, due to the interference of many rays reflected in different ways, and to be independent of the height above the earth.

Because, at the considered frequency range, the base of the ionosphere is at altitudes of only a few wavelengths and the transition region, in which the electron density and hence the conductivity increases rapidly with height, has a thickness on the order of a wavelength or less, the D-region does not absorb energy from incident very low or low-frequency radiation, but it affects its direction of flow and thus it contributes to long-distance propagation.

The interference of the various waves in the earth-ionosphere waveguide, according to experimental evidence quoted by Al'pert [1973], even at short distances from the transmitter, produces a characteristic quasiperiodic oscillatory pattern during both the day and the night. At greater distances, there is very little separation between the ground wave and the wave reflected from the ionosphere, so that at a distance of over 2000 to 3000 km only a single wave, representing the "zero mode" of the excited wave spectrum, will be detected. At shorter distances, it is justifiable to consider the field to be the result of the superposition of the ground and reflected waves. Because the reflectivity of the ionosphere increases with the distance-dependent angle of incidence, the contribution of the reflected wave increases with distance. Then, as the magnitude of the reflected wave exceeds that of the ground wave by a factor of 100 to 1000 or more, the amplitude and frequency of the field-strength oscillations diminishes with distance from the transmitter. At short distances where a very large number of modes must be considered, a resolution into rays (i.e., the geometrical optics approach to the waveguide problem) is applicable.

The rate of attenuation of very low-frequency signals received from the west was found to be lower than that of signals received from the east, and the difference in the rate of attenuation was found to decrease with an increasing frequency. It disappears above about 20 kHz. Crombie [1960] attributes the phenomenon to the interaction of the earth's transverse horizontal magnetic field with free electrons contained in the elliptically polarized electric field in the vertical plane of propagation of very low-frequency waves propagating by the TM mode in the earth-ionosphere waveguide. The refractive index of the region containing the free electrons depends on the direction of propagation and so does the propagation constant of the waveguide [Stratton, 1941].

According to this theory, the propagation in the direction of the earth's horizontal magnetic field is reciprocal, but it is not reciprocal in directions inclined to this field. Crombie [1960] shows that, because the ratio of the gyrofrequency of the electrons in the ionosphere and the critical frequency is negative for west to east and positive for east to west propagation, the coefficients of reflection differ for these two directions so that the earth-ionosphere waveguide exhibits nonreciprocal propagation characteristics.

Taylor found that the attenuation of very low frequencies of about 6 kHz is about 7 to 9 dB/1000 km and about 1 to 3 dB/1000 km for frequencies above 10 kHz. The difference in attenuation rates for frequencies below 8 kHz propagating in the west-to-east direction is about 3 dB/1000 km lower than for waves propagating in the opposite direction, whereas it is only 1 dB/1000 km lower for frequencies above 10 kHz.

5.1.1 Very Low-Frequency Propagation Phenomena

Although very low frequency waves exhibit great stability, they are affected to a certain extent by some ionospheric phenomena: short-duration field-strength variations, time of day or night, season, 11-year solar-activity period, and ionospheric disturbances.

5.1.1.1 Short-Duration Field-Strength Variations

The field at a location far removed from the transmitter, as described earlier, is due to the interference from a number of waves propagating over various paths. The charge carrier density at the area of reflection in the ionosphere varies continuously due to ionospheric inhomogeneities, as described in Chapter 4. This causes variations in the height of reflection, which in turn affects the phase relation of the interfering rays and thus affects the field-strength at the receiver. A phase shift of π rad requires a half-wavelength variation. This would be 15 km at a wavelength of 30000 m and 1.5 km at 3000 m. Short-term height variations causing path-length variations on this order do not occur in the ionosphere, hence fades of field strength at very low frequencies are not too deep nor too rapid. They cannot be detected by ear and are noticeable only on field-strength records. In general, field-strength variations, according to Doluchanow [1956], on the order of 10–30%, and their time intervals may last from several times ten minutes to several hours.

5.1.1.2 Diurnal Field-Strength Variations

Recordings of the relative time of arrival of very low-frequency signals show considerable changes at about local sunrise and sunset. These are attributed to the change in the height of reflection because nighttime reflection is taken to occur from the *E*-region, whereas the daytime one is from the *D*-region. Variations between sunrise and sunset and *vice versa* are relatively small when normal propagation conditions prevail.

The occurrence of the large variations may be of importance when using very low-frequency transmissions for navigational purposes; their time of occurrence may be predicted by using the formulas [Rohan, 1963a]:

$$D = 12\left\{1 \pm \frac{L}{180}[\cos(L - n/A) - \cos(L + n/A)]\right\}$$

$$D = 12\left\{1 \pm \frac{L}{180}\left[\cos\left(L - \frac{A-n}{A}\right) - \cos\left(L + \frac{A-n}{A}\right)\right]\right\} \tag{5.2}$$

where D is the duration of direct daylight in hours (direct daylight meaning direct illumination by the sun), L is the geographic latitude of the considered location, A is the number of days from the preceding solstice or equinox to the following one, whichever is applicable for the period containing the day in question; that is,

$$
\begin{aligned}
A &= 92 \text{ between March 22 and June 21,} \\
&= 92 \text{ between June 22 and September 21,} \\
&= 91 \text{ between September 22 and December 21,} \\
&= 90 \text{ between December 22 and March 21,}
\end{aligned}
$$

and n is the count of days from the nearest preceding day as just given.

The first expression in (5.2) applies to the first and third of these periods, and the second to the others. The negative sign is used for locations in the southern hemisphere for days between March 22 and September 21 inclusive; and, *vice versa,* the positive one applies to the northern hemisphere. Assuming equal intervals of direct daylight between sunrise and noon, at 1200 o'clock, and between noon and sunset, the time of sunrise and sunset may be determined.

Figure 5.1 shows a characteristic plot of the relative arrival time variations. The bottom part shows the relative daytime pathlength variations. This is obtained by drawing a line of a length of the calculated direct daylight above the local time scale of one terminal and a similar line displaced by the difference in local times and the great circle distance between the two considered terminals. The pathlength in direct daylight may be read from the fully drawn trapezium. However, because ionospheric activity above a location on the earth starts earlier and ceases later than the local sunrise or sunset, we should consider the geometric instead of the geographic horizon for the calculation of the sunlit interval. This is approximated by increasing the pathlength by $2d$, where

$$d = r\cos^{-1}[r/(r + h)] \tag{5.3}$$

with r representing the earth's radius and h the ionospheric height. The so-obtained trapezium is shown by the dashed lines.

The intersections of the sides of the trapezium with the extended parallel sides of the original one give the approximate times of the start and cessation of E-region activity and the consequent variations in the relative times of arrival of the signals at the considered location.

5.1.1.3 *Seasonal Influences*

The weak field-strength variations from seasonal influences cause about 20–50% higher daytime and lower nighttime values in summer than in winter. This probably is due to the shorter nights, too short for a sufficient disappearance of the absorbing charge carriers under the reflecting regions, whereas the winter's daytime minimum is a twilight-like effect due to selective absorption.

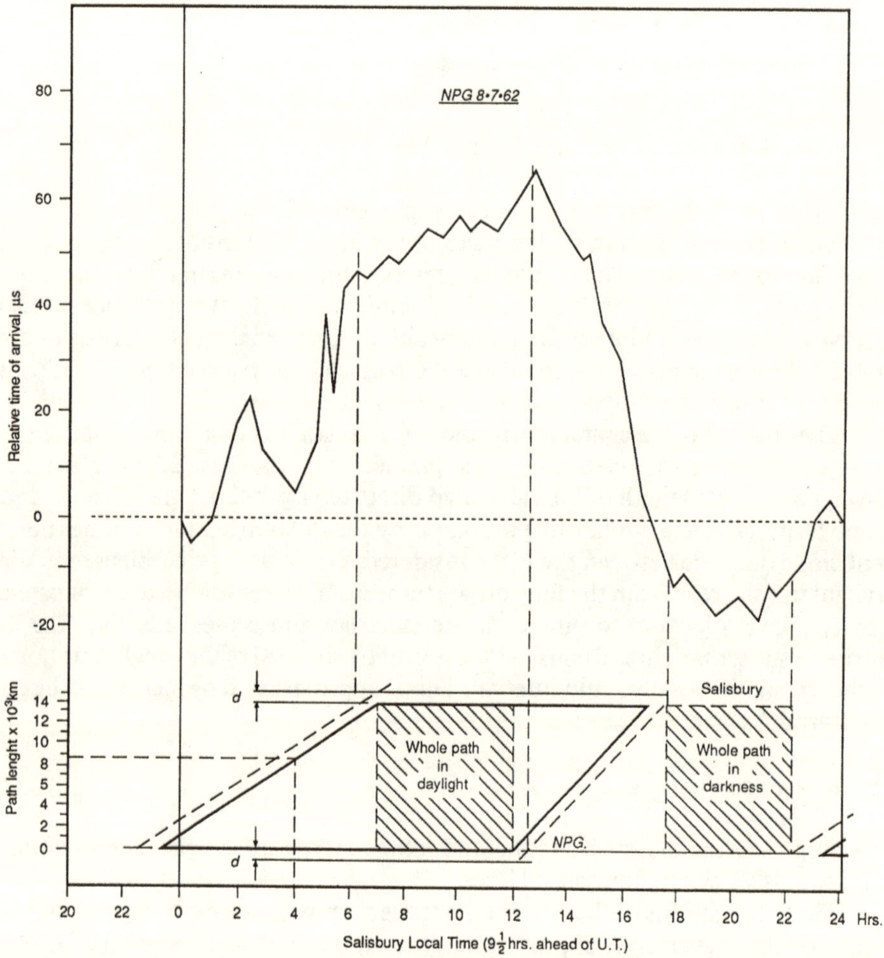

Figure 5.1 The determination of the pathlength in daylight and its correlation with changes in the relative times of arrival (after Rohan, 1963a, © DRCS).

5.1.1.4 The Influence of the 11-Year Period of Solar Activity

Although of no importance in very low frequency propagation, this effect manifests itself in a slight rise of daytime field strength with an increasing number of sunspots. A doubling of field strength at sunspot maxima from that at sunspot minima sometimes occurs. An increase in the daytime *D*-region charge carrier density and hence conductivity and reflection coefficient accounts for this.

5.1.1.5 *The Influence of Ionospheric Disturbances*

Because disturbances ascribed to corpuscular penetrations affect mainly the higher ionospheric regions that do not participate in very low-frequency propagation, ionospheric disturbances of this type are of no concern in the frequency range used for communication. Short-duration ultraviolet emission during the sun's eruptions may cause short-time, but rapidly disappearing, improvement of very low-frequency propagation due to stronger *D*-region ionization. Magnetic disturbances, however, often improve propagation conditions though their influence usually is perceivable only a few days after their occurrence.

5.1.2 Field-Strength Calculations

A theoretical approach to field-strength calculations may proceed, according to Dolchanow [1956], as follows. Denoting by P the radiated power and remembering that the antennas used at very low frequencies of necessity are omnidirectional, the energy flow is given by the Poynting vector:

$$S = P/A \tag{5.4}$$

where A is the area over which the energy is distributed. This area is the area of a rotational cone of a mean base radius of $(r + h/2)\sin\xi$, where r is the earth's radius and h the ionospheric height. Then,

$$A = 2\pi(r + h/2)h\sin\xi \tag{5.5}$$

with ξ the angle subtended at the earth's center by the lines connecting it with the transmitter and receiver locations, respectively.

Inserting (5.5) into (5.4) and using expression (3.15), we arrive at

$$S = \frac{P}{2\pi(r + h/2)h\sin\xi} = \frac{E_{\text{eff}}^2}{120\pi} \tag{5.6}$$

from which the theoretical value of the field strength

$$E = \sqrt{\frac{60P}{(r + h/2)h\sin\xi}} \tag{5.7}$$

with all quantities expressed in the same system of units.

The denominator of this expression grows with increasing distance from the transmitter, and the field strength falls accordingly. At a distance equal to half of the earth's circumference, however, where the field is due to the superposition of two waves diverging in opposite directions from the source, the denominator becomes zero, which would indicate an infinitely large field strength. This is caused by our neglecting the losses on reflection both at the ground and ionosphere and the strong absorption by partly ionized gasses below the reflecting ionospheric region. Additionally, consideration should be given to the asymmetry of the earth-ionosphere waveguide caused by the different electric properties of the earth's surface and those of the ionosphere and by the day-night transitions. Even with these considerations, a slight increase in the field strength is observed at the antipodes, often referred to as the *antipodal effect*.

An empirical formula, the Austin-Cohen formula, generally is accepted for field-strength calculations at the considered frequencies. It determines the daytime field strength for propagation over sea and is applicable for over land propagation when the distances exceed 2000 to 3000 km. Its validity has been proved by numerous measurements. It reads

$$E_{\text{eff}} = \frac{298 \times 10^3 \sqrt{P}}{d} (\xi/\sin\xi)^{1/2} \exp(-\alpha d/\lambda^{\beta}) \; \mu\text{V/m} \tag{5.8}$$

with P the radiated power in kW, d the distance between the transmitter and receiver in km, ξ the angle subtended at the earth's center by the transmitter's and receiver's location in rad, α is the attenuation factor, and λ the wavelength in km.

Early workers who investigated the field intensity of remote transmitters differed as to the values of α and β. Table 5.2 gives the values considered by different authors [Morecroft, 1933].

Table 5.2
Values of α and β in the Austin-Cohen Formula

Author	α	β
Sommerfeld	0.0017	0.33
Austin-Cohen	0.0015 over sea 0.003 to 0.01 over land	0.50
Fuller	0.0045	1.4
Espenschied	0.005	1.25

The attenuation factor α depends on the height and conductivity of the ionosphere and the conductivity of the ground from which reflections occur.

As the electron density and conductivity of the ionosphere decreases during the night, the Austin-Cohen formula, as given earlier, will not apply for night conditions.

Example 5.1

Communication with submerged submarines was proven to be possible in shore-to-ship very high-power transmissions at very low frequencies. (a) Compute the field strength of a 500 kW, 15 kHz transmitter at Nauen, Germany, at New York. (b) What will be the field strength at the receiving antenna of a submarine submerged at a depth of 10 m?

Solution: (a) By using the well-proven Austin-Cohen formula and assuming propagation over sea having, at very low frequencies, a relative permittivity $\epsilon' = 80$ and a conductivity $\sigma = 6.5$ mS/m and taking the approximate distance between the transmitter and receiver sites to be $d = 6600$ km, the angle subtended at the earth's center, when using the average earth radius $r = 6378$ km, becomes

$$\frac{6600}{2\pi 6378} = \frac{\xi}{2\pi}, \quad \xi = 1.03 \text{ rad}, \quad \xi = 59.29°$$

The field strength at New York then, with Austin-Cohen's values for α and β, is

$$E_{\text{eff}} = \frac{298 \times 10^3 \times \sqrt{500}}{6600} \sqrt{\frac{1.03}{0.86}} \exp\left(-\frac{0.0015 \times 6600}{\sqrt{20}}\right) \quad \mu\text{V/m}$$

$$= 120.28 \quad \mu\text{V/m}$$

(b) the absorption coefficient, on the basis of expression (2.38), is

$$\delta = \frac{2\pi}{\lambda}\kappa$$

where κ is given by (2.41). Using the given seawater parameters yields $\kappa = 62.77$.

The absorption coefficient then is $\delta = 0.02$/m, and the magnitude of the electric field, as given by (2.37) becomes

$$E_{\text{eff}} = 120.28 \exp(-0.2) = 98.48 \quad \mu\text{V/m}$$

Several formulas have been given for the more exact computation of the field intensity of very low-frequency transmissions. Wait [1960], accepting Sommerfeld's suggestion that the field is the sum of the contributions from a number of propagation modes, calculates the field of a vertical dipole source by the following simplified expression:

$$E = E_0 \sqrt{\frac{\xi}{\sin\xi} \frac{\sqrt{d/\lambda}}{h/\lambda}} \, e^{j[(2\pi d/\lambda)-\pi/4]} \sum_{n=0}^{\infty} \delta_n S_n \, e^{-j2\pi S_n d/\lambda}$$

The unknown symbols have the following meaning: E_0 is the field of the radiator at a distance d over a flat, perfectly conducting earth that for $d/\lambda \gg 1$ is given by

$$E_0 = j\frac{120\pi}{d} X \, ds \, e^{-j2\pi d/\lambda}$$

and usually normalized to a value $E_0 = 100$ mV/m at a distance of 1 mile, $d = r\xi$ is the distance between the source and the receiver, λ is the free-space wavelength, $S_n = \sqrt{1 - C_n^2}$ is the sine of the complex angle of incidence of the rays in the guide with C_n the cosine of this angle. The factor X is a function of the considered modes, and δ_n is approximately equal to 0.5 for $n = 0$ and to unity for $n = 1$.

Wait gives a similar expression for the vertical electric field of a horizontal dipole. Field intensities calculated for a 16.6 kHz transmission over the path between San Diego and Hawaii and for an 18.6 kHz transmission from San Diego to a distance of about 600 km toward Tokyo have been found to be in good agreement with measured fields.

The mode theory of very low-frequency propagation was found to offer an accurate enough method of determining field intensity for frequencies below about 25 kHz, where the ionosphere may be considered to be a conducting sheet. A stratified ionosphere has to be considered at higher frequencies. Bremmer [1949] approached the problem on the basis of geometrical optics to produce a rather involved formula incorporating the length of the zig-zag ray path within the earth-ionosphere waveguide as well as the reflection coefficients of the ground and the ionosphere.

It is evident then that the calculation of the field of a very low-frequency transmitter is rather involved. Nevertheless, because in the end the signal-to-noise ratio is the determining factor for successful communication, we recommend computing the field strength for a number of frequencies within the very low-frequency range, taking the dependency of the antenna's efficiency on frequency into consideration, and calculating the signal-to-noise ratio on the basis of a known noise-level dependence on the frequency. The chosen frequency should be the one yielding the maximum signal-to-noise ratio. Figure 5.2 shows the frequency dependency of the noise level.

5.2 LONG- AND MEDIUMWAVE PROPAGATION

Similar to the previous section, where the propagation of part of the low-frequency range was considered together with the very low frequency one, the properties of the

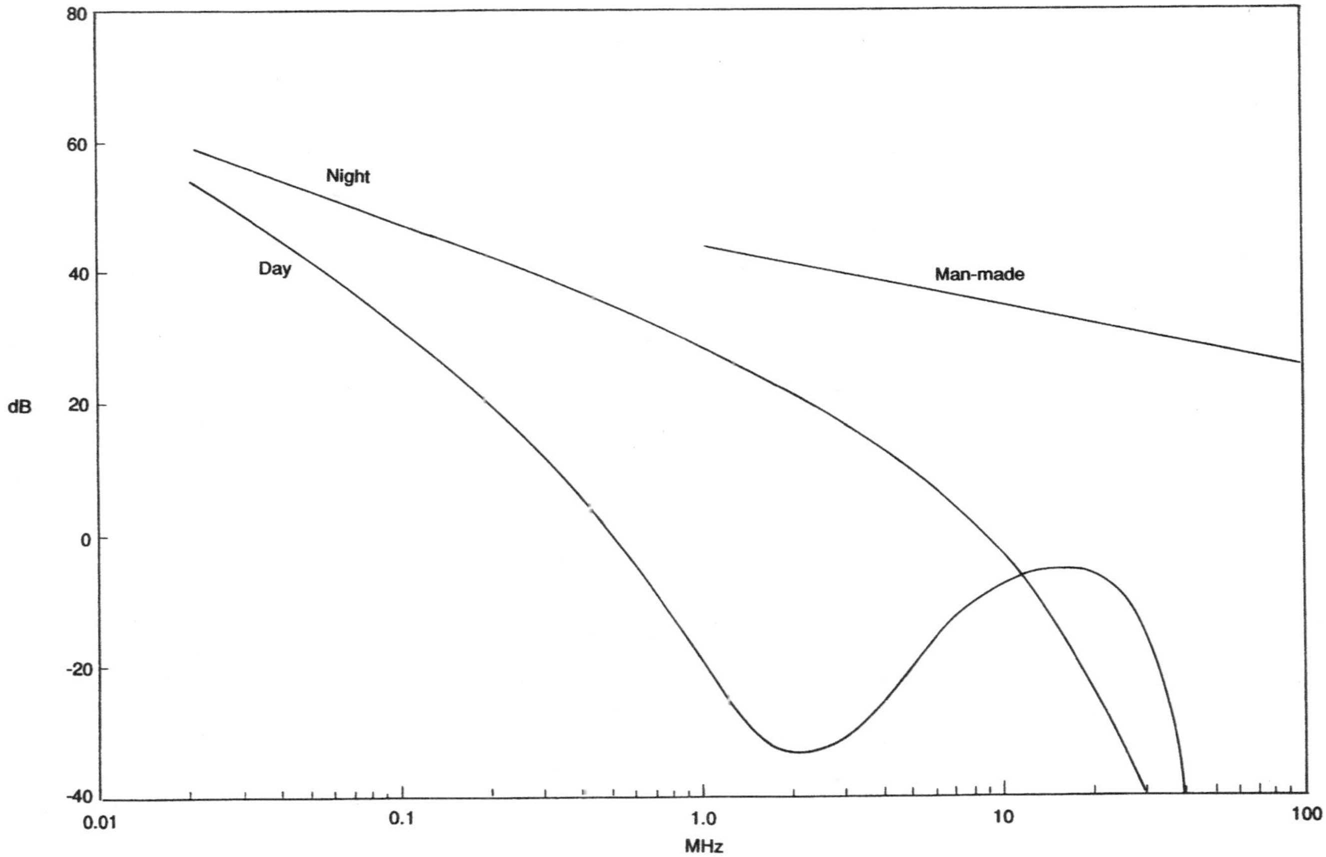

Figure 5.2 Frequency dependency of the median value of atmospheric and artificial noise levels for a receiver bandwidth of 10 kHz (0dB = 1 μV/m).

remaining low frequencies will be studied together with those of waves at frequencies up to 1.5 MHz, or a wavelength of 200 m. Waves in the considered frequency range of 100 kHz to 1.5 MHz again propagate as ground and sky waves. Ground-wave propagation was dealt with in Chapter 3; here, we treat the propagation of sky waves.

The nature of the *D*- and *E*-regions of the ionosphere at 100 kHz was found earlier. At 1.5 MHz the *E*-region, by using the charge carrier density and collision frequency of Section 5.1, yields a ratio of displacement to conduction current of well over 5000, indicating that it behaves like a dielectric. The necessary charge carrier density for reflection at oblique and normal incidence is given by (4.32) or (4.33). For the *E*-region at a height of 120 km expression (4.32) permits the determination of a charge carrier density requirement of $N \approx 10^9$ electrons/m^3 for reflection of waves leaving the transmitter antenna horizontally at a frequency of 1.5 MHz, whereas for the *D*-region at a height of 90 km it is $N \approx 8.4 \times 10^8$ electrons/m^3. At the lower frequency of 100 kHz the two charge carrier densities are 5×10^6 and 3.7×10^6 electrons/m^3, respectively. Higher frequencies then require higher charge carrier densities for reflection.

The lower boundary of neither of these ionospheric regions is sharply defined. The gradient of the charge carrier density is small, so that the lower level of neither region can reflect electromagnetic waves; however, it contributes to a large extent to their absorption.

Ionospheric absorption processes are not adequately understood so far; they are attributed to the fuzzy structure of the lower ionosphere regions. Due to an inadequate knowledge of the electron-to-ion ratio and the collision frequencies and those of the considered frequency range being of the same order, the computation of the absorption is rather difficult. The application of (2.38) and (2.43) yields values of absorption far too large for both the lowest and highest frequencies of the considered range. Measurements show that maximum attenuation occurs at a frequency of about 400 kHz, a frequency accepted to be the *D*-region limiting frequency, in the vicinity of which absorption reaches values that prevents daytime communication at these frequencies. The absorption becomes greatly reduced at night when it is so low that the sky-wave field strength exceeds that of the ground wave.

The ionospheric reflectivity exhibits great variations with the time of the day. It is about unity after sunset but drops, according to Al'pert [1973], to about 10^{-3} to 10^{-2} during daytime. The daytime field then is due only to the ground wave, although generally it is the result of the ground component and those arriving at the receiver site after reflections from the ionosphere. The nighttime field is higher but may exhibit considerable variability. Irregular field-strength fluctuations about the daily mean value are experienced at twilight, even at distances as short as 150 to 200 km from the transmitter. Prominent shallow variations appear at night.

The fluctuations result from multipath propagation, causing interaction of several waves arriving at the receiver by different paths. The individual waves are affected by irregular variations of the ionosphere. Most intense fluctuations occur at ranges where the amplitudes of the ground wave and the first reflected wave are comparable. Multiply reflected waves are very much weaker than the singly reflected one. Remote broadcast transmitters receivable only at night, often produce a more stable field than closer ones, because often only the singly reflected wave is received after the others are scattered by the ground's irregularities [Al'pert, 1973].

Periodic amplitude variations also may be observed, with the depth and frequency of the variations increasing with frequency. The properties of the fluctuations change with the separation of the location of the receivers.

5.2.1 Long- and Mediumwave Propagation Phenomena

In the considered frequency range, field strength may be influenced by several phenomena: time of day, season, ionospheric disturbances and solar activity, and ionospheric cross modulation.

5.2.1.1 Diurnal Field-Strength Variations and Fading

Because the ground wave of electromagnetic radiations in the considered frequency range predominates during the day and the sky wave at night, we must assume that the wave reaching a receiver will depend on the time of the day and the distance from the transmitter. Three cases may be considered, as shown by Doluchanow [1956].

If the receiver's location is close to the transmitter, the ground wave will be received during the day and the night. At intermediate ranges the transmitter's field is determined by the ground wave during the day and the sky wave at night, when both waves are equally strong or the sky wave even stronger. The two waves interfere then and cause deep fluctuations in the received signal's strength; fluctuations exhibiting a statistical character are designated by the name *fading.*

Fading may be caused as well by drifts of the cloudlike ionosphere resulting in a continuous charge density and hence reflection height variation. A Δd change in path length results in a $\Delta\varphi = 2\pi\Delta d/\lambda$ phase change. This is more pronounced at higher frequencies, where the phase variation compared to a wavelength is greater; however, its duration diminishes with a rising frequency of the propagating electromagnetic wave. Field-strength variations due to fading may reach more than 10 dB and last from about a second to many minutes.

At very long distances, only the sky wave may be received at night because the highly attenuated ground wave reaches the receiver neither by day nor by night. No daytime communication therefore is possible in the considered frequency range between the two locations. At night, despite the absence of the ground wave, fading may occur due to the interference between sky waves propagating over various ionospheric paths and the ionospheric drift. The dividing lines between these above three zones depend on frequency. They lie closer to the transmitter at higher frequencies than at lower ones.

Improved nighttime propagation conditions necessitate an at least 1000 km separation between transmitters operating on identical frequencies to reduce possible mutual interference. The ground-wave field strength should be at least three times that of the sky wave for fading-free reception. Various fade-combatting receiver circuits have been developed, but they are beyond the scope of the current study.

5.2.1.2 Seasonal Field-Strength Variations

Figure 4.12 intimates that, although the nighttime *E*-region charge carrier density varies but little over the year, the daytime one is higher in summer than in winter. A similar variation may be observed in the *D*-region. This increases the daytime absorption of electromagnetic waves with little change in the nighttime one. Further, because atmospherics caused by thunderstorms are more frequent in summer, the signal-to-noise ratio is reduced during this season. It is worst at times of the highest daytime absorption.

5.2.1.3 The Influence of Ionospheric Disturbances and Solar Activity

A reduction of nighttime sky wave radiation accompanied by strong scatter echoes, and polarization variations are noticeable during corpuscular penetrations in auroral zones. However, because reflection probably occurs at heights below those reached by corpuscular penetrations, no daytime influence of ionospheric disturbances have been detected so far at the considered frequency range. The influence of sunspot number variations manifests itself in reflection index and field-strength variations attributable to magnetic activity.

5.2.1.4 Ionospheric Cross-Modulation (Luxembourg Effect)

In 1933 it was observed in the Netherlands that a receiver tuned to the Beromünster, Switzerland, mediumwave transmitter reproduced the program of the Luxembourg longwave transmittter during lulls in the Beromünster transmitter's modulation, although the two transmitters were operating at widely different frequencies. This, explained Bailey and Martyn [1934], is caused by a cross-modulation effect due to nonlinear properties of the ionosphere in the region of cross modulation. Because the first observation occurred on the Luxembourg transmitter, the phenomenon was named the *Luxembourg effect*.

A strong amplitude-modulated transmitter operating in an area near the point of ionospheric reflection of the waves of a remote transmitter may influence the absorption coefficient of the ionosphere around the point of reflection in sympathy with its modulation. The modulation of the interfering transmitter thus may be superimposed on the carrier wave of the remote one.

Experimental evidence shows that cross modulation is most marked when the wanted wave is received by ionospheric transmission from a mediumwave broadcasting station and the sender of the disturbing wave is a high-power longwave station situated somewhere near the midpoint of the transmission path of the wanted wave.

5.2.2 Field-Strength Calculations

Ground-wave field-strength calculations were described in Section 3.5. This section concerns sky-wave field strengths, which are inversely proportional to the distance from the transmitter.

The length of the ray's path *ACB* in Figure 5.3, with the simplifying relations

$$\xi = d/2r \text{ and } \cos\xi \approx 1 - d^2/8r^2$$

acceptable due to the small value of the angle ξ, follow from the law of cosines and is

$$1 \approx 2\sqrt{d^2/4 + h^2} \tag{5.9}$$

The vertical component of the electric field $E \cos\epsilon$ at B, due to a transmitter at A, is

$$E_v = 2\,\frac{300\sqrt{P_{\text{kW}}}\cos\epsilon}{l_{\text{km}}}\,\varphi(\epsilon)\ \text{mV/m} \tag{5.10}$$

with $\varphi(\epsilon)$ determining the directional characteristic of the antenna and the factor 2 based on the assumption that complete reflection occurs at an ideally perfectly conducting earth.

The vertical coverage diagram of a vertical antenna of height $\lambda/3$ or less, is a half circle, so that $\varphi(\epsilon) = \cos\epsilon$ may be used in (5.10). The vertical component of the electric field becomes then

$$E_v = \frac{600\sqrt{P_{\text{kW}}}}{l_{\text{km}}}\cos^2\epsilon\ \text{mV/m} \tag{5.11}$$

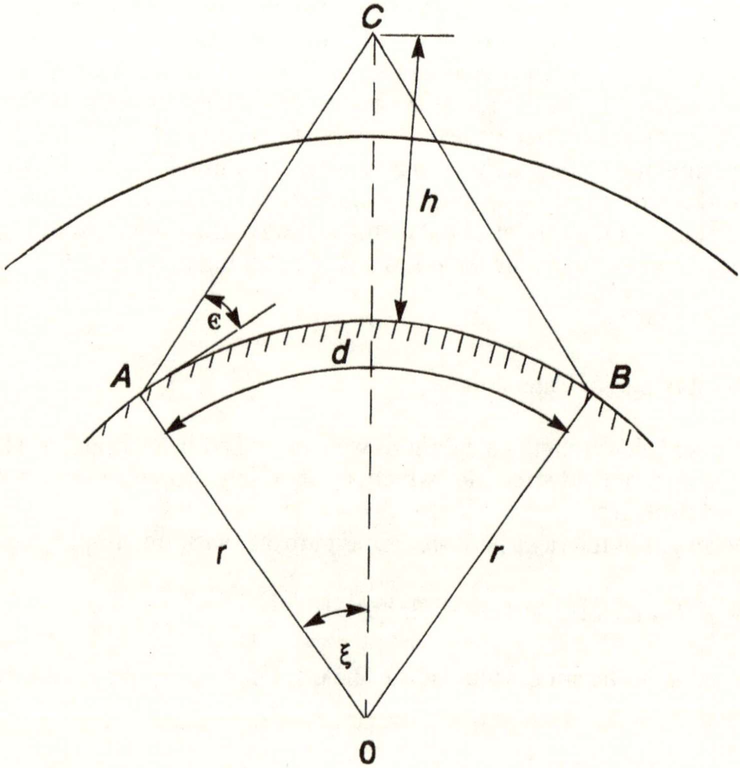

Figure 5.3 The length of the ray path.

Applying the law of sines to the triangle *OAC*, permits expressing $\cos\epsilon$ in terms of known quantities by

$$\cos\epsilon = \frac{d/2}{\sqrt{d^2/4 + h^2}}$$

Then, neglecting the curvature of the earth and using (5.9), the vertical component of the electric field is given by

$$E_v = \frac{300\sqrt{P_{\text{kW}}}}{\left[\dfrac{d_{\text{km}}^2}{4} + h_{\text{km}}^2\right]^{3/2}} \frac{d_{\text{km}}^2}{4} \text{ mV/m} \tag{5.12}$$

The maximum of the field at $d = 2\sqrt{2}h$ was confirmed experimentally.

Example 5.2

Compare the values of the vertical components of the electrical field strength of a 100 kW transmitter at a distance of 100 km when reflection from the *D*- or *E*-region of the ionosphere occurs.

Solution: The dependency of the vertical component of the electrical field on ionospheric height is given by expression (3.12). Taking then the height of the two ionospheric regions to be

$$h_D = 80 \text{ km and } h_E = 120 \text{ km}$$

the field strengths of interest become, respectively,

$$E_{vD} = 8.93 \text{ mV/m and } E_{vE} = 3.41 \text{ mV/m}$$

The computation of the field strength at ranges exceeding about 1000 km is complicated by the contribution of waves reflected several times from the earth and the ionosphere. Experimental evidence [Al'pert, 1973] shows that at nightfall the field strength may rise by factors of 200 or more, with amplitudes increasing 1000 to 10000 times at distances beyond 1000 to 2000 km from the transmitter. The average nighttime field intensity was found not to depend significantly on frequency and to be characterized not only by its average value but also by a "quasi-maximum" one defined as that value which is exceeded by the instantaneous values of the field for only 5% of the total measurement time.

Curves assembled by the CCIR and given in Figure 5.4 (a–c) show in μV/m the averaged nighttime field strength of a 1 kW transmitter for propagation over land and sea. Curves *A* refer to propagation over paths far removed from polar regions, and curves *B* refer to propagation over these areas. Data for these figures were obtained by a large number of measurements over different seasons. The dot-dashed curves in the first two figures show ground-wave field strengths over land and sea; the curves annotated by $3 \times 10^5/d$ show free-space field-strength values.

The vertical component of the electric field of an *n*-times reflected wave is

$$E_v = \frac{300\sqrt{P_{\text{kW}}}}{[(d_{\text{km}}/2n)^2 + h_{\text{km}}^2]^{3/2}} \left(\frac{d_{\text{km}}}{2n}\right)^2 \quad \text{mV/m} \tag{5.13}$$

5.3 SHORTWAVE PROPAGATION

The frequencies to be considered in this section include electromagnetic waves within the range of 1.5 to 30 MHz. Ground waves at these frequencies are greatly attenuated so that they may be neglected at ranges beyond about 100 km. Sky waves

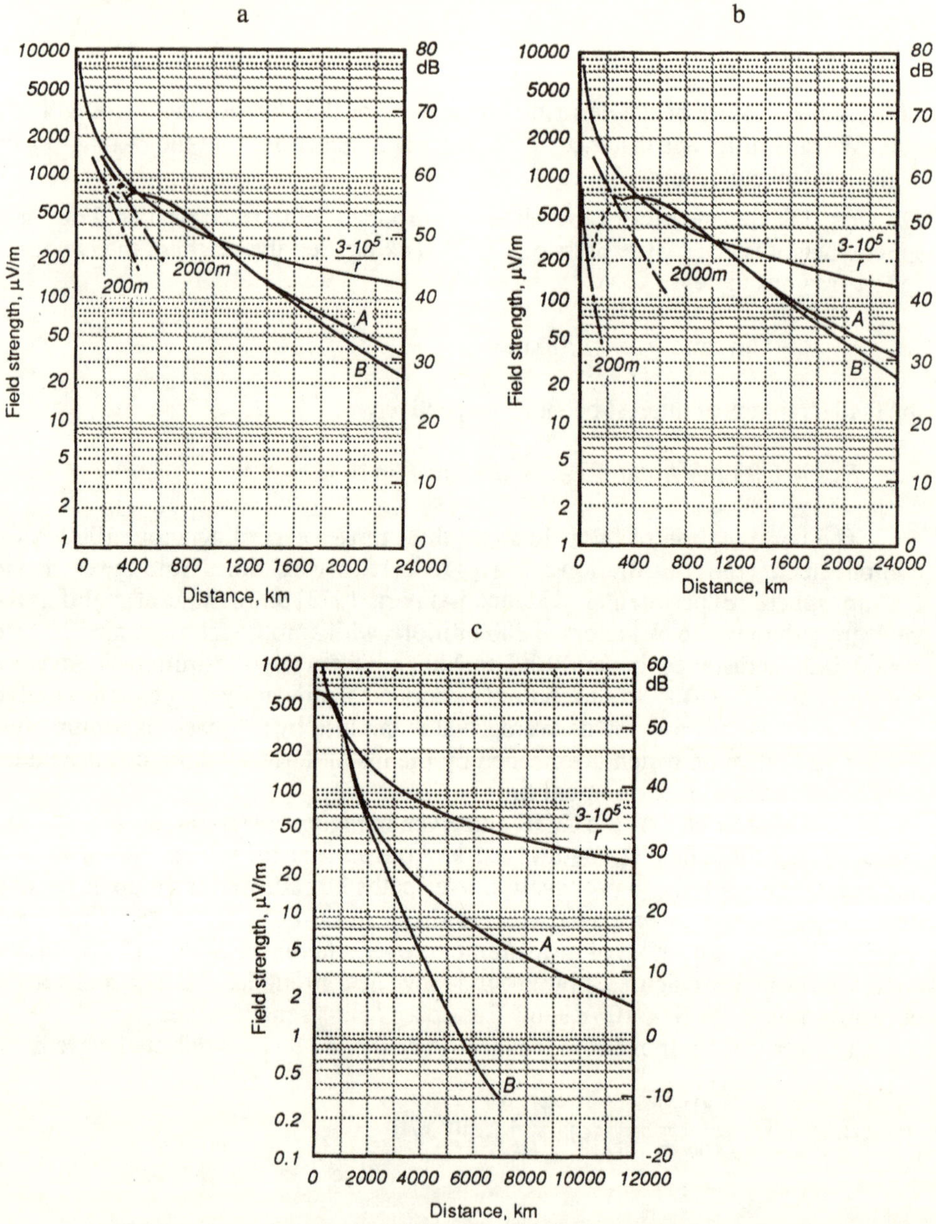

Figure 5.4 Mean "quasi-maximum" field-strength of a 1 kW transmitter for propagation (a) over land, (b) over sea, and (c) at long distances (after Doluchanow, 1956, © VEB Verlag).

propagate in single or multiple hops between the earth and the F-region of the ionosphere. Because the ionospheric attenuation is very small, propagation by ionospheric reflections is possible to very great distances; shortwaves are known to circle the earth several times in this manner.

The distance covered in a single hop, as computable on the basis of simple geometric considerations of a spherical earth and ionosphere, is approximately

$$d = 2r\cos^{-1}\left\{\frac{1}{r+h}\left[r\cos^2\epsilon - \sin\epsilon\sqrt{r^2\sin^2\epsilon + h(2r+h)}\right]\right\} \tag{5.14}$$

where ϵ is the angle of radiation with respect to the local horizontal, r is the earth's radius, and h the considered ionospheric height. It is important to note that, as the ray's trajectory in the ionosphere depends to a great extent on the distribution of electron concentration, the maximum distance covered by a single hop does not occur when $\epsilon = 0$, but at small values of $\epsilon > 0$, depending on the ratio of the operating and critical frequencies. At an ionospheric height of 400 km the distance d exceeds 4000 km.

The F_2 ionospheric region supports shortwave propagation, whereas the D- and E-regions attenuate short electromagnetic waves passing through them.

The mid-latitude noon F-region charge carrier density, according to Doluchanow [1956], is on the order of 10^{12} electrons/m^3 and the collision frequency is about 10^3/s. Shortwave reflections occur then at higher levels of the F_2-region, and the lower levels of the F-region contribute to the attenuation of waves passing through them. The magnitude of attenuation may be determined from expression (2.38) when the values computed from (4.23) for the relative permittivity and conductivity are used in expression (2.43) for κ.

The nature of the ionosphere at shortwaves causes the displacement current density to be much larger than the conduction one, so that the attenuation coefficient per unit length of the path, expression (2.38), reduces to

$$\begin{aligned} \delta &= 60\pi\frac{\sigma}{\sqrt{\epsilon'}} \\ \epsilon' &= 1 - 80.5N/f^2, \end{aligned} \tag{5.15}$$

as given by (4.30).

Because the dielectric constant ϵ' differs but little from unity and the frequencies of the waves in the considered range are much higher than the collision ones, the absorption coefficient per unit length of the path becomes

$$\delta \approx 1.35 \times 10^{-7} N\nu/f^2 \tag{5.16}$$

showing that the attenuation, unlike the behavior of other dielectrics, is inversely proportional to the square of the frequency and directly proportional to the product of the charge carrier density and collision frequency.

In the lower ionospheric regions, where the collision frequency is comparable with that of the electromagnetic waves, it may not be neglected in expressions (4.23) for the two ionospheric parameters. The process of absorption becomes more complicated, however, though absorption still diminishes with a rising frequency of the electromagnetic wave. Due to the inhomogeneity of the ionosphere, the absorption coefficient changes from place to place. We therefore must integrate it along the propagation path of the waves within the ionosphere to obtain the total absorption.

In accordance with Section 4.3, where it was stated that the maximum usable frequency is the highest frequency reflected by the ionosphere to be returned to earth at the required area, the frequency chosen for communication over a specified distance should be lower than the maximum usable one because no signal, disregarding the transmitter's power and the antenna's efficiency, will reach the receiver at frequencies exceeding the maximum usable one. Due to the diurnal dependency of the charge carrier density, several frequencies will be required for uninterrupted 24-hour communication over this distance.

Experience has shown that frequencies within the range of about 12 to 30 MHz are most suited for daytime communication. At nighttime, when the charge carrier density of the *F*-region diminishes, we must resort to lower frequencies in the range of about 1.5 to 8 MHz, as shown by Figure 4.1. The reduction of charge carrier density reduces absorption and thus counteracts its expected increase due to the frequency reduction. Frequencies between these two ranges proved to be most suitable for use during twilight. This frequency division is not rigid and is subject to seasonal effects, solar activity, and geographic location of the link's terminals.

At times of high relative sunspot numbers the *E*-region ionization may reach values adequate for the reflection of shortwaves arriving at low grazing angles whereas those at high grazing angles continue toward the *F*-region, thus causing double transmission. Transmission based on reflection from the sporadic *E*-region, when transmission over the *F*-region fails, may result in greatly reduced ranges due to the lower heights of reflection. Reflection of waves arriving at the upper boundary of the E_s-region from the *F*-region leads to an *M*-shaped trajectory mentioned in Section 4.8.

5.3.1 Shortwave Propagation Phenomena

5.3.1.1 Skip Zones

The area between the limits of the ground-wave range and the shortest distances reached by waves reflected by the ionosphere is called the *skip zone.* It surrounds the shortwave transmitter and, due to irregularities of the propagation in the various directions, generally is irregular in shape.

The distance to the inner skip-zone boundary depends on ground-wave attenuation and, although not dependent on the time of the day, exhibits a frequency dependency and seasonal variations. The distance to the outer boundary that depends on ionospheric propagation conditions, depends on the time of day because the ionosphere's charge carrier density varies with the sun's position. The charge carrier density required for the reflection of waves at a given frequency occurs at night at higher ionospheric levels than it does during daylight. The outer skip-zone boundary therefore will be more remote at night than during the day. This also explains the frequency dependency of the outer boundary's distance from the transmitter.

Al'pert computes the above boundary, the skip distance, by using the formula

$$d \approx 2 \int_{r}^{r+h} \frac{r^2 \sin\epsilon_0 d(r + h)}{(r + h)\sqrt{n^2(h)(r + h)^2 - r^2 \sin^2\epsilon_0}} \tag{5.17}$$

with ϵ_0 the initial radiation angle and $n(h)$ the refractive index at the considered ionospheric height. When the inequality $h_m/h_0 \ll 1$ applies, where h_0 is the altitude of the ionospheric region's lower boundary and h_m is the height of maximum electron concentration above h_0, the distance to the outer boundary of the skip zone, d_m, is given by

$$d_m \approx \frac{2h_0\sqrt{(f/f_c)^2 - 1}}{\sqrt{1 + [(f/f_c)^2 - 1]h_0/r}} \tag{5.18}$$

and the maximum usable frequency, f_{max}, at which a wave launched at an elevation angle ϵ_0 still will be reflected, by

$$f_{max} = f_c \sqrt{\frac{4\left(h_0 + \dfrac{d^2}{8r}\right) + d^2}{4\left(h_0 + \dfrac{d^2}{8r}\right)^2}} \tag{5.19}$$

Figure 5.5 shows the propagation range's dependency on the angle ϵ_0 for several values of the ratio of the used frequency to the critical frequency. A definite value of the distance d_m and the ray's emergence angle ϵ_m corresponds to each value of the frequency ratio. Figure 5.6, based on the previous one, shows the dependency of the extent of the skip zone d_m on the ratio of the maximum frequency to critical frequency for the case $h_m/h_0 = 0.4$ and $h_0 = 240$ km. An examination of Figure 5.5 shows that a ray may arrive at a location at a distance d_0 for a specified value of f/f_c

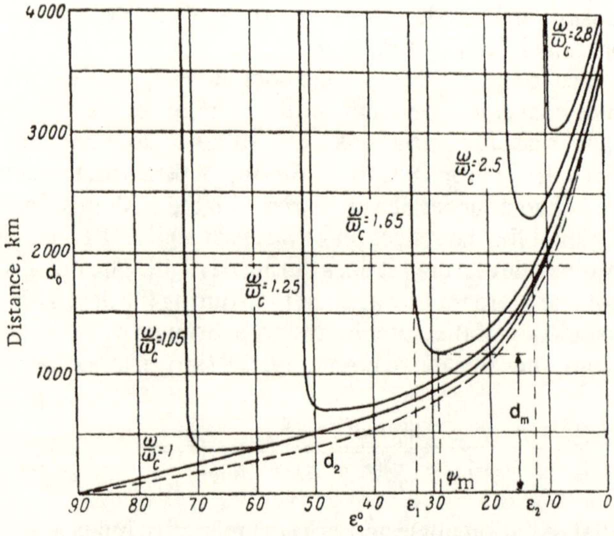

Figure 5.5 The dependency of the propagation range on the ray's elevation angle with ω/ω_c as a parameter for h_m/h_0 = 0.4 and h_0 = 240 km (after Al'pert, 1973, © Vsesojuznoje Agentstvo).

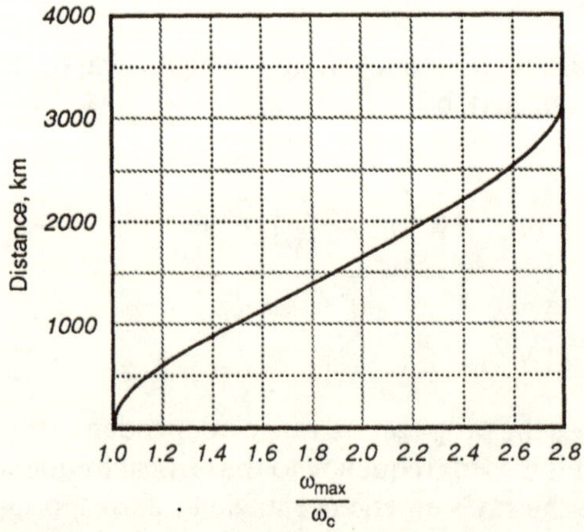

Figure 5.6 The skip distance as a function of f_{max}/f_c, for h_m/h_0 = 0.4 and h_0 = 240 km (after Al'pert, 1973, © Vsesojuznoje Agentstvo).

over two different paths determined by the two values, ϵ_1 and ϵ_2, of the elevation angle ϵ_0. The "Pedersen ray" follows the path with the higher elevation angle. This is longer than the path appropriate to the lower elevation angle.

Example 5.3

Assuming that the initial radiation angle is 52.69°, compute the distance to the outer boundary of the skip zone when $h_0 = 240$ km. Compare your result with that obtained from Figure 5.6 when the used frequency is 10 MHz.

Solution: From expression (4.35)

$$\sin^2\phi_0 = 1 - (f_c/f)^2$$

$$0.63 = 1 - (f_c/f)^2 \cdots f_c/f = 0.61$$

$$f/f_c = 1.64$$

$$f_c = 10/1.64 = 6.1 \text{ MHz}$$

from expression (5.18)

$$d_m = 604.99 \text{ km}$$

from (5.19)

$$f_{max} \text{ for the preceding } d = d_m$$

$$f_{max} = 7.69 \text{ MHz}$$

Then $f_{max}/f_c = 1.26$ and from Figure 5.6 for $f_{max}/f_c = 1.26$, the distance $d = 650$ km.

5.3.1.2 *Echoes*

Under favorable conditions signals of a transmitter may reach a remote receiver over various paths. The first signal to arrive there propagates over the shortest path; subsequent ones traverse longer paths [Weiss, 1936]. All travel in the earth-ionosphere waveguide in the previously described zig-zag manner. Shortwave signals are known to circle the earth, often several times, and appear as echoes of the first one received. They take about 0.1378 second to circle the earth; they are perceivable by the ear and differ very little in intensity from the first signal received. Echoes may cause errors in telegraphic communication, in telephonic communication they appear as protracted echoes and multiple superimposed pictures shifted relative to one another in television reception. The frequency of occurrence of round-the-world echoes and their signal strength depends on the time of day, the season, and the state of the ionosphere.

Because all of these signals are at the same frequency and most arrive from the same direction, they cannot be eliminated. However, a judicious choice of the operating frequency, close to the maximum usable one, may reduce the occurrence of multiple echoes by preventing the reflection of low-angle incidence waves from the ionosphere.

5.3.1.3 Fading Phenomena and Other Effects

Two types of fading are known to occur: interference fading and polarization fading. The first type of fading at this frequency range is due either to the interference of two or more sky waves arriving at the receiving antenna in different phases or to reflection from an uneven ionosphere. It manifests itself as a random, often quasi-periodic, variation of the signal strength capable of reaching values of 20 to 40 dB with periods of between about a second to several times 10 seconds. Fading also may be irregular in character. Both types of fading often are superimposed on fadings of longer periods [Doluchanow, 1956].

Interfering waves may arrive at the receiving antenna after one or more reflections, hence, exhibiting phase differences due to the difference in the traversed path-lengths. At these wavelengths only small changes in the height of reflection may cause considerable phase variations.

Ionospheric irregularities may diffuse the reflection. The incident ray splits into bundles of divergent rays. Rays originating in different bundles may interfere and cause an uninterrupted random variation of the field.

Magneto-ionic splitting, investigated in Section 4.5, causing the arrival of an ordinary and an extraordinary wave at the receiver also may cause fading. Both waves usually are elliptically polarized. The resulting elliptically polarized field has the character of a highly elongated ellipse with its direction of polarization continually changing due to the fluctuation of the ionospheric charge density. This changes the direction of the ellipse's large axis without undue variations in its dimensions. Fading in this case is due to the directional variation of the polarization plane. Maximum signal strength occurs when the polarization ellipse's large axis is parallel to the receiving antenna.

Fading may cause dropout of telegraphic signal elements. Although this is unwelcome, it is not as serious as dropout in signal elements in teletype communication, which may result in printing wrong letters as an experienced telegraphist usually can fill in spaces caused by dropout in telegraphic signals. In amplitude modulated voice communication fading may cause volume and pitch variations.

The expanded expression of an amplitude modulated signal

$$E = E_0 \cos\omega t + \frac{mE_0}{2}\cos(\omega + \Omega)t + \frac{mE_0}{2}\cos(\omega - \Omega)t$$

shows that the field consists of three elements, the carrier wave of angular frequency, ω, and two sidebands of frequencies $\omega + \Omega$ and $\omega - \Omega$, where Ω is the instantaneous modulating frequency and m is the modulation index. Because the ionospheric refractive index is frequency dependent, each of these fields will propagate over a different path and be reflected at a different height. Fading will affect them in different ways, maxima and minima of the three components will not coincide—selective fading results.

Relative phase variations, due to these pathlength variations, vary the amplitudes of the three field components. Although in an undistorted amplitude modulated wave the amplitude of the sidebands is about half that of the carrier, a change of this ratio has the same effect as a change of the modulation index. This leads to distortion, which may manifest itself either as amplitude distortion (i.e., a change in the relative amplitudes) or harmonic distortion (caused by the appearance of modulation harmonics not present in the original signal and phase distortion due to shifts of the modulation's phase).

5.3.2 The Influence of Ionospheric Disturbances and Solar Activity

The highest known ionospheric layer and the one most important for shortwave propagation, the F_2-layer, is the most susceptible to the influences of ionospheric disturbances. These may manifest themselves in various ways.

Disrupted regions of the reflecting layer suffer changes in charge carrier densities resulting in changes of the virtual height of reflection and necessitating corresponding changes of the used frequencies. Ionospheric disruptions increase losses; they reduce signal strength and therefore the signal-to-noise ratio. Shortwave communication may cease completely during periods of strong ionospheric disturbances.

The frequency and intensity of ionospheric disturbances is known to depend on the geomagnetic latitude; auroral zones have been found to be the ones most affected. The seasonal frequency of ionospheric disruptions exhibits weak maxima at the equinoxes.

Ionospheric disruptions clearly depend on the 11-year cycle of solar activity. Weak and medium strength disruptions appearing at times of increasing solar activity exhibit a 27-day period coinciding with the duration of a solar day. Observations show that solar eruptions are followed 15 to 20 hours later by ionospheric storms.

Short-term forecasts of such storms may be based on the observation of certain geomagnetic effects.

Corpuscular penetrations occurring during ionospheric storms may further disrupt shortwave communication. They are more frequent during periods of sunspot maxima and may be completely absent during years of minima. Individual disturbances that generally appear as fadeouts lasting no longer than a few minutes cannot be predicted so far.

5.3.3 Field-Strength Computations

The determination of the exact values of the received field strength is complicated by our inadequate knowledge of the lower ionospheric layers' absorption of waves at the frequencies of interest. The computation of the received field strength may proceed on the same basis as that in Section 3.2.1. Then by using expression (3.19) we can determine the free-space field strength at a distance of R_{km} from the transmitter. This expression has to be extended to take into consideration attenuation of the waves, if more realistic results are to be gained.

Most of the attenuation is contributed by absorption in the ionized atmospheric layers. Hence, it depends on the pathlength of the wave in the ionosphere, which, in turn, depends on the separation of the terminals of the considered communication link because this determines the number of hops the wave takes to cover the required distance.

The pathlength in the ionosphere for a single-hop trajectory is given by Al'pert as

$$L = 2\int_{r}^{r+h} \frac{n(h)(r+h)\,\mathrm{d}(r+h)}{\sqrt{n^2(h)(r+h)^2 - r^2\sin^2\epsilon_0}} \tag{5.20}$$

with the symbols having the earlier given meanings.

Due to the inhomogeneity of the ionosphere, the absorption of electromagnetic waves varies from place to place. To obtain the total absorption, the variable absorption must be integrated over the whole ionospheric path, as indicated earlier.

Writing then expression (3.19) modified to account for absorption, we obtain

$$E_{\text{eff}} = \frac{173\sqrt{P_{kW}G}}{R_{km}}\,10^{-0.5A}\ \text{mV/m} \tag{5.21}$$

where P_{kW} is the radiated power in kW,
G is the directional gain of the antenna,
R_{km} is the separation of the link's terminals, and
A is the ionospheric absorption coefficient consisting of four terms and expressed by

$$A = JQ\overline{K}d \tag{5.22}$$

with J, given in Table 5.3, accounting for seasonal effects,

Q = $1 + 0.0037\overline{R}$, a solar cycle variation factor, where $\overline{R}$ is the 12-month average number of sunspots over a time span of 6 months preceeding and 6 months following the date of interest,
$\overline{K}$ = $(\cos 0.881\chi)^{1.3}$, an average solar zenith angle factor,
d = the illuminated pathlength in thousands of km, which may be taken to be unity at pathlengths shorter than 3200 km.

The product $Q\overline{K}$ usually is called *absorption index I.*

Table 5.3
Seasonal Variation Factor J

Month	*Both Terminals in the Same Hemisphere:* *Northern*	*Southern*	*Terminals in Different Hemispheres*
Jan., Feb., Nov., Dec.	1.3	1.0	1.15
Mar., Apr., Sept., Oct.	1.15	1.15	1.15
May, June, July, Aug.	1.0	1.3	1.15

The top part of Figure 5.7 shows the dependency of the zenith angle for the month of June on the geographic latitude and local time, and the bottom one shows that of the average solar zenith angle factor $\overline{K}$. This factor becomes equal to unity for a zero zenith angle and falls to zero when the sun is about 10° below the horizon. Maps like those in Figure 5.7 are available in some of the publications listed in the references to this chapter (e.g., N.B.S. [1948], Davies [1966], Picquenard [1974]) for each month of the year.

Example 5.4

Compute the field strength of a 10 kW transmitter at 18 h local time in June 1970 at a location 10° south when the link's terminals are in different hemispheres and 4000 km apart and the antenna gain is 10 dB.

Solution: The number of sunspots for 1970, as read from Figure 4.10, is 91. Then $Q = 1 + 0.0037 \times 91 = 1.34$. The zenith angle for the specified time and latitude, as read from Figure 5.7, is $\chi = 95°$ so that the average solar zenith angle factor $\overline{K} = (\cos 0.881 \times 95)^{1.3} = 0.06$, as may be read as well from the bottom map of this figure.
Then $I = Q\overline{K} = 0.08$, and from Table 5.3, $J = 1.15$.

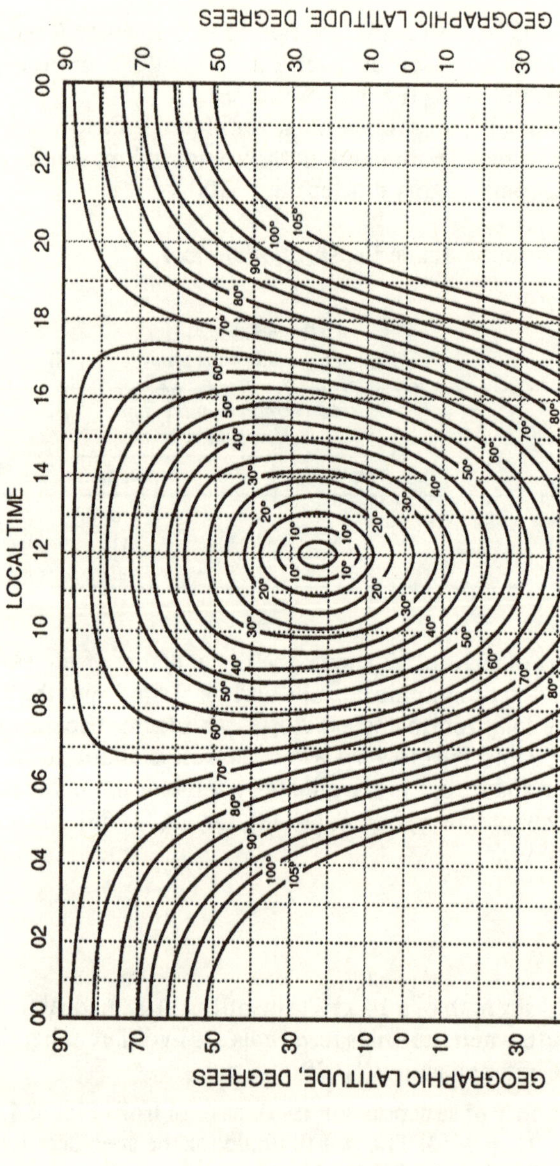
LOCAL TIME
GEOGRAPHIC LATITUDE, DEGREES
GEOGRAPHIC LATITUDE, DEGREES

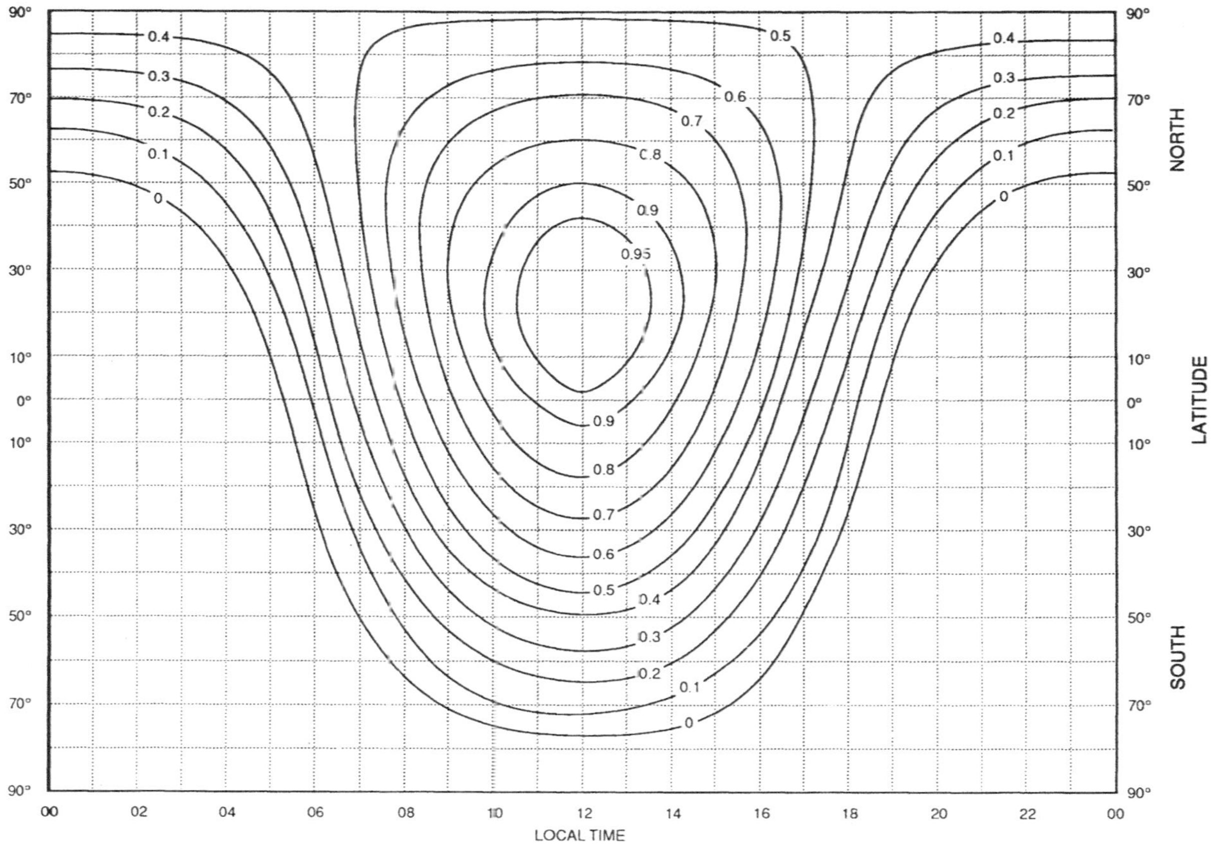

Figure 5.7 Zenith-angle of the sun and the ionospheric absorption factor for June: (a): after Davies, 1966, Dover Publications; (b) after NBS Circular 462, Ionospheric Radio Propagation, 1948.

The ionospheric absorption coefficient, with $d = 4$, is $Ad = 1.15 \times 1.34 \times 0.06 \times 4 = 0.37$ and expression (5.21) yields the field strength:

$$E_{eff} = \frac{175\sqrt{10 \times 10}}{4000} 10^{-0.5 \times 0.37} = 286 \ \mu V/m$$

A method of field determination described in the earlier publications considers the ionospheric conditions at intermediate locations between the terminals and may provide closer values of field strength to those experienced in practice. The required radiated power for maintaining reliable communication depending on the necessary signal-to-noise ratio for the considered type of communication may be computed from (5.21).

Because the available signal power at the receiver decreases and the noise power increases with a decreasing operating frequency, the signal-to-noise ratio varies in sympathy with the operating frequency; there will be a minimum frequency at which the signal-to-noise ratio still assures that reliable communication is possible. This frequency is called the *lowest usable frequency* (LUF). It depends on the transmitted power and the parameters determining the path loss and ambient noise level at the receiver's location.

5.3.4 Ionospheric Predictions

A number of ionospheric observatories publish, three to four months in advance, data on solar activity and ionization of the ionospheric regions on the basis of which we can determine the maximum usable frequency, absorption, and noise levels likely to occur and affect communication circuits using ionospheric wave trajectories.

Although predictions concerning the *E*-region are rather accurate, those for the *F*-region usually are within about 10% of values observed in practice. However, as we have to expect some variations in the ionospheric conditions, it is recommended to be conservative in the estimation of the frequencies to be used for the various communication circuits.

Procedures applicable to the calculation of quantities of interest on the basis of the predictions are beyond the scope of this book; the reader is referred to the publications named earlier for detailed instruction.

5.4 VERY AND ULTRA HIGH-FREQUENCY PROPAGATION

Very and ultra high-frequency waves extend in wavelength from 10 m down to millimeters; they include all usable frequencies above 30 MHz. Lately it is more accepted to talk about meter waves in the range of 10 m to 1 m (30 to 300 MHz), which could be termed very high-frequency waves, decimeter waves between 1 m and 10 cm (300 MHz to 3 GHz), centimeter waves 10 cm and 1 cm (3 to 30 GHz), and millimeter waves in the frequency range above 30 GHz. All of these are at ultra-high frequency waves or microwaves.

The reason for dividing the very and ultra high-frequency range into these categories is due not so much to the differences in their propagation characteristics as to the differences in their generation and the construction of the appropriate equipment. The apparatus for the generation, amplification, and radiation of meter waves differs very much from that found to be efficient in centimeter waves, where the dimensions of the various parts of the equipment are comparable with the wavelength.

Ionospheric reflections occur only at frequencies lower than the *maximum usable frequency* (MUF) for the prevailing composition of the appropriate ionospheric region. The maximum daytime usable frequency over the 11-year solar activity period was found to be, on the average, near 30 MHz although, at times of high solar activity, much higher frequencies occasionally may be reflected from the F_2-region. The sporadic E-region, which was found to be independent of solar activity, occasionally may reflect waves at frequencies of up to about 80 MHz.

Communication at the frequencies considered in this section then cannot rely on regular ionospheric reflections, and indeed, these waves are used mainly for communication over line-of-sight paths.

5.4.1 Some Theoretical Considerations

Propagation at very and ultra high frequencies usually is dealt with by the principles of geometrical optics, for which the scalar wave equation is required. If the field E is subjected to sinusoidal time variations expressed by the exponential form

$$E = E\,e^{j\omega t}$$

then

$$\frac{\partial^2 E}{\partial t^2} = -\omega^2 E \tag{5.23}$$

so that expression (2.7) becomes

$$\nabla^2 E + \omega^2 \mu\epsilon \mathbf{E} = 0 \tag{5.24}$$

Considering only the E_y component, the preceding expression becomes the scalar wave equation, which may be written [Rohan, 1981] as

$$\nabla^2 E + k^2 n^2 E = 0 \tag{5.25}$$

with the wavelength constant $k = 2\pi/\lambda$ and n, a modified index of refraction, equal to $(c/v_0)^2$, where c is the velocity of propagation of light. One possible solution of this equation has the form

$$E = Q\, \mathrm{e}^{-jkS} \tag{5.26}$$

where Q and S are real functions of position. Inserting expression (5.26) into (5.25) and performing the indicated operations, we arrive at

$$\nabla^2 Q - 2jk\nabla \mathbf{Q} \cdot \nabla S - jkQ\, \nabla^2 S - k^2 Q(\nabla S)^2 + k^2 n^2 Q = 0 \tag{5.27}$$

The scalar wave equation is satisfied if the real and imaginary parts of (5.27) are equal to zero, or if

$$\frac{\nabla^2 Q}{k^2 Q} - (\nabla S)^2 + n^2 = 0 \text{ and } \nabla^2 S + 2\frac{\nabla Q \cdot \nabla S}{Q} = 0 \tag{5.28}$$

If $(\nabla^2 Q/k^2 Q) \ll n^2$, the first of these expressions becomes

$$(\nabla S)^2 = n^2 \tag{5.29}$$

That is, the differential equation of the "eikonal" of S that, with increasing frequency and hence k, leads to increasingly closer approximations as long as the ratio $\nabla^2 Q/Q$ remains finite [Kline and Kay, 1965]. Adoption of the geometrical optics approach permits the consideration of rays instead of wavefronts, thus simplifying the solution of propagation problems.

Due to expression (5.26), the surfaces S = const. are surfaces of constant phase or wavefronts of E and their normals, given by the gradients of S, determine the direction of the rays that may be regarded as the orthogonal trajectories of the constant S surfaces and that, for a position-dependent n, are curved. Integration of expression (5.27) permits the determination of the wavefronts and hence the directions of the rays.

The radius of curvature of the ray's trajectory may be computed after Picquenard [1974] considering a straight-line approximation of the trajectory and the wavefront AB when A is at a tropospheric level $h = AB \cos\epsilon$ above that of B, Figure 5.8. Due to the negative gradient of the refractive index, point A will travel faster than point B and hence, although B covers the distance

$$BB' = \frac{c}{n}\,\mathrm{d}t$$

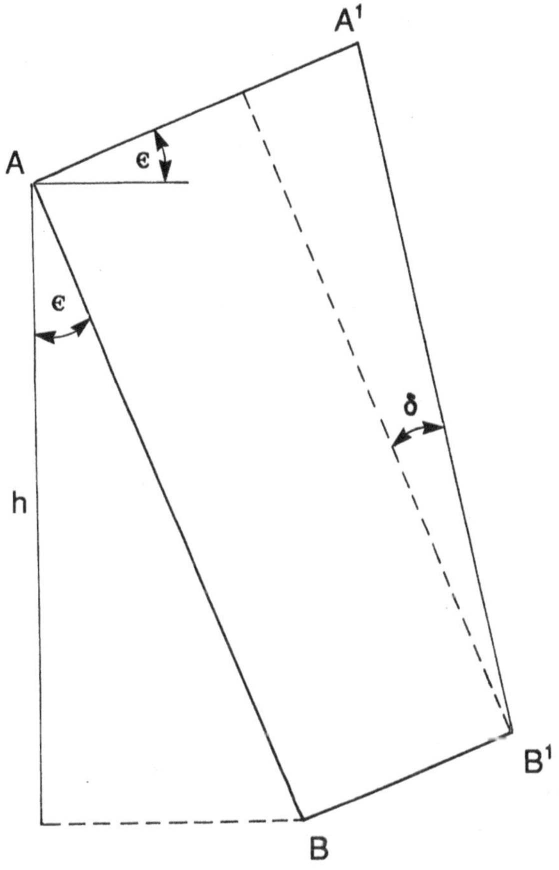

Figure 5.8 Ray trajectory (after Picquenard, 1974, Macmillan).

in time dt, point A traverses the distance

$$AA' = \frac{c}{n + h\dfrac{\mathrm{d}n}{\mathrm{d}h}}\mathrm{d}t = BB' + AB\tan\delta$$

The new distance between the two ends of the wavefront is

$$A'B' = AB/\cos\delta$$

when the wavefront, due to the curvature of the trajectory, deviates from the direction of the original by angle δ.

The radius of curvature is

$$\rho = \frac{AA'}{\delta} = \frac{AA' \times AB}{AA' - BB'}$$

or with $\delta \approx \tan\delta$ and the preceding expressions for the required distances

$$\rho = \frac{\dfrac{h}{\cos\epsilon}\,\dfrac{c}{n + h(\mathrm{d}n/\mathrm{d}h)}}{\dfrac{c}{n + h(\mathrm{d}n/\mathrm{d}h)} - \dfrac{c}{n}} = -\frac{n}{\cos\epsilon}\frac{\mathrm{d}h}{\mathrm{d}n} \tag{5.30}$$

which, due to $\mathrm{d}n/\mathrm{d}h < 0$ in the considered direction, will be negative; that is, the trajectory will bend toward the earth.

The validity of the geometric optics approach to problems of propagation is restricted to cases where the relative change of the modified index of refraction and the distance of neighboring rays over a pathlength of a wavelength is much smaller than $2\pi n$. Caution is to be exerted in applying a geometric optics approach to cases where the rays converge or diverge; that is, where neighboring rays cross to form foci or caustics. The first restriction depends on the frequency of the radiation and the structure of the medium of propagation, whereas the second one depends on the type of wave the ray pattern is to represent. Both restrictions are irrelevant in conventional propagation problems and the use of the geometric optics approach for their solution is generally accepted.

5.4.2 The Various Propagation Modes of VHF and UHF Waves

The antennas used at this frequency range are, as a rule, elevated. It is necessary therefore to consider the ray reflected from the ground with all of its accompanying phenomena, as described earlier in Section 2.5 on reflection and Section 3.5. The use of elevated antennas is necessitated by the rapid decrease in the field strength beyond the horizon, due to which visibility between the transmitting and receiving antennas is mandatory. Propagation may take place over short distances at which the earth's surface may be taken to be plane, over longer distances within the optical range where the curvature of the earth and atmospheric refraction cannot be neglected, and beyond the horizon.

5.4.2.1 *Propagation over a Plane Earth*

As it was pointed out in Sections 2.5 and 3.5, the resultant field at the receiving antenna is the vector sum of the field due to a ray propagating directly between the transmitting and receiving antenna, the so-called direct ray, and a ray that includes a reflection from the ground on its path between the two antennas. Due to the short distances involved, we may neglect atmospheric refraction.

The field strength from the direct ray is

$$E_d = E_0 \cos\omega t \tag{5.31}$$

where E_0, given by (3.19), reads

$$E_0 = \frac{245\sqrt{P_{\mathrm{kW}}G(\epsilon)}}{R_{\mathrm{km}}}\ \mathrm{mV/m}$$

$G(\epsilon)$ is the antenna gain in the direction of the direct ray between the transmitting and receiving antenna, as shown in Figure 5.9 and R is the slant range between the antennas.

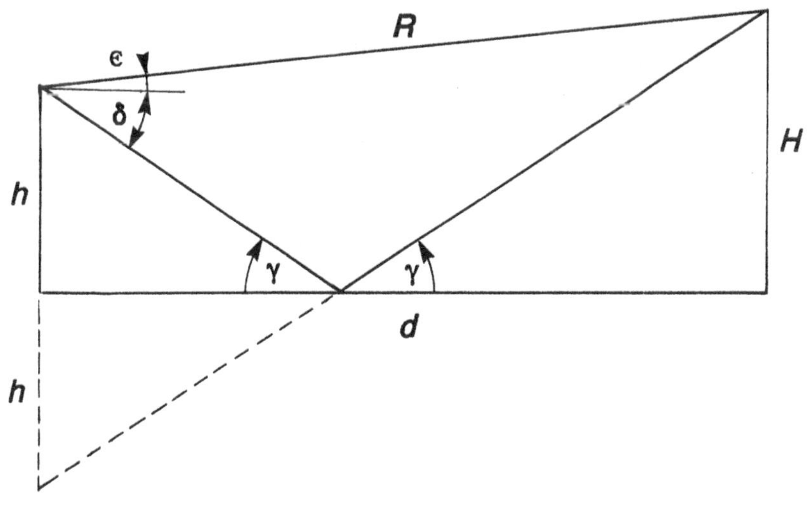

Figure 5.9 Propagation over a plane earth.

The field strength from the reflected ray is

$$E_r = \frac{245\sqrt{P_{kW}G(\delta)}}{R + \Delta R} \rho \cos\left(\omega t - \mu - \frac{2\pi}{\lambda}\Delta R\right) \text{ mV/m} \tag{5.32}$$

with $G(\delta)$ the antenna gain in the direction δ shown in Figure 5.9, μ is the phase shift on reflection given by (2.65), (2.69), or (2.72), ρ is the reflection coefficient given by (2.65), (2.69), (2.70), or (2.71), both of these quantities depending on the used polarization, and ΔR is the difference between the direct and reflected ray pathlengths. The slant range is

$$R = \sqrt{d^2 + (H - h)^2}$$

and the reflected ray length is

$$R' = \sqrt{d^2 + (H + h)^2}$$

as may be deduced from simple geometric considerations. Then, due to $(H + h) \ll d$ and also $(H - h) \ll d$, we arrive with these approximations:

$$R \approx d\left[1 + \frac{1}{2}\left(\frac{H - h}{d}\right)^2\right]$$

$$R' \approx d\left[1 + \frac{1}{2}\left(\frac{H + h}{d}\right)^2\right]$$

at

$$\Delta R \approx 2\frac{Hh}{d} \tag{5.33}$$

or at large distances where the direct and reflected rays may be assumed to be parallel, as shown by Figure 3.8, we arrive at

$$\Delta R \approx 2h \sin\epsilon \tag{5.34}$$

The distance d is the ground range between the transmitter and the receiver. A more detailed transmitter-receiver geometry has been shown by Figure 3.17.

The dimensions of the various quantities of interest are given by formulas (3.64) to (3.70). The angle of incidence and reflection is

$$\gamma = \pi/2 - |\delta - \alpha|$$

The resulting field is the vector sum of E_d and E_r, or

$$E = \sqrt{E_d^2 + E_r^2 + 2E_dE_r\cos(\varphi_1 - \varphi_2)} \tag{5.35}$$

where $\varphi_1 - \varphi_2 = \mu + 2\pi\Delta R/\lambda$ is the phase difference between the two rays.

There are no known guidelines on how far we can consider the earth to be plane, and the use of this outlined approach depends on the required accuracy of the computed quantities. Generally, we restrict the plane earth approach for distances of only a few kilometers between the two antennas.

Example 5.5

Determine the field of a horizontally polarized 100 W, 50 MHz transmitter at a distance of 20 km when the transmitting antenna, having a gain $G(\epsilon)$ of 27 dB, is at a height of 20 m, and its $G(\delta)$ gain is 16 dB. The receiving antenna is at a height of 12 m.

Solution: Because, according to expression (3.13), the earth may be considered to be plane when the distance between the transmitter and the receiver is smaller than

$$R_{\text{km}} = \frac{80}{\sqrt[3]{f_{\text{MHz}}}} = \frac{80}{\sqrt[3]{50}} = 21.72 \text{ km}$$

the geometry becomes similar to that depicted in Figure 5.9.

The grazing angle may be determined from the similarity of the triangles constituted by the two antenna heights and the lines from the top of the antennas to the point of reflection. Writing

$$r_1 + r_2 = d$$

$$\frac{h}{r_1} = \frac{H}{r_2}$$

where r_1 and r_2 are the distances of the point of reflection from the feet of the antennas, we obtain $r_1 = 12.5$ km and $r_2 = 7.5$ km, respectively. The grazing angle follows from $\tan\gamma = h/r_1 = H/r_2$, and it is 5.25°. The coefficient of reflection at horizontal polarization may be taken to be unity, and the phase shift on reflection is $\mu = \pi$ rad. The antenna gains specified in dB, converted to multipliers, become $G(\epsilon) = 501.19$ and $G(\delta) = 39.81$.

Inserting the required parameters into expression (5.31) leads to the field of the direct ray $E_d = 86.72\cos\omega t$, whereas expression (5.32), with $\Delta R = 0.02$ obtained from (5.33), yields the field of the reflected ray as $E_r = 24.44\cos(\omega t - \pi - 0.04\pi/$

λ) or, because the last term in the bracket is negligibly small, $E_r = 24.44 \cos(\omega t - \pi)$.

The result then, from (5.35), when $\varphi_1 - \varphi_2 = \pi$, is

$$E = \sqrt{E_d^2 + E_r^2 - 2E_dE_r} = 56.38 \text{ mV/m}$$

5.4.2.2 Tropospheric Propagation over a Spherical Earth

The curvature of the earth affects the propagation of electromagnetic waves in various ways, most of which have been dealt with earlier but are repeated here in part to give a coherent picture of the occurring phenomena. The optical horizon for antennas at height h and H in meters, as given by (3.47) above, is

$$R_{\text{km}} = 4.12(\sqrt{h} + \sqrt{H})$$

Beyond this range the propagation is complicated by diffraction at the earth's hump, due to which the signal strength falls off very fast. Further, due to reflection from a convex surface, the plane earth reflection coefficient must be modified to account for the divergence of rays on reflection, as indicated in Section 2.5.5 by expression (2.77).

Because the pathlengths are large, atmospheric refraction and the accompanying bending of rays cannot be neglected. Very and ultra high-frequency propagation takes place mostly in the lowest atmospheric layer, the troposphere, which varies in height from about 6 km above the poles to about 18 km above the equator. As the troposphere is not homogeneous, it will affect the propagation of electromagnetic waves. The varying density of the troposphere and varying amount of moisture vapor in it causes refraction.

A dipole moment induced by a passing electromagnetic wave in each gas molecule increases the dielectric permittivity proportionally to the number of induced dipoles per unit volume. Any spatial variation of the density implies a corresponding spatial variation in the velocity of propagation. The air may be considered to be a perfect gas in which the density and hence the refractive index variation is directly proportional to the pressure P and inversely proportional to the temperature T.

The electromagnetic wave causes the alignment of the permanent dipole moments of the water molecules. They contribute then to the polarization of the medium, whereas random molecular motion, due to rising temperatures, depolarizes it. The refractive index of a medium rises with a rising partial pressure e of its water vapor and is inversely proportional to its temperature.

The components of a mixture of gases affect the refractive index in proportion to their partial pressures. The atmospheric refractive index, given by $n = \sqrt{\epsilon\mu}$, where ϵ and μ are permittivity and permeability, depends on the atmospheric pressure, the temperature, and humidity. It is related by

$$n = 1 + N \cdot 10^{-6} \tag{5.36}$$

to the refractivity N given by

$$N = \frac{77.6}{T}\left(P + \frac{4810e}{T}\right) \tag{5.37}$$

where the pressure P and the partial pressure of water vapor e are to be inserted in millibars and the absolute temperature T in Kelvins or $t + 273$, where t is in degrees Celsius. The first term of (5.37) is the "dry" term; the second one is the "wet" term. The height dependency of the three parameters of 5.37 is 1.2 mb/10 m for the pressure P, 1°C/200 m for T and 1 mb/300 m for the humidity e.

The magnitude of the N-units follows from the average vertical gradient of the refractive index

$$\frac{dn}{dh} = -0.039 \times 10^{-6} N\text{-units/m} = -39 N\text{-units/km} \tag{5.38}$$

Various approximations to the height dependency of the refractive index have been proposed. The one most often used, the so-called standard atmosphere, assumes the atmosphere to consist of thin shells of constant values of the refractive index with a linear height dependence given by $n + dn$ at heights $h + dh$. This artificial atmosphere offers a good likeness of real atmospheric conditions for the first few km above the earth. Another approximation is provided by the "fundamental reference" atmosphere, in which the height dependency of the refractive index is

$$n(h) = 1 + 289 \times 10^{-6}\, e^{-0.136h} \tag{5.39}$$

with h, the height above sea level, in km. This type of artificial atmosphere, although equivalent over the first km to the "standard" one, offers a reasonable approximation to the real atmosphere over a more extensive region, where propagation of electromagnetic waves occurs.

Neither of the artificial atmospheres represents true atmospheric conditions, especially below about 500 m, where convection currents due to humidity and temperature variations on the ground cause exchanges between air and ground and thus determine the air temperature and create turbulence and hence vertical air movement accompanied by compression and expansion without the exchange of heat. Additionally, the temperature and humidity distribution of the atmosphere is affected by the horizontal movement of air masses of different heat properties.

A ray traveling in the earth's atmosphere meets regions of various refractive indices and is forced to bend its path. The refractive index is slightly higher than unity at the earth's surface but approaches this value with increasing altitude. Vertical refraction, causing downward curvature of the ray path, is proportional to the negative vertical gradient of the refractivity.

The height dependency of the true refractive index, as mentioned earlier, is rather complicated so that we adopt, for theoretical considerations, those described earlier. The solution of propagation problems is made very much easier by considering straight-line instead of curved-ray trajectories. However, this requires a modification of the earth's radius to compensate for the change of the ray trajectory's curvature.

Accepting then the "standard" atmosphere with its linear height-dependent variation on the refractive index, we can determine a modifying factor for the true earth radius in the following manner. Denoting the angle of the ray's incidence into the lowest boundary of the refractive layer by $\varphi = \pi/2 - \epsilon$, expression (5.30) for the ray's radius of curvature becomes

$$\rho = \frac{n}{-\dfrac{dn}{dh}\sin\varphi} \tag{5.40}$$

whereas that of the earth's curvature is r, the true earth radius. The difference of the two curvatures

$$1/r - 1/\rho = 1/r + \frac{dn \sin\varphi}{n\, dh} \tag{5.41}$$

must be equal to the curvature of an imaginary earth having such a radius as to make the curvature of rays' trajectories zero or their radius of curvature infinite. Then,

$$1/r + \frac{dn \sin\varphi}{n\, dh} = 1/kr$$

and

$$k = \frac{1}{1 + \dfrac{r}{n}\dfrac{dn \sin\varphi}{dh}} \tag{5.42}$$

or, as the refractive index of air is approximately equal to unity, the preceding correction factor of the earth's radius becomes

$$k = \frac{1}{1 + r\dfrac{dn}{dh}\sin\varphi} \tag{5.43}$$

Further, because our interest is primarily in waves propagating almost horizontally that reach the refractive layers approximately at an angle of $\pi/2$ rad, this expression reduces to

$$k = \frac{1}{1 + r\dfrac{\mathrm{d}n}{\mathrm{d}h}} \tag{5.44}$$

This yields an earth-radius modifying factor permitting the consideration of straight propagation paths and hence simplifies our computations. Although permitting the straightening of the propagation path, the new geometry retains the relative position of the earth's surface and ray trajectory.

Because the gradient of the refractive index $\mathrm{d}n/\mathrm{d}h$ is not readily available, we usually resort to the expression [Bean and Dutton, 1966] for the evaluation of this factor

$$k = \frac{1}{1 - \dfrac{r}{n_s} c_e N_s 10^{-6}} \tag{5.45}$$

where n_s, given by (5.36) for N_s the surface refractivity, is the refractive index at the earth's surface where it is approximately equal to unity. Values of c_e are computable from

$$c_e = \ln \frac{N_s}{N_s + \Delta N}$$

with ΔN the difference between N_s and the value of N at a height 1 km above the earth.

With $c_e = 0.1357887/\text{km}$ and $N_s = 289.03627$, obtained by interpolation between values given by Bean and Dutton, the factor $k = 4/3$ for a "standard" atmosphere.

The introduction of a modified refractive index [Kerr, 1951]

$$n' = 1 + M \times 10^{-6}$$

where M, called the *refractive modulus,*

$$M = [(n - 1) + h/a] \cdot 10^{6}$$

with h, the height above a perfectly conducting ground, and $a = kr$, the effective earth radius, permits the consideration of propagation through a plane instead of a spherically stratified region. The height dependency of this modified refractive index is

$$\frac{dn'}{dh} = \frac{1}{a} - \kappa$$

where κ is the lapse rate of the refractive index with an approximate value of 3×10^{-5}/km under normal turbulent conditions.

Atmospheric influences on the ray's trajectory were described earlier. Another effect that, however, influences the energy content of electromagnetic waves in all weather conditions is attenuation due to atmospheric oxygen and water vapor. Absorption of electromagnetic waves by atmospheric gases was studied, among others, by Van Vleck [1947a and 1947b].

Energy loss from a wave traveling through the atmosphere may be ascribed to two phenomena. Absorption and the accompanying transformation into other forms of energy, mainly heat, accounts for one; whereas scattering by particles excited into oscillations is the other one.

Although oxygen molecules are electrically nonpolar, they possess a permanent magnetic moment of two Bohr magnetons [Joos, 1944]. Because Maxwell's equations apply symmetrically to electric and magnetic fields, absorption by oxygen is due to the interaction of the magnetically polarized molecules and the traveling electromagnetic wave. The oxygen molecule's magnetic dipole moment is caused by the unpaired spins of electrons. It interacts with the "end-over-end" rotation of the oxygen molecule and forms a "rho-type" triplet yielding microwave resonances. According to the selection rules of quantum mechanics [Joos, 1944], the rotational quantum number cannot change by more than one unit, so that transition between components removed by two units cannot occur and therefore they cannot be magnetically active. Otherwise, strong absorption could occur even at frequencies as high as 3 GHz.

The interaction of the molecules and the wave is very much weaker than it would be if oxygen molecules possessed electric instead of magnetic dipole moments because a Bohr magneton, given by $\beta = eh/4\pi mc$, with e and m the charge and mass of an electron, $h = 6.625 \times 10^{-27}$ ergsec Planck's constant, and c the velocity of electromagnetic wave propagation in free space is only about one-hundredth of a Debye unit of electric polarity.

Mutual combination of closely spaced energy levels by electric dipole radiation causes absorption by uncondensed water vapor. Transitions between nearby states obey the previously mentioned selection rules. The most important rules for dipole radiation are Laporte's parity rule and the restriction of the total rotational quantum number [Aller, 1953]. The rotational energy levels of the water molecule are of four different symmetry types and the rotational quantum number is permitted to change only by one unit with restrictions on the permitted combinations.

Only one pair of levels of the water molecule, excited at normal temperatures, may combine and due to the close spacing resonate at microwave frequencies. Such a resonance occurs at a frequency of 22.22 GHz, where absorption reaches a value of almost 0.2 dB/km.

Figure 5.10 shows the dependency of this specific attenuation, in dB/km, on frequency at an atmospheric pressure of 760 mm Hg, temperature of 20°C, and water vapor density of 7.5 g/m³. The curve for attenuation by molecular oxygen shows this to be lower than attenuation by water vapor up to a frequency of about 43 GHz. It exhibits attenuation of about 15 dB/km due to a series of very close spectral lines between about 50 and 70 GHz and an isolated absorption line at 118.74 GHz, where the attenuation reaches a value of about 1.5 dB/km. The absorption due to water vapor has absorption lines at 22.22, 183.3, and 323.8 GHz with attenuations of about 0.16, 40, and 40 dB/km, respectively. The calculation of absorption due to either of these atmospheric gases is rather complex. Approximate values may be obtained by the following empirical formulas (CCIR).

For oxygen attenuation at a temperature of 20°C the attenuation in dB/km is

$$\begin{aligned}
\gamma_o &= \left[\frac{6.6}{f^2 + 0.33} + \frac{9}{(f-57)^2 + 1.96}\right] f^2 10^{-3} \text{ dB/km} \quad \text{for } f < 57 \text{ GHz} \\
\gamma_o &= 14.9 \text{ dB/km} \quad \text{for } 57 < f < 63 \text{ GHz} \\
\gamma_o &= \left[\frac{4.13}{(f-63)^2 + 1.1} + \frac{0.19}{(f-118.7)^2 + 2}\right] f^2 10^{-3} \text{ dB/km} \\
&\quad \text{for } 63 < f < 350 \text{ GHz}
\end{aligned} \tag{5.46}$$

whereas that for water vapor is

$$\begin{aligned}
\gamma_W = &\left[0.067 + \frac{2.4}{(f-22.3)^2 + 6.6} + \frac{7.33}{(f-183.5)^2 + 5}\right. \\
&\left. + \frac{4.4}{(f-323.8)^2 + 10}\right] f^2 \rho 10^{-4} \text{ dB/km} \quad \text{for } f < 350 \text{ GHz}
\end{aligned} \tag{5.47}$$

The frequency is to be inserted in GHz whereas the water vapor density ρ is in g/m³.

The total attenuation by atmospheric gasses over a terrestrial or slightly inclined path d_{km} long is

$$A_a = \int_0^d [\gamma_o(d) + \gamma_W(d)]dd \quad \text{dB} \tag{5.48}$$

The formula for slant ranges inclined by angles $\epsilon < 10°$, the total atmospheric attenuation, according to CCIR, is given in dB by

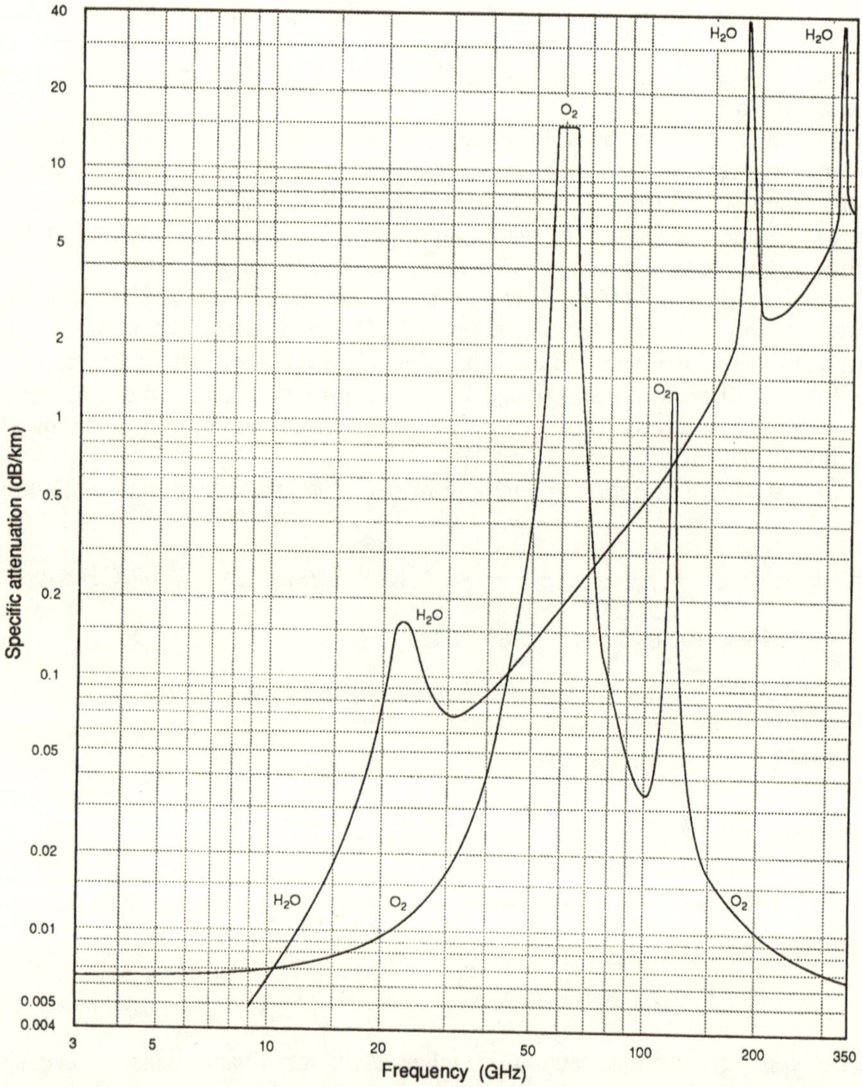

Figure 5.10 Specific attenuation by atmospheric gases: pressure 1 atmosphere; temperature 20°C, water vapor content 7.5 g/m^3 (after CCIR, 1978–1982).

$$A_a = \frac{16\gamma_o}{\sqrt{\sin^2\epsilon + 16/a} + \sin\epsilon} + \frac{4\gamma_W}{\sqrt{\sin^2\epsilon + 4/a} + \sin\epsilon} \tag{5.49}$$

where $a = kr$ is the effective earth radius.

For elevation angles $\epsilon > 10°$ this formula reduces to

$$A_a = \frac{8\gamma_o + 2\gamma_W}{\sin\epsilon} \text{ dB} \tag{5.50}$$

It is interesting to note that, although the troposphere is known to extend in height to about 9 to 15 km above the earth, the effective distance of absorption by oxygen, according to CCIR, is approximately the distance the electromagnetic wave travels in a constant density atmosphere extending in case of absorption by oxygen to a height of only 8 km, and to only 2 km for the absorption by water vapor, with a vacuum above.

Example 5.6

Evaluate the attenuation of a 50 GHz wave by oxygen and water vapor when the water vapor content is 7.5 g/m^3 and compare your results with values read from Figure 5.10. Plot the total attenuation for ray elevation angles from 0.1° to 30° in a "standard" atmosphere.

Solution: Because the given frequency is less than 57 GHz, the first expression of (5.46) for the attenuation by oxygen yields

$$\gamma_o = 0.45 \text{ dB/km}$$

and expression (5.47) shows that the attenuation by water vapor is

$$\gamma_W = 0.13 \text{ dB/km}$$

There is a good agreement between the computed values and those obtained from the graph.

Using the average earth radius of 6378 km and $k = 4/3$, makes a, the modified earth radius in a "standard" atmosphere $a = 8504$ km. Formulas (5.49) and (5.50) provide data for the construction of the graph of Figure 5.11.

Attenuation of electromagnetic energy propagating through precipitation is due partly to scattering by drops of water and partly to conversion of the absorbed energy to heat. Precipitation includes rain, hail, and snow.

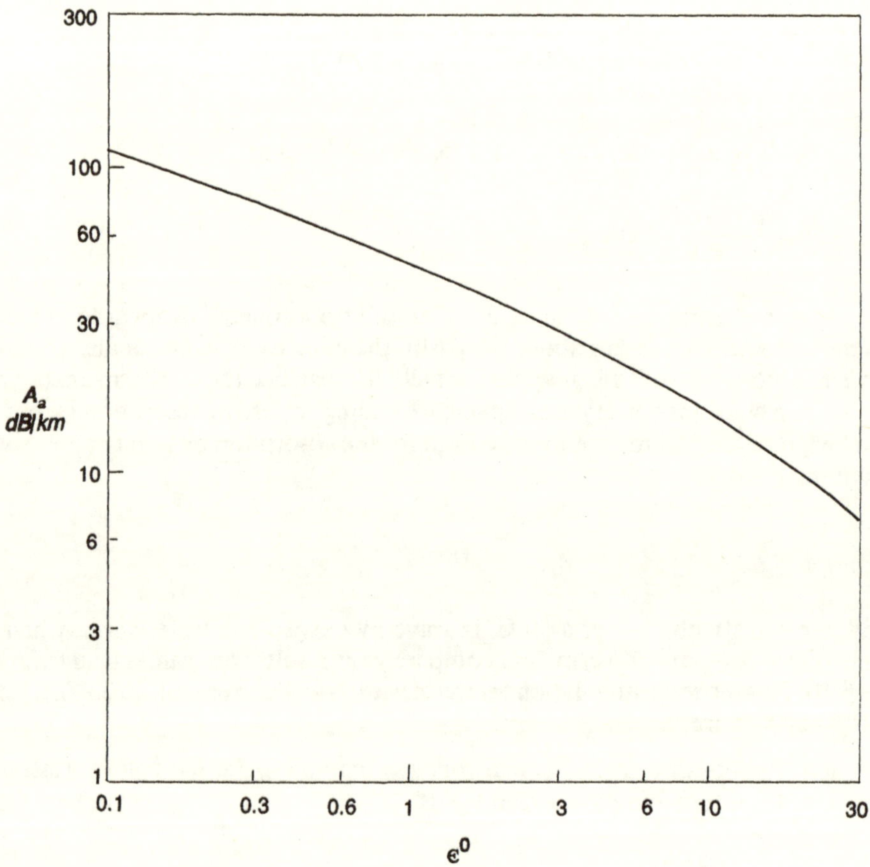

Figure 5.11 Graph for Example 5.6.

Although attenuation by precipitation is negligible at frequencies below about 3 GHz, at higher frequencies it may exceed the combined attenuation by oxygen and water vapor. Attenuation by rain depends on the rainfall rate and the pathlength of the electromagnetic waves in rain. Rainfall is not uniform: both the rate and the density of the rain along its path are functions of position and time.

The total energy removed from a propagating wave of length λ by a spherical particle of radius a, according to Goldstein [in Kerr, 1951], is

$$W = SQ(a, \lambda)$$

where S is the incident Poynting vector and $Q(a, \lambda)$, the total cross section of the raindrops (i.e., the sum of the absorption and scattering cross section), is given by

$$Q(a, \lambda) \approx \lambda^2(2\pi a/\lambda)^2[1 + (2\pi a/\lambda)^3] \tag{5.51}$$

For n drops of radius a in a range ds and layer thickness l, the attenuation, in dB/km, is

$$S = S_0 \exp\left[-434 \int_0^l \int_0^\infty n(a)Q(a, \lambda)\, da\, dl\right] \tag{5.52}$$

Precipitation rates vary to a great extent, even between places only a few hundred meters apart. Further, measurements of drop size are very inaccurate, and there is no conclusive evidence that rainfalls of a given rate have a unique drop size distribution, although a probable distribution usually is assigned to various rainfall rates. Additionally, larger raindrops may break up while falling into smaller ones due to the strength of the air currents. Most drops are in the range of 0.01 to 0.6 cm. Because temperature influences the dielectric properties of water, this parameter must be considered in the computation of attenuation by rain, fog, or clouds.

The attenuation of an electromagnetic wave traveling in a volume of rain of extent L [Ippolito, 1981] is

$$A = \int_0^L \alpha\, dx$$

with α the specific attenuation of the rain volume in dB/km. The attenuation constant may be expressed as

$$\kappa = \rho Q_a$$

with ρ denoting the drop density, or the number of raindrops per unit volume, and Q_a the attenuation cross section of a raindrop of radius a.

The attenuation by rain at a temperature of 18°C for frequencies between 3 and 30 GHz, according to Ryde [1946], is given by

$$\alpha_{\mathrm{dB/km}} = kp \tag{5.53}$$

where p is the precipitation rate in mm/h; and the coefficient k, expressed analytically by Rohan [1983], is

$$k = 0.002747\lambda^2 - 0.042408\lambda + 0.149679 \tag{5.54}$$

where λ is the wavelength in cm. This multiplicative coefficient is unity at 18°C but may double its value at 0°C.

The specific attenuation at the surface of the earth, as a function of the rain rate p in mm/h, is

$$\alpha_{\mathrm{dB/km}} = ap^b$$

when a and b are frequency- and temperature-dependent constants, given, at a temperature of 0°C, in Table 5.4.

Table 5.4
Rain Attenuation Coefficients at 0°C Rain Temperature (after Ippolito, 1981)

Frequency (GHz)	*Coefficient*	
	a	*b*
2	3.45×10^{-4}	0.891
4	1.47×10^{-3}	1.016
6	3.71×10^{-3}	1.124
12	0.0215	1.136
15	0.0368	1.118
20	0.0719	1.097
30	0.186	1.043
40	0.362	0.972
94	1.402	0.744

The total attenuation over a slant path at an elevation angle θ and specific attenuation $\alpha_{\mathrm{dB/km}}$ is

$$A_\theta = L\alpha/\sin\theta$$

Attenuation by rain of various rates is shown by Figure 5.12. Attenuation by hail generally is lower than that due to rain. For attenuation by hail, the parameter p in expression (5.53) indicates the rate of precipitation, in mm/h, of liquid water produced after melting. Although p is less than 500 even in the severest thunderstorms, a value of 100 may cause an attenuation of several dB/km. The attenuation diminishes with a falling temperature but rises inversely with the stone diameter. At a wavelength of 1 cm attenuation by hail is only about one-tenth of that produced by rain; at λ = 3 cm it is only one-hundredth; and only about eight-hundredths at λ = 10 cm [Ryde, 1946].

According to Ryde, clouds of ice crystals and snow do not cause appreciable attenuation, even if the rate of fall exceeds 125 mm/h. Attenuation by concentric

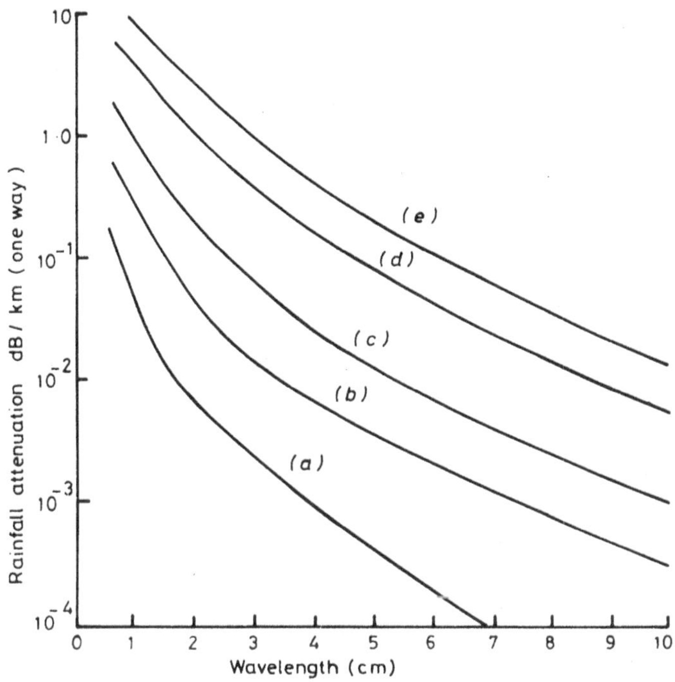

Figure 5.12 One-way attenuation (dB/km) in rain at a temperature of 18°C (after Skolnik, 1974, © McGraw-Hill).

spheres of different dielectric constant, for example, melting ice spheres, may approach that of rain.

Fog and clouds consist of such fine liquid water droplets that no perceptible precipitation takes place. Droplet diameters are assumed to be less than 0.01 cm. The density of the fog, which is judged by the distance in meters over which a prominent dark object may be identified in daylight by the naked eye against the sky at the horizon or over which a moderately unfocused light source may just be seen by night [Bean, Dutton, and Warner, 1970] determines electromagnetic wave attenuation.

Eldridge, as quoted by Altshuler [1984], derived the empirical formula

$$V = 0.024\rho^{-0.65} \text{ km}$$

for the visibility V in km. The symbol ρ denotes density in g/m^3. The formula is claimed to be valid for drop diameters between 0.3 and 10 μ. In an abundance of very small drops, as in a dense haze, the coefficient of 0.024 changes to 0.017.

The liquid water content of clouds varies from about 0.05 to 0.25 g/m^3 for stratiform to 0.3 to 1.3 g/m^3 for stratocumulus clouds [Ippolito, 1987]. Large cumulus clouds associated with thunderstorms may exhibit a peak water content of 5 g/m^3; however, fair weather cumulus clouds average less than 1 g/m^3, and cumulonimbus clouds about 2.5 g/m^3. Typical cumulus clouds extend to 6 to 8 km in range both horizontally and vertically.

Attenuation by clouds depends on their liquid water content

$$M = \frac{4\pi\rho}{3} \sum_{i=1}^{N} r_i^3 \text{ g/m}^3$$

with ρ the density of water in g/m^3, N the number of drops of radii r_i per unit volume.

The attenuation in dB/km, according to Gunn and East, as quoted by Bean *et al.* [1970], is

$$K = 0.4343 \frac{6\pi}{\lambda} \operatorname{Im}\left(-\frac{m^2 - 1}{m^2 + 2}\right) \tag{5.55}$$

where m is the water's complex index of refraction and Im indicates the imaginary part of the ratio. The attenuation is inversely proportional to the wavelength and falls exponentially with visibility.

Fog, consisting of suspended water droplets from the condensation of atmospheric water vapor, may reach a water content density from 0.4 to 1 g/m^3.

Altshuler [1984] computes attenuation by fog, in dB/km/g/m^3, from the formula

$$A = -1.347 + 0.0372\lambda + \frac{18.0}{\lambda} - 0.022T$$

which is claimed to be valid for 3 mm $< \lambda <$ 3 cm and -8°C $< T <$ 25°C. The wavelength is to be in mm and the temperature T in C. The standard error of the computed attenuation is allegedly 0.146 dB. Figure 5.13 shows the attenuation by fog at a temperature of 18°C.

5.4.3 Other Meteorological Influences

5.4.3.1 Indirect Influences of Precipitation

We showed earlier that the field at a remote location is due partly to the field component propagating directly between the transmitting and receiving antenna and

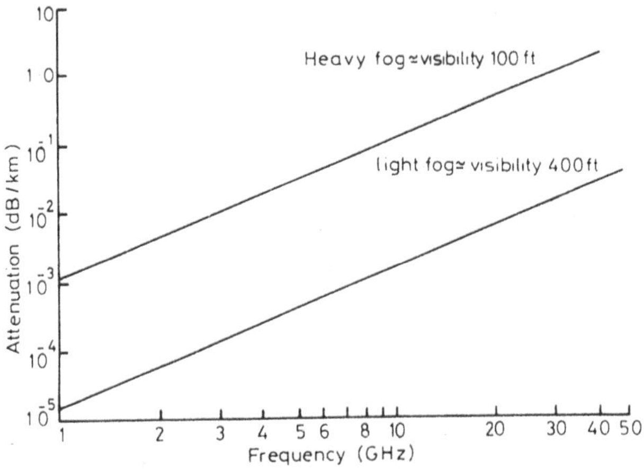

Figure 5.13 One-way attenuation (dB/km) of electromagnetic waves in fog at a temperature of 18°C (after Skolnik, 1974, © McGraw-Hill).

partly to the one reflected from the ground and other objects. Because this latter component depends on the reflection coefficient, which is a function of the complex dielectric constant of the reflecting medium determined by its permittivity and conductivity, both of which change with humidity, weather conditions will influence the magnitude and phase of the reflected field component.

The ground's permittivity and conductivity exhibit seasonal variations. Josephson and Blomquist [1958] recorded seasonal field-strength variations of up to 14 dB with vertically polarized very high-frequency transmissions over the same path. This change, related to the dielectric constant of the reflecting ground, corresponds to a variation of 3 to 30. They found that the effective dielectric constant depends mainly on the water content of the ground and is relatively independent of the type of the ground.

An empirical formula

$$\epsilon = 0.78W + 2.5 \tag{5.56}$$

for the dielectric constant was obtained by them from very high-frequency field-strength measurements.

A more general formula

$$\epsilon = kW + 2.0 \tag{5.57}$$

where

$$k = \begin{cases} 0.5 \text{ for sandlike soil} \\ 1.0 \text{ for average soil} \\ 1.5 \text{ for clay} \end{cases}$$

and W is the percentage of weight of water in a given volume to the weight of water that could occupy the same volume, has been proposed by Biggs [1970].

The dependency of ground conductivity on moisture and temperature t°C, as given by Albrecht [1962], is

$$\sigma = 7.7 \cdot 10^{-5}(0.73W^2 + 1)(1 + 0.03t) \text{ S/m} \tag{5.58}$$

Although the last expressions show the dependency of the complex dielectric constant and hence the reflection coefficient on the moisture content of the ground, in practice the ground's roughness was found to have much more influence on the reflected field than its electrical constants.

Various authors have measured values of ground constants. However, comparison is difficult as no strict measure of the soil's moisture content or degree of dryness is given and their frequencies of measurement often differ as well. With this in mind, Table 5.5 presents ground constants based on those given by (1) Bashkuyev and Tsydypov [1976], (2) Doluchanow [1956], and (3) Kerr [1951]. The values given by (1) are average effective values within the frequency range of 0.044 to 0.088 GHz, except for undergrowth where it is in the range of 0.066 to 0.088 GHz. Doluchanow does not reveal the frequencies of measurement; and those given by Kerr are shown in the table.

The constants shown in Table 5.5 are to be taken as approximate because they depend on factors that may differ from those prevailing at the time of measurement.

Seawater constants at frequencies below about 1.5 GHz are close to $\epsilon' = 80$ and $\sigma = 4.3$ S/m. The approximate values at frequencies above these may be computed from the following expressions [Rohan, 1983]

$$\epsilon' = 80 - 0.00733(f_{\text{MHz}} - 1500)$$

$$\sigma = 4.3 + 0.001467(f_{\text{MHz}} - 1500)$$

where 1500 MHz $< f <$ 3000 MHz, and

$$\epsilon' = 69 - 0.000628(f_{\text{MHz}} - 3000)$$

$$\sigma = 6.5 + 0.001492(f_{\text{MHz}} - 3000)$$

Table 5.5
Ground Constants of Different Terrains

Source	*Type of Ground*	*Frequency (GHz)*	*Relative Permittivity*	*Conductivity (mho/m)*
3	Sea water:	0.1–1.5	80	4.3
	20–25°C	3	69	6.5
	28°C	9.375	65	16
2			80	1–4.3
1	Fresh water in lakes and rivers	0.044–0.088	69	0.07
2			80	$10^{-3} - 2.4 \times 10^{-2}$
3		0.3	80	$10^{-3} - 10^{-2}$
1	Alluvial-boggy light loam of average moisture content	0.044–0.088	11	0.017
2	Moist ground		10–30	$3 \times 10^{-3} - 3 \times 10^{-2}$
3		0.3	30	10^{-2}
1	Alluvial-boggy soil on gravel	0.044–0.088	13	0.039
1	Alkaline black soil on gravel	0.044–0.088	10	0.066
1	Alluvial-boggy light loam on gravel	0.044–0.088	10	0.028
1	Alluvial-boggy light loam	0.044–0.088	14	0.012
1	Sandy loam, chestnut brown, dry soil	0.044–0.088	8	0.009
3	Very wet sandy loam	3.33	24	0.6
2	Dry soil		3–4	$1.1 \times 10^{-5} - 2 \times 10^{-3}$
3	Very dry sandy loam	3.33	2	0.03
3	Very dry ground	0.3	4	10^{-4}
1	Pine forest, average tree height 10 m	0.044–0.088	38	0.045
1	Gray sandy forest soil	0.044–0.088	14	0.123
2	Forest			10^{-3}
1	Undergrowth 4–5 m average height	0.066–0.088	18	0.022
1	Wind-borne sand	0.044–0.088	4	0.007
1	Corn field, 0.6 m high	0.044–0.088	33	0.027
2	Mountainous country			7.5×10^{-4}
2	Large city			7.5×10^{-4}

Note: (1) Bashkuyev and Tsydypov [1976]; (2) Doluchanow [1956]; and (3) Kerr [1951]. See references at the end of Chapter 5.

where $f > 3000$ MHz. These expressions were obtained by linear approximation of Kerr's values.

Expressions

$$\epsilon' = 50 + 31\cos(\pi f/30) \quad 0 < f_{GHz} < 15$$

$$\epsilon' = 50 + 22.54\cos(\pi f/30) \quad 15 < f_{GHz} < 30$$

fitting the curve $\epsilon' = f(f_{GHz})$, given by Al'pert [1973] as based on the work of Ford and Oliver, yield approximate values of seawater permittivity at frequencies up to 30 GHz. The inaccuracy of published seawater permittivity figures is shown by the difference between the values obtained from these two sets of formulas.

5.4.3.2 The Effect of Thunder- and Dust Storms

Sengupta and Das Gupta [1984] observed the influence of thunderstorms on a 6.7 GHz microwave signal propagating over a 35-km long line-of-sight path. Although thunderstorms occur occasionally in moderate climates, they are very frequent in tropical ones. They develop within one air mass as wind perturbations when low-level inflow of warm humid air triggers convection storms. Thunderstorms are characterized by extensive, black rolling clouds associated with very severe up- and downdrafts, very severe turbulence, lightning, and heavy rain. The downdraft air spreads over the ground in the lowest levels of the atmosphere and modifies the pressure, temperature, humidity, and wind.

Air pressure increases sharply followed by a dip then a further increase followed by another dip and yet another increase to high wind velocities and a 15 to 20 dB increase in signal level; this heralds the onset of the storm. The observed signal level exhibits a negative correlation with wind velocity and a positive one with pressure variations. It fluctuates greatly during the storm.

In a sandstorm, the diameter of sand grains exceeds 0.2 mm. The grains are heavy and, driven by wind, they produce low-flying clouds below about 2 m above the ground. Dust particles may produce dense clouds rising to a height of 1 km or more. Sand clouds, being below the average height of microwave antennas, are not likely to affect propagation. However, dust clouds, being in the propagation path, will attenuate microwaves [Chu, 1979].

The uncertainty about the value of the permittivity and size distribution of dust particles makes the analytical determination of attenuation difficult. Chu uses data on optical visibility to compute the attenuation coefficient and phase shift at centimeter waves as

$$\alpha_{\text{dB/km}} = \frac{3k}{13}\, 8.68\alpha_0 a \left[\text{Im}\left(\frac{\epsilon - 1}{\epsilon + 1}\right)\right]$$

$$\varphi^\circ = \frac{3k}{13}\, \alpha_0 a \left[\text{Re}\left(\frac{\epsilon - 1}{\epsilon + 1}\right)\right] \frac{180}{\pi}$$

with α_0 the optical attenuation coefficient; a is the radius of the sand or dust particles; ϵ is their permittivity; and k is the free-space phase constant.

The equations show a linear dependence of the sought quantities on particle size. Because particle size and density in sand or dust clouds is larger at heights closer to the ground, antennas of microwave links should be elevated. Satellite links are not likely to suffer greatly from this cause of attenuation.

The values of attenuation and phase shift computed from the preceding formulas agree well with measured ones. At a relatively poor visibility of 0.1 km in a uniform duststorm of particle sizes 0.01 and 0.1 mm, the calculated attenuation at a frequency of 11 GHz is less than 0.03 dB/km. The phase shift ranges from 1.5 to 30°/km. Beam displacement and broadening may occur in the presence of strong density gradients in the dust cloud.

5.4.4 Depolarization by Precipitation

The great demand for frequencies, even in the microwave range, causes overcrowding of frequency bands allocated to the various services. Frequency sharing therefore is inevitable. Mutual interference between stations operating on the same or adjacent frequency is reduced by polarization discrimination.

Absorption and scattering caused by precipitation reduces the signal amplitude. Depolarization changes the plane of polarization and hence causes interference with stations designed to operate at different polarizations.

Theoretical investigations of electromagnetic wave attenuation by rain initially assume that all raindrops are spherical and the same size. The results do not depend on polarization because the scattering cross section of spheres is independent of polarization.

Surface tension tends to cause raindrops to be spherical; however, upwardly acting aerodynamic forces flatten falling raindrops and cause them to acquire the shape of oblate ellipsoids [Cox, 1981]. Ananasso [1980] has observed that wind contributes to this process. Photographs of falling raindrops show that drops exceeding about 1 mm in diameter become asymmetrical oblate spheroids in shape.

Microwaves propagating through rain were found to be depolarized by the scattering of the incident waves by the oblate spheroidal drops. Part of the scattered wave is polarized in a plane orthogonal to the polarization plane of the incident wave. The amount of orthogonal polarization is determined by the rain rate, the shape, size, and inclination of the raindrops, and the size of the rain cell. Circular polarization was found to become elliptical.

Oguchi [1960, 1983] determined the difference in the attenuation of waves polarized along the two main axes of assumed ellipsoidal raindrops. According to Chu [1974], the attenuation of vertically polarized waves is 3 to 5 dB lower than that of horizontally polarized ones. These differential changes in attenuation and phase shift cause depolarization [Crane, 1984].

Snowflakes cause little attenuation. Their depolarization effect is similar to that caused by smaller particles. Battesti [1983] found that dry snowflakes may produce low attenuation of long duration; moist snow attenuation may be rather high. The ratio of snow and hail attenuation of vertically and horizontally polarized waves always exceeds 0.9 even when the attenuation is larger than 30 dB.

Ice crystal depolarization in regions above the ice's melting layer, originally called *anomalous* due to its unknown cause, was observed on earth-space paths. Ice crystals produce nearly pure differential phase shifts accompanied by very low copolarized attenuation. The ice-crystal depolarization frequency of occurrence and severity depends on the geographical location and season of the year.

Depolarization reduces the efficiency of antennas because they exhibit maximum gain at the design polarization. Their reduced response to depolarized waves adds to the attenuation suffered in propagation.

5.4.5 Electromagnetic Wave Propagation Beyond the Horizon

Electromagnetic waves may produce fields beyond the radio horizon either due to diffraction or to scattering.

5.4.5.1 Diffraction

Electromagnetic waves generally travel away from their source along radial lines perpendicular to the surface of uniform phase, the phasefront, and deviate from this direction when the phasefront becomes irregular either in amplitude or in phase. Changes in the direction of travel may be caused by reflection and refraction that, as shown in the preceding sections, may cause sudden or gradual changes in the direction of travel.

Another phenomenon, diffraction, occurs whenever an electromagnetic wave encounters an obstructing object around the edge of which energy bends, thus reducing the shadowing effect of the obstruction. The extent of the obstruction depends on the size of the obstacle in terms of the used wavelength and the distance from the source and hence the angle subtended by the obstruction at the source. If the distance to a small obstacle is large, so that the angle subtended at the source is small, the radiated energy may penetrate the region further behind the obstacle where the shadowing effect is negligible. Shadowing is observed, however, close behind the obstruction. The obstructions most often encountered are the hump of the earth at the horizon, hills, mountains, buildings, trees, and so on.

At very low frequencies where the wavelength is comparable to the earth's radius and much longer than the extent of possible obstructions, the shadowing by the earth's hump is negligible. At high frequencies, where the opposite applies and the earth's surface cannot be considered smooth as even small height variations of the earth's surface (e.g., waves at sea) may be much larger than the wavelength, however, they may exert considerable effect on electromagnetic wave propagation.

The diffraction phenomena of interest in electromagnetic wave propagation occur when electromagnetic energy propagates between antennas at locations below the radio horizon. Shadowing then is caused by the earth's surface at the horizon or obstacles of the kind mentioned. Because the antennas are beyond the range of mutual visibility, the electromagnetic waves propagating between them will be attenuated so that, in addition to the attenuations mentioned earlier, attenuation due to diffraction has to be considered in field-strength computations. This loss increases with distance and frequency and depends to some extent on the antenna heights. Formulas for its computation are rather involved and not particularly well suited to practical application; therefore, at frequencies in excess of 30 MHz, we resort to Bullington's nomograms reproduced in Figures 5.14 and 5.15. At these frequencies both antennas may be considered elevated above the earth and the point-to-point variation of the dielectric constant has little influence on the propagation's characteristics.

The initial quantities required for entering the nomograms are the distances to the radio horizon given by

$$d = \sqrt{2kr}\,(\sqrt{h} + \sqrt{H}) \tag{5.59}$$

where h and H are the transmitting and receiving antenna heights, kr is the effective earth radius, and the limiting antenna height, according to Burrows, is

$$h_0 = 30\lambda^{2/3} \tag{5.60}$$

Two distinct cases may arise. Either both antenna heights are below h_0, or at least one of them is lower than the limiting antenna height h_0. In the first case we use the nomogram of Figure 5.14 to determine the combined effect of refraction, accounted for by the factor k (= 4/3 for a "standard atmosphere") and diffraction around the curvature of the earth.

When either antenna is as much as twice the limiting antenna height, figures obtained from Figure 5.14 for loss were found to be about 2 dB in error, which increases with increasing antenna height. In these cases the nomogram of Figure 5.15 is applied.

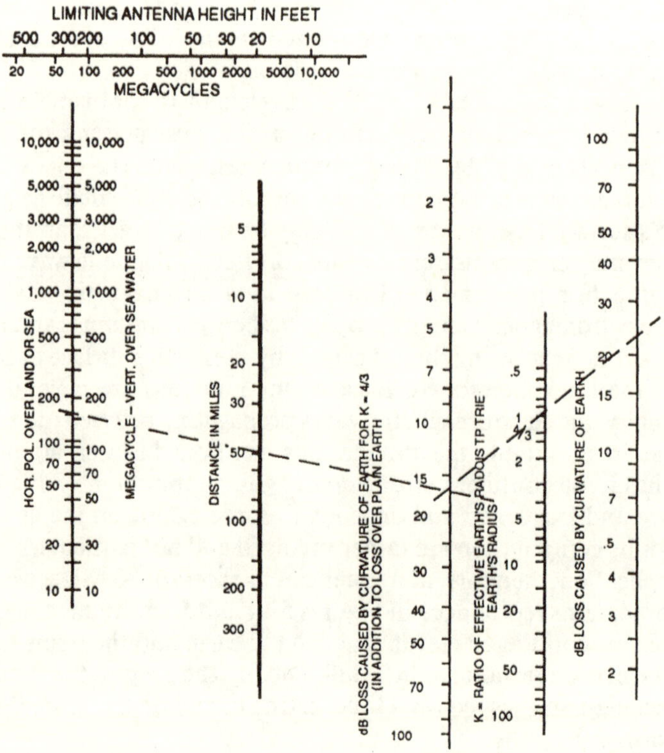

Figure 5.14 Diffraction loss due to the earth's hump when both antennae are shorter than h_0 (from Bullington, 1947, © 1947 IRE).

The loss shown here consists of losses over three distances; namely,

$d_1 = \sqrt{2krh}$, the distance of the first antenna's radio horizon;

$d_2 = \sqrt{2krH}$, the similar distance of the second antenna; and

$d_3 =$ the distance between the lines of sight of the two antennas (see Figure 5.15), or

$d_3 = D - d$, where D is the separation of the antennas and d is given by (5.59)

For more accurate results of the field strength or received power at, or beyond, the horizon, we should increase the figures obtained from Figure 5.15 by

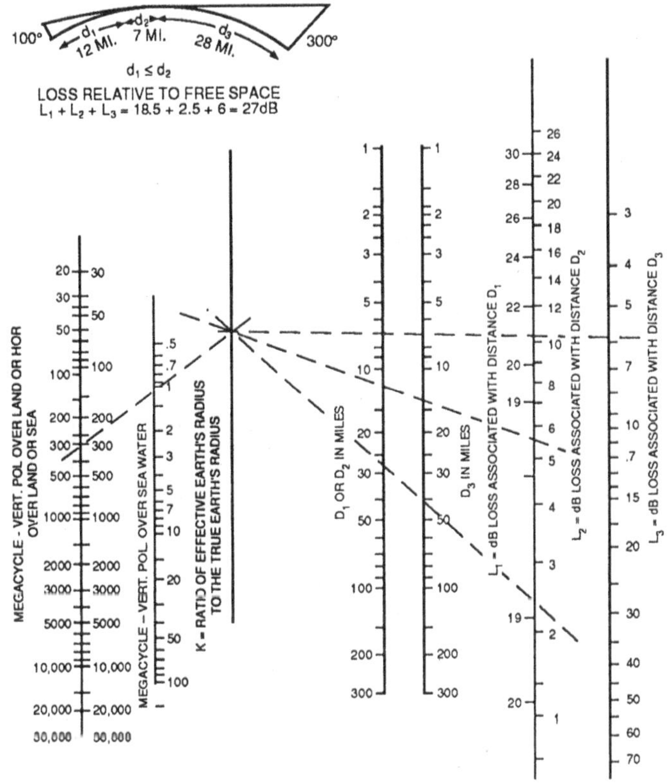

Figure 5.15 Propagation loss relative to free-space transmission over smooth earth paths exceeding the total line of sight of both antennas (from Bullington, 1947, © 1947 IRE).

$$10 \log \left[\frac{1}{\sqrt{2}} \left(1 + \frac{d_1}{d_2} + \frac{d_3}{d_2} \right) \right] \tag{5.61}$$

as long as one antenna is higher than h_0.

Example 5.7

Determine the diffraction loss of a 60 MHz vertically polarized wave traveling in a "standard" atmosphere over land, when the separation of a 15-m high transmitting and an 8-m high receiving antenna is 400 km.

Solution: The limiting height, as determined from expression (5.60), is $h_0 = 87.72$ m. Because both antennas are lower than h_0, we read the sought value of the diffraction loss from Figure 5.14.

By drawing an extended straight line from the 60 MHz point on the left-most scale to that designating the distance of 400 km, or 249.22 miles on the distance scale, yields the loss due to the earth's curvature of about 100 dB in addition to the loss due to propagation over a plane earth. The intersection of the extended line from this point to the $4/3 = 1.33$ mark appropriate to a "standard" atmosphere, on the fifth scale from the left, leads to a loss of 118.7 dB, read from the extreme right scale, for the loss caused by the curvature of the earth.

The influence of hills, buildings, trees, *et cetera* on propagation is difficult to compute. For theoretical considerations aimed at evaluating losses due to their presence within or near the propagation path, they are replaced by the extreme case of a perfectly absorbing knife edge of height H above the line joining the two antennas, as shown by the sketch in Figure 5.16. The investigations are based on the application of the Kirchhoff-Huyghens principle, as shown by Stratton [1941].

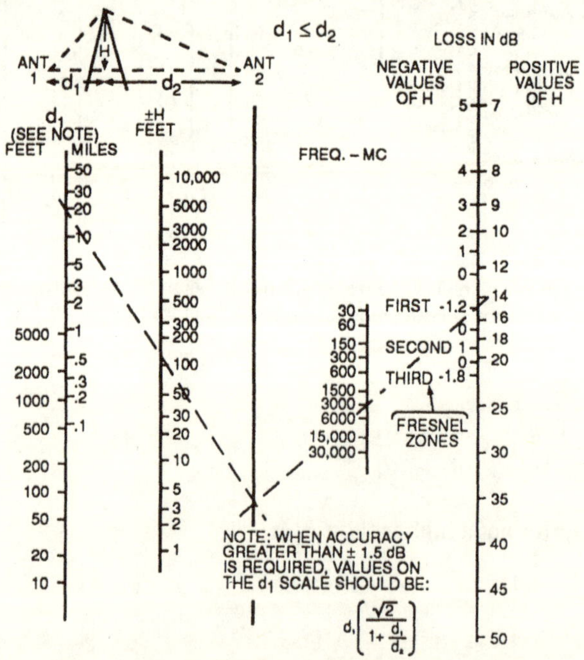

Figure 5.16 Shadow loss relative to free space for diffraction over a ridge (from Bullington, 1947, © 1947 IRE).

Bullington's nomogram in Figure 5.16 presents the magnitude of the shadow loss as a function of the ridge's height and the frequency used. At grazing incidence, when the height H of the ridge approaches zero, the shadow loss approaches 6 dB and increases with rising values of H. When the propagation path clears the ridge, the shadow loss oscillates around the 0 dB value and tends to remain so when the clearance increases. Free-space transmission occurs when the clearance is large.

Optimum clearance occurs when the path from the first antenna to the top of the ridge and from there to the second antenna is longer by $\lambda/2$ than the direct path between the antennas, or

$$\delta = R_1 + R_2 - (d_1 + d_2) = \lambda/2 \tag{5.62}$$

that is, the phase difference between the waves propagating over these two paths is π rad. A surface satisfying these conditions is an ellipsoid of revolution having the two antennas at its foci, or as encountered earlier, the first Fresnel zone.

The pathlength difference indicates which Fresnel zones are cleared by the propagation path. The location of the first three zones are shown in the nomogram. A phase difference of 2π rad indicates that the direct path between the antennas clears the first two Fresnel zones; whereas it clears the first three zones when the phase difference is 3π rad.

The virtual height and location of an imaginary single ridge, replacing, for purposes of loss computation, two successive ridges, may be obtained by the intersection of the extended lines from the two antennas to the top of their adjacent ridges [Starr, 1953].

The optimum clearance height, as obtained from (5.62) when writing

$$R_i = \sqrt{d_i^2 + h^2} \approx d_i\left(1 + \frac{h^2}{2d_i^2}\right)$$

is

$$h \approx \sqrt{\lambda \frac{d_1 d_2}{d_1 + d_2}} \tag{5.63}$$

Figure 5.17 yields the shadow loss for propagation over a terrain similar to that shown in the sketch. The obtained shadow loss is to be added to the normal loss over a smooth earth for the antenna heights and distances involved. Measured losses in the frequency range of 30 to 150 MHz were found to be within 5 to 6 dB of those given by the nomogram.

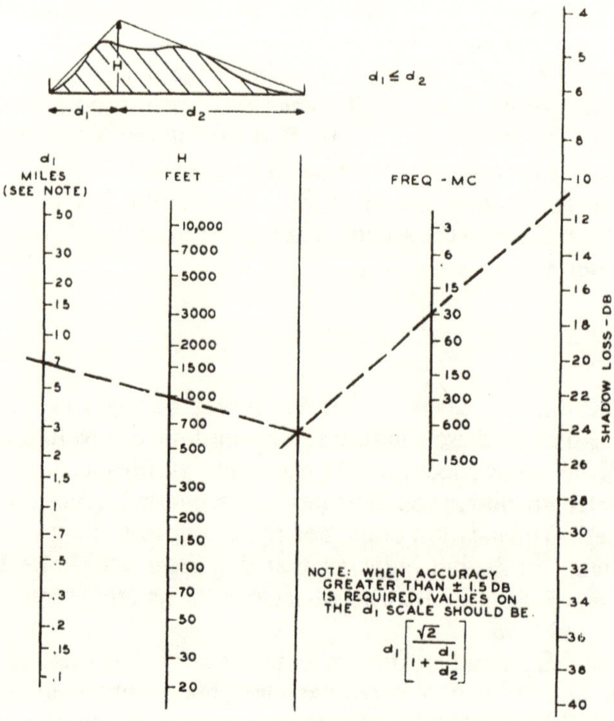

Figure 5.17 Shadow loss relative to smooth earth (from Bullington, 1947, © 1947 IRE).

Remember that all the latter losses (i.e., losses due to the earth's hump, diffraction, and shadowing) where applicable, must be added to the plane earth propagation loss, and the total loss must be used for field-strength computations.

5.4.5.2 Scattering

The field of transmitters operating at very and ultra high frequencies at ranges beyond the radio horizon, if due only to diffraction over the hump of the earth, diminish very fast with distance. Experiments have shown that the field at these ranges, although subjected to deep fading, often is higher than that attributable to diffraction. It was suggested, therefore, that the presence of a field at the experienced levels must be due to some phenomenon other than diffraction. Some incidental meteorological or special ionospheric conditions were suspected of causing the increase in signal strength. However, because it occurred more frequently and for longer periods than could be ascribed to the previous special conditions, it was concluded that some scattering of electromagnetic energy into all possible directions must occur in the lower troposphere.

The phenomenon received a lot of attention since about 1950 and several theoretical proposals attempted its explanation. The most plausible proposal assumes the scattering of electromagnetic energy to be caused by inhomogeneities in the atmosphere. These are due to regions of fluctuation, turbulence occurring when disturbances to the smooth flow (i.e., eddies) develop in the presence of large-scale motions accompanied by a continuous spectrum of fine-scale irregularities.

Turbulence then is irregular motion causing the instantaneous speed and direction of the air at particular regions in the troposphere to differ from the mean values of these quantities. Although little is known of the causes of turbulence and the exact nature of the eddies, some experimental data indicate that atmospheric turbulent velocities vary from a few times ten meters per second to a few centimeters per second. The scale of turbulence (i.e., the dimension of eddies) varies from about 10 m under stable conditions to 10 cm under unstable ones. The time constants of the turbulent fluctuations were found to vary from a fraction of a second to several minutes, and the intensity was found to be normally distributed. The described irregularities represent fluctuations of the dielectric constant ϵ' or, from expressions (4.27) and (4.25), fluctuations of the refractive index. They vary continuously with time.

An estimate of the fluctuations of the index of refraction may be obtained from the measured values of the standard deviation of the temperature from the mean of its instantaneous value. The expression

$$\Delta n \approx 2 \times 10^{-6} \Delta T \tag{5.64}$$

is based on the adiabatic law, which offers a good approximation for eddy motion in the atmosphere. It shows that a 1°C temperature variation causes a 2 M-unit deviation from the mean. The definition of the M-unit was given earlier [Booker and Gordon, 1950]. These experimentally determined figures apply to the atmosphere near ground level. At higher levels, the magnitude and linear extent of the fluctuations probably decreases.

The physical basis of dielectric constant variations assumes ϵ' to be dependent on the position P of the investigated point in the troposphere and on the time t, or

$$\epsilon_p' = \epsilon'(P, t) \tag{5.65}$$

The time variation of the permittivity affects only slow signal variations so that we can write

$$\epsilon_p' = \epsilon'(P) = \overline{\epsilon'} + \Delta\epsilon'(P) \tag{5.66}$$

where $\overline{\epsilon'}$ is the mean value of the permittivity at point P.

The departure of the permittivity from its average value at a point P', $\Delta\epsilon'(P')$, differs from that at point P unless both points coincide. The variance of the permittivity in a volume V becomes then

$$\overline{(\Delta\epsilon')^2} = \frac{1}{V}\int_V (\Delta\epsilon')^2 \, dV \tag{5.67}$$

whereas the autocorrelation function of $\Delta\epsilon'$ is given by

$$\rho = \frac{1}{V\overline{(\Delta\epsilon')^2}}\int_V \Delta\epsilon'(P)\Delta\epsilon'^*(P') \, dV \tag{5.68}$$

where the asterisk indicates the complex conjugate value and assumes that the complex permittivity is used to account for losses in the medium, if significant. The magnitude of this expression depends on the relative position of the points P and P': it tends to unity when the two points coincide and to zero for their infinite separation. The autocorrelation function depends only on the separation $r = P' - P$ of the considered locations if the medium is isotropic. Then, assuming an isotropic radiation, we can write arbitrarily

$$\rho = e^{-r/l} \tag{5.69}$$

where l is the scale of the turbulence. Remember though that the statistical distribution of l affects the frequency of the scattered field.

Assuming that the conductivity of the considered medium is $\sigma = 0$ and inserting expression (5.66) into the Ampère-Maxwell equation (1.11) when the electric field is given by $\mathbf{E} = \mathbf{E}_0 + \mathbf{E}_1$, where $\mathbf{E}_0$ is the primary field and $\mathbf{E}_1$ the scattered field, yields [Picquenard, 1974]:

$$\begin{aligned} \nabla \times \mathbf{H} &= \epsilon_0 \frac{\partial}{\partial t}[(\overline{\epsilon'} + \Delta\epsilon')(\mathbf{E}_0 + \mathbf{E}_1)] \\ &= \epsilon_0(\overline{\epsilon'} + \Delta\epsilon')\left[\frac{\partial \mathbf{E}_0}{\partial t} + \frac{\partial \mathbf{E}_1}{\partial t}\right] + (\mathbf{E}_0 + \mathbf{E}_1)\epsilon_0 \frac{\partial \Delta\epsilon'}{\partial t} \end{aligned} \tag{5.70}$$

Neglecting the last term of this expression, due to the earlier mentioned slow variation of $\Delta\epsilon'$, and disregarding the field $\mathbf{E}_1$ due to multiple scattering, which is negligible when compared with the primary field $\mathbf{E}_0$, leads to the approximation:

$$\nabla \times \mathbf{H} = \epsilon_0\epsilon' \frac{\partial \mathbf{E}_0}{\partial t} + \epsilon_0\Delta\epsilon' \frac{\partial \mathbf{E}_0}{\partial t} \tag{5.71a}$$

or, for a periodically varying primary field, to

$$\nabla \times \mathbf{H} = j\omega\epsilon_0\epsilon'\mathbf{E} + j\omega\epsilon_0\Delta\epsilon'\mathbf{E} \tag{5.71b}$$

Comparing this expression with (1.11) shows that the effect of inhomogeneities of the permittivity may be compared with that of currents of density:

$$\mathbf{i} = \sigma\mathbf{E} = j\omega\epsilon_0\Delta\epsilon'\mathbf{E} \tag{5.72}$$

where σ stands for the conductivity of the medium.

The field from scatterers other than those on a straight line between the transmitter and the receiver is due to these currents; they represent randomly distributed elementary dipoles. Their fields, due to the various distances of the individual dipoles from the transmitter, are not in phase. Radiation is maximum in the direction and sense of the incident radiation from the transmitter. It decreases quickly at places removed from this direction.

Scattering may be assumed, for theoretical considerations, to occur from spherical bodies of radius a, tightly packed into a volume V. The permittivity of the spheres differs by $\Delta\epsilon'$ from ϵ', the permittivity of the surroundings.

At a distance of r_1 from the transmitter, the field is

$$E_1 \, e^{j\omega(t-r_1/c)} \tag{5.73}$$

so that the electric dipole moment due to the displacement current in the mentioned spheres becomes

$$\Delta\epsilon' E_1 \, e^{j\omega(t-r_1/c)} \, dv \tag{5.74}$$

where dv is the volume of one sphere.

The potential at the receiver at a distance r becomes

$$V = \frac{1}{4\pi\epsilon'} \int_V \frac{\Delta\epsilon' E_1 \, e^{j\omega(t-r_1/c)}}{r} \, dv \tag{5.75}$$

If α, the angle between the direction of the radiation from the transmitter to the scattering mass and the direction from here to the receiver, is about $\pi/2$ rad, an assumption valid for most cases encountered in practice, the field strength at the receiver at a distance $r \gg \lambda/2\pi$ may be expressed by [Doluchanow, 1956]:

$$E_2 = (2\pi/\lambda)^2 V \text{ V/m} \tag{5.76}$$

which, with expression (3.14), permits the determination of the rate of energy flow per unit area produced by the joint action of all m spheres in a unit volume of the scattering mass, as

$$S = \frac{1}{m}\frac{E_2^2}{120\pi} = \frac{1}{m}\frac{E_1^2}{120\pi}(2\pi/\lambda)^4\left(\frac{\frac{4}{3}\pi a^3\Delta\epsilon'}{4\pi\epsilon' r}\right)^2 \text{ W/m}^2 \tag{5.77}$$

This shows that the rate of energy flow depends on the ratio $\Delta\epsilon'/\epsilon'$ that, due to

$$\begin{aligned} \epsilon' &= n^2 \\ \Delta\epsilon' &\approx 2\Delta n \\ \Delta n &= 10^{-6}\Delta M \end{aligned} \tag{5.78}$$

for $n \approx 1$, makes that ratio

$$\left(\frac{\Delta\epsilon'}{\epsilon'}\right)^2 = 4 \times 10^{-12}(\Delta M)^2 \tag{5.79}$$

Denoting further the effective scattering surface of the spheres in a unit volume by σ reduces (5.77) at the receiver's location to

$$S = \frac{E_1^2}{120\pi}\frac{\sigma}{4\pi r^2} \tag{5.80}$$

that equated to (5.77) yields an expression

$$\begin{aligned} \sigma &= \frac{\pi}{m}\left(\frac{2a^3}{3}\right)^2\left(\frac{2\pi}{\lambda}\right)^4\left(\frac{\Delta\epsilon'}{\epsilon'}\right)^2 \text{ m}^2/\text{m}^3 \\ &\approx \frac{\pi}{m}\left(\frac{2a^3}{3}\right)^2\left(\frac{2\pi}{\lambda}\right)^4 (\Delta M)^2 4 \times 10^{-12} \text{ m}^2/\text{m}^3 \end{aligned} \tag{5.81}$$

for the scattering surface per unit volume.

The atmospheric turbulence, although proven not to be isotropic, often is assumed isotropic for theoretical considerations. Anisotropy complicates calculations greatly.

Expressions (5.81) show that the scattering surface depends on the ratio $\Delta\epsilon'/\epsilon'$ and the frequency of radiation used.

Measurements over distances up to 300 km show that power decreases with approximately the fifth power of the distance between the transmitter and the receiver if the axes of both antennas are horizontal. An additional attenuation, equal to the square of the mean altitude of the scattering volume, has to be considered if the antennas are above the horizon. Clear antenna horizons are recommended for the reduction of this latter attenuation. Further, because the expression for the scattering surface depends on the permittivity of the atmosphere and hence the atmospheric refraction, tropospheric scatter propagation depends on the climate and meteorological conditions.

Figure 5.18 depicts the geometry of an over-the-horizon tropospheric scatter communication link. The lines $\overline{AC}$ and $\overline{AD}$ and lines $\overline{BC}$ and $\overline{BD}$ represent the 3 dB beamwidths of the transmitting and receiving antennas. The scattering volume, the volume of space common to both beams, is represented in a single plane by the shaded area. Its dimensions may be determined from purely geometrical considerations.

If the power radiated by the transmitter equipped by an antenna having a directional gain G is P, the energy scattered by the scattering spheres in volume V is

$$W = \frac{PG}{4\pi r^2} V\sigma \tag{5.82}$$

and the energy flow per unit area at B, the location of the receiver, according to (5.80) becomes

$$S = \frac{PGV\sigma}{16\pi^2 r^2 r_1^2} \text{ W/m}^2 \tag{5.83}$$

The field strength at the receiver is then

$$E_2 = \sqrt{120\pi S} = 1.55 \frac{\sqrt{PGV\sigma}}{rr_1} \text{ V/m} \tag{5.84}$$

for P in W and all lengths in m. The directional gain of the receiving antenna is implicit in this expression because it determines the scattering volume V.

The total received energy consists of contributions traveling over various paths and therefore arriving at the receiver with various delays. If Δt is the maximum time difference between contributions traversing the longest and shortest paths, the available bandwidth (i.e., the bandwidth that the system can handle without distortion) becomes

$$B = 1/\Delta t$$

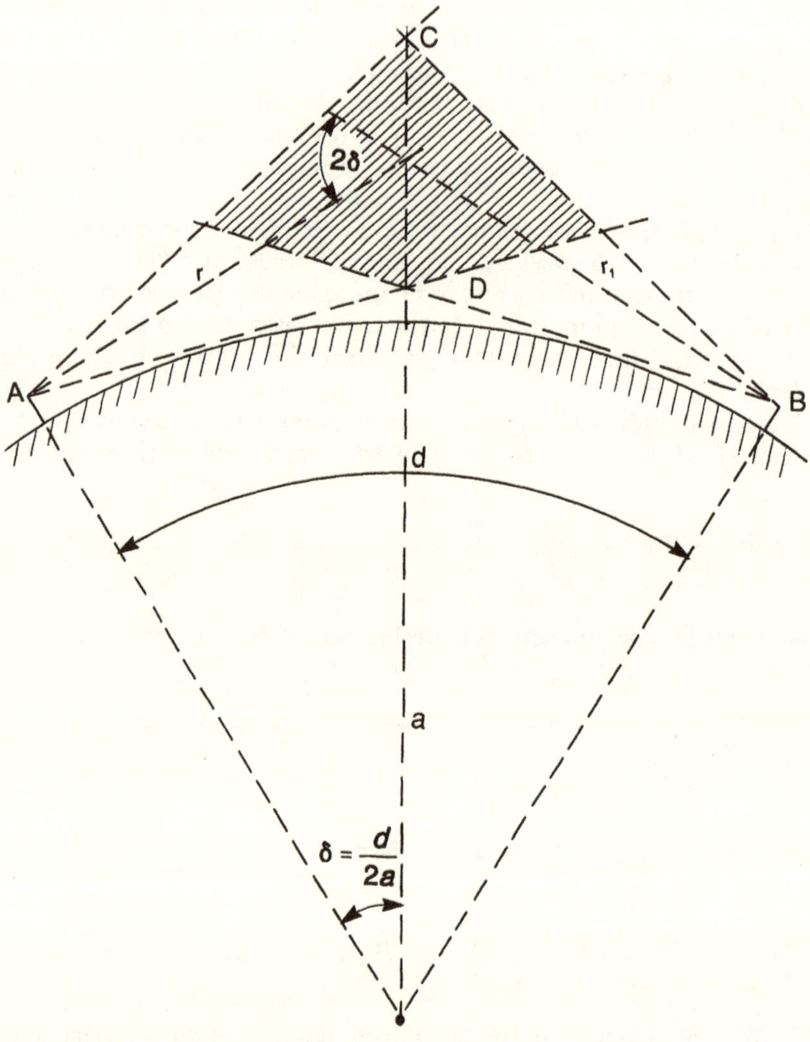

Figure 5.18 Over-the-horizon tropospheric scatter communication link geometry.

Various levels of distortion are permissible for various types of service. Narrowbeam antennas reduce the range of path delays and are suitable for communicating wide-band signals.

The generally used frequencies for tropospheric scatter communication links are between about 150 MHz and 4 GHz because high-gain antennas for lower frequencies would be physically rather large and attenuation of electromagnetic waves increases sharply at higher ones.

The strength of signals reflected by tropospheric discontinuities varies in time and space due to the random motion of the atmosphere. Because, according to the doppler effect, the frequency of a signal reflected by moving elements differs from that of the incident radiation, the beating between the frequencies of the received signal reflected by various scattering elements causes signal strength variations. The time variations of the signal strength manifest themselves as fading.

The doppler frequency, when v is the speed of the reflecting elements and c that of electromagnetic waves, is given by

$$f_D = 2fv/c \tag{5.85}$$

or by

$$f_D = 2f\frac{v}{c}\sin\frac{\theta}{2} \approx f\frac{v}{c}\theta$$

if scattering through a small angle θ is considered, so that the "period" of fading becomes

$$\lambda/v\theta \tag{5.86}$$

It is directly proportional to the wavelength of the incident wave and inversely proportional to the velocity of the reflecting elements and the scattering angle [Booker and Gordon, 1950].

The speed of fading, on the other hand, is directly proportional to the carrier frequency, and the frequency difference yielding correlated fields is inversely proportional to the scattering angle and hence the dimensions of the scattering volume.

The signal strengths over short-period (1–5 min) fading experienced in practice were found to have a Rayleigh probability distribution, whereas those of the randomly varying phase exhibit a constant probability density distribution.

The correlation of short-period fading signal strength extends over longer distances in the direction of propagation than over vertical distances. This intimates that vertically spaced antennas, in a plane perpendicular to the direction of propagation, are to be used for space diversity operation. Long-period fading field strength observes an approximately log-normal probability distribution with hourly mean values extending over large regions and a standard deviation depending on distance and climate.

Maxima of field strength in temperate climates occur in summer, and minima in winter. The mean value difference between these two seasons may reach 8 to 10 dB. Minima of signal strength, for propagation over paths of about 150 km, usually are observed during summer afternoons; at pathlengths of about 600 km, maxima are observed during daytime.

The interest in receiving broadcast transmissions by mobile receivers, for police, fire-fighting, taxicab, and other mobile telephone communication, led to the investigation of electromagnetic wave propagation in built-up areas. The proximity of structures, large in comparison with the wavelength causing scattering, and multiple reflections changing not only with the position of the mobile terminal, but also with that of other mobile objects in the vicinity makes it impossible to determine the field strength by a single simple mathematical expression. The field includes too many random contributions. Several investigators attacked the problem on a statistical basis; others sought expressions fitted to data obtained by measurement.

Suzuki [1977] assumed the urban radio channel to be multipath due to reflection, refraction, and scattering by buildings that cause energy to reach the receiver over various paths at various strengths and phases. The field strengths and phases of the constituting components of the field at a given point within the built-up area depend on the geometry of the transmitter-receiver system and on the frequency and length of delay. The interfering components produce a spatial standing-wave pattern, so that the signal received by a moving receiver exhibits fluctuations. Reflections often produce changes in the plane of polarization as well, so that waves originally radiated with one linear polarization during propagation acquire components of another one.

Ossanna's [1964] theoretical model considered waves reflected by the flat sides of buildings; and Clarke's [1968] model assumed scattered waves. Both accept a Rayleigh-distributed path strength. Young [1952] has shown that this probability distribution of path strength at any delay yields an excellent fit to the observed amplitude fluctuations in mobile radio reception at very and ultra high-frequencies in New York City, in a sample area of less than about 1 km^2, whereas in suburbs it fits measured data over distances of 5 to 9 km from the transmitter.

The probability distribution was found to change to a Ricean or even a Gaussian one when a strong direct, not reflected, component raises the average signal level; for example, at the intersection of the mobile receiver's path with a street pointing radially toward the transmitter.

Davis and Bogner [1985], although accepting a Rayleigh-distributed path strength for a large number of scatterers where the central limit theorem applies, assume fast fading, other than Rayleigh-distributed, multipath statistics in a small number of scatterers. Further, they state that the field-strength variation along a street running radially from the transmitter may differ very much from that along one oriented transversely. In the latter case, it may vary from one side of the street to the other because shielding by buildings will be greater on the side closer to the transmitter.

On the basis of measurements at frequencies of 150, 450, 900, and 3700 MHz, Young [1952] found frequencies around 500 MHz to be best suited for mobile radio communication. Transmissions at 3700 MHz were found to suffer from audible rate fluctuations in the received carrier level while the mobile unit moved at normal speed.

Allenbrook sums plane earth and building diffraction losses to obtain small mean prediction and low rms errors compared with measured values of losses in "flat" and "sloping" cities, both at very-high and ultra-high frequencies.

Causebrook investigated ground-wave propagation at frequencies of 908, 1214, and 1457 kHz along radial lines at azimuth angles of 169° and 181° from a transmitter at Brookman Park, 25 km north of central London, to locations about 50 km away. Both paths crossed the densely built-up areas of central London.

A loop antenna mounted on the roof of a van was manually steered to maximum signal reception. The recordings of signal strength were used for the determination of the magnetic field strength H.

An empirical formula

$$\eta = (1 - B)^{95/\lambda}\eta_1 + \left[\sqrt{\frac{\lambda B}{206}} - 1.23B + 0.35B^{1.5}\right] j\frac{2\pi h}{\lambda}$$

where B is the fraction of the considered area covered by buildings, η_1 the surface impedance with the buildings absent, and h the height of the buildings in meters, yields the surface impedance relative to $120\pi\Omega$ used to compute the ground wave attenuation factor

$$g(R) = 1 - \int_0^R (\eta + \psi)\exp(-j2\pi\xi/\lambda)g(r)\sqrt{\frac{jR}{\lambda r(R - r)}}\,dr$$

relative to a perfectly conducting plane. Here, R (in meters) is the distance for which the attenuation is required, ψ the angle (in radians) between the line from the receiver to the surface at range r and the tangent to the surface at this point, and ξ the distance between the terminals subtracted from the sum of distances from the terminals to the surface at r.

Figure 5.19 shows the comparison of the calculated and measured magnetic field strength dependence on distance. The curves exhibit a rapid decay with distance from the transmitter in densely built-up areas with an increase in less built-up, though more remote, ones. The steep decrease in field strength occurs in each case near the center of London even though this phenomenon depends on the used frequency. The second dip in figures 5.19 e and f is due to high buildings in the Croydon area.

The validity of Causebrook's method is proved by the good agreement in the shape and field strength values of the measured and computed curves.

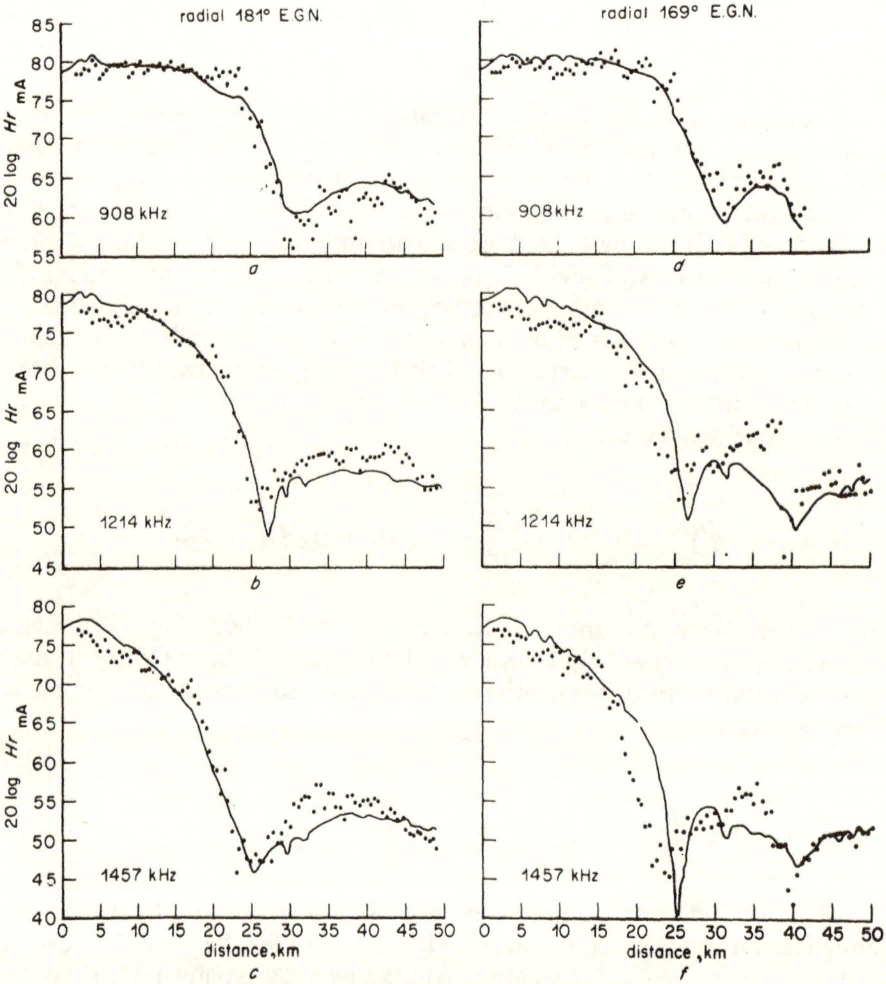

Figure 5.19 Magnetic field-strength dependency on distance from the transmitter; calculations compared with measurements (after Causebrook, 1978, © IEE).

The field strength at any point within a built-up area depends on many factors and changes from instant to instant, even for a stationary observer, since the movement of other vehicles will affect it.

While methods for field strength calculation in built-up areas exist, it is always recommended to perform a set of measurements in order to ascertain that the chosen method of calculation fits the prevailing conditions.

5.4.6 Anomalous Propagation

Operators of 600 MHz radars stationed during the second World War on the Mediterranean island of Malta observed returns from Mt. Etna and other terrain features in Sicily at ranges of about 600 km (Booker, 1946). Similarly, operators of 200 MHz radars in Bombay observed returns from the far side of the Arabian peninsula at distances of about 2500 km. Because propagation to such large distances at the above frequencies cannot be due to either diffraction or scatter, some other mechanism, present at favorable meteorological conditions, must account for it.

This phenomenon attracted considerable attention and several theoretical studies were devoted to it. As a result, it was concluded that subject to certain meteorological conditions the paths of very- and ultra-high frequency waves originating at the earth's surface are refracted to such an extent that they are forced back to the earth or to travel in paths parallel to the earth's surface. The lower atmosphere is then super-refractive and the waves become ducted.

On the other hand, meteorological conditions may reduce the atmosphere's refractivity so that wave paths bend less than they do under "normal" conditions. They may deviate upwards and tend towards the ionosphere when sub-refraction prevails.

Super-refraction and sub-refraction cause anomalous propagation.

5.4.6.1 Meteorological Conditions Conducive to Super-Refraction

As was demonstrated earlier, the path of a ray in the atmosphere depends on the gradient of the refractive index along the path. Since the atmosphere is not uniform, both vertical and horizontal variations of the refractive index are experienced; however, the latter ones are so small as to be normally negligible. The atmosphere may be assumed to be horizontally homogeneous, and the vertical gradient of the index of refraction is determined by the distribution of atmospheric temperature, pressure, and humidity.

Atmospheric temperature and humidity decrease with height under normal conditions since turbulence prevents any great changes in the consistency. It often happens, however, that the air becomes so calm that temperature inversions and humidity lapses may develop and both of these phenomena may cause ducting of very- and ultra-high frequency electromagnetic waves.

Due to the temperature differences between night and day, and due to the surface temperature falling faster than the air temperature, the surface heat loss may lead to temperature inversions at heights below about 300 m and produce conditions favorable to ducting.

Similar conditions may be caused by advection or a horizontal transport of air when a mass of warm, dry air moves from the hot land over a cooler sea, where it causes evaporation with a consequent increase in humidity and reduction of temperature near the surface. Advective ducts may develop when an excess of unmodified air temperature is above that of water or water vapor pressure differs in the modified and unmodified air. Advective duct formation is more frequent above the sea than above land though it may occur over land when warm, dry air shifts over wet ground.

In a subsidence inversion the air mass from a high pressure center of an anticyclone descends from higher levels. The descending dry air is heated by adiabatic compression and spreads out to form a layer well above the earth's surface. Water vapor from the moist surface may ascend to the base of the inversion by convection or turbulence and intensify the humidity contrast.

Ducting over land is less pronounced and smaller in extent because terrain and other features, like hills, forests, and buildings, disturb conditions conducive to duct formation. Over sea, however, they are frequent and may extend over vast areas. Duct propagation over the sea was observed over very long distances, as mentioned earlier.

Two types of ducts may develop. The boundaries of a surface duct are an elevated layer of stratified air on one side and the surface of the earth on the other; stratified air layers are those of an elevated duct. Propagation within a duct occurs in a waveguide manner; here, however, one or both waveguide boundaries are soft so that the plane of reflection depends on the wave's angle of incidence into the boundary. Only waves of lengths less than or equal to

$$\lambda_{max} = 2.296 \left(\frac{\Delta n}{\Delta h} d^3 \right)^{1/2}$$

[Kerr, 1951] with all lengths in meters, may propagate within a duct of width d when $\Delta n/\Delta h$ is the gradient of the refractive index.

Assuming that the refractive index is approximately equal to unity, the ray curvature, as given by (5.30), becomes

$$\rho = -\mathrm{d}h/\mathrm{d}n$$

from where, due to 5.36, the gradient of the refractive index becomes

$$\frac{N}{h} = -\frac{1}{\rho} 10^6$$

For ducting to occur, the radius of curvature of the ray must be equal to or smaller than that of the earth of a radius r, or

$$\frac{\Delta N}{\Delta h} \leq -\frac{10^6}{r} \tag{5.87}$$

For an average value of $r = 6371.228$ km, this expression yields

$$\frac{\Delta N}{\Delta h} \leq -157 N\text{-units/km} \tag{5.88}$$

as a condition for ducting.

Because N differs from unity by only about three parts in 10^4, for many purposes it is preferable to introduce the refractive modulus M, defined earlier in Section 5.4.2.2. Figure 5.20 shows the variation of this modulus with height. It indicates the negative slope required for ducting.

5.4.6.2 Propagation in Ducts

The theoretical aspects of propagation in ducts with various refractive index profiles was studied by Budden [1961b], among others, for both horizontal and vertical polarization.

Although the path bending investigation was based on the assumption of a linearly decreasing refractive index with height, it must be appreciated that such a refractive index profile cannot exist because the refractive index may not assume values below unity. The previously mentioned hypothetical "standard" atmosphere consists of three layers. The lowest one extending to an altitude of about 1 km, the troposphere, was assigned a linear refraction index gradient. The second one, the stratosphere, extending over the region of 1–9 km altitude is assumed to have an exponentially falling refractive index; and in the third one, the ozonopause, the decrease of the refractive index follows a different exponential law. Because ducts occur at heights below 1 km, the linearly decreasing refractive index is acceptable for initial theoretical studies.

The modified refractive index introduced in Section 5.4.2.2 simplifies the following investigation of horizontal polarization. For a positive and constant lapse rate κ, the modified index of refraction is

$$n' = n'_0(1 - \kappa h) \tag{5.89}$$

with n'_0 the value of n' at ground level, where $h = 0$. The values of κh encountered in practice do not exceed 10^{-4} so that we introduce a negligible error when writing

$$n'^2 = n'^2_0(1 - 2\kappa h) \tag{5.90}$$

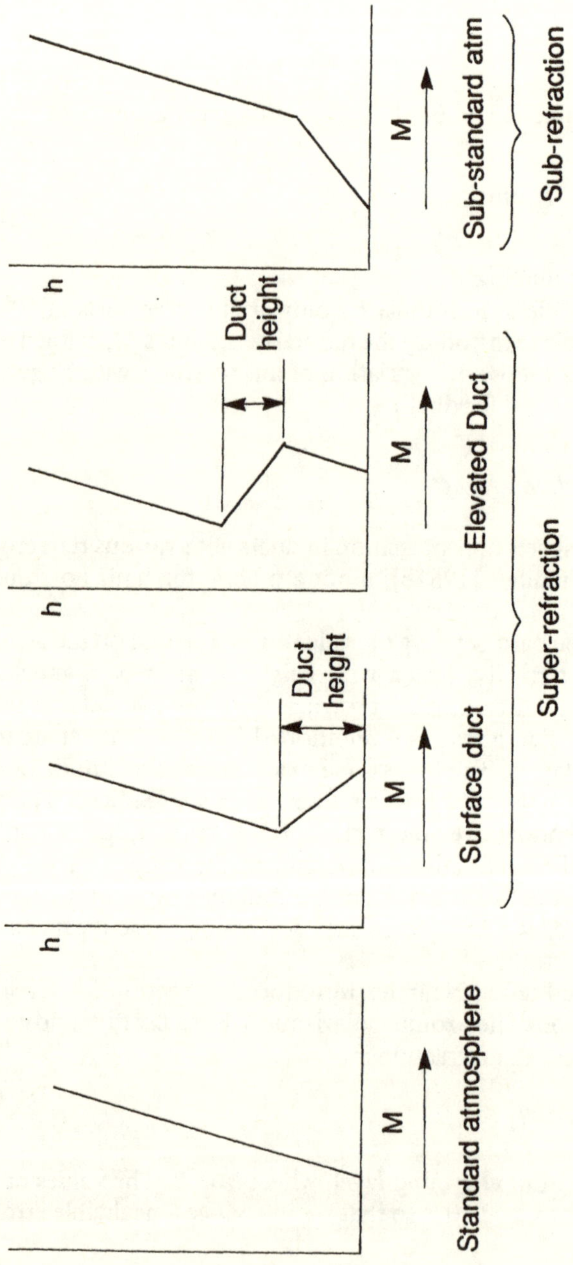

Figure 5.20 Modified refractivity variations with height.

The last two expressions yield acceptable values of the modified refractive index to heights of several hundred meters for cases where temperature inversion exists; however, because at great enough altitude the curvature of the earth always yields a negative lapse rate of the refractive index, they cannot exist in reality.

The phase shift suffered by a wave leaving the earth with a normal at an angle θ and entering an elevated stratified layer at a height h and distance d at an angle ψ with the horizontal, is

$$-j\frac{2\pi}{\lambda}\,n'(d\cos\psi + h\sin\psi)$$

which with Snell's law

$$n'\cos\psi = n_0'\cos\theta \tag{5.91}$$

becomes

$$-j\frac{2\pi}{\lambda}\,(n_0' d\cos\theta + n'h\sin\psi) \tag{5.92}$$

where the second term is a function of the altitude. This expression may then be written as

$$-j\frac{2\pi}{\lambda}\left(n_0' d\cos\theta + \int_0^h n'\sin\psi\, dh\right) \tag{5.93}$$

Applying Snell's law (5.91), to (5.90) yields

$$n'^2\sin^2\psi = n_0'^2(\sin^2\theta - 2\kappa h)$$

or, for $n' \approx 1$ at ground level,

$$n'^2\sin^2\psi = \sin^2\theta - 2\kappa h \tag{5.94}$$

thus leading to an expression for the virtual height of reflection

$$h_r = \frac{\sin^2\theta}{2\kappa} \tag{5.95}$$

an altitude where the vertical component of n', $n'\sin\psi = 0$.

Because the ray travels from the ground to height h_r and back, the total phase shift will be twice the value given by (5.93), where the first term, which is independent of n', may be neglected. The total phase shift then is

$$-2j\frac{2\pi}{\lambda}\int_0^{h_r} n' \sin\psi \, dh$$

or substituting (5.94) for the term under the integral sign, it becomes

$$-2j\frac{2\pi}{\lambda}\int_0^{h_r} (\sin^2\theta - 2\kappa h)^{1/2} \, dh \tag{5.96}$$

Budden solved the integral to arrive at a reflection coefficient referred to level h_1:

$$\rho = j \exp\left(-\frac{2}{3}j\frac{2\pi}{\lambda}\frac{\sin^3\theta}{\kappa}\right) \tag{5.97}$$

which, for horizontal polarization, is the ratio of the electric field in the ascending and descending waves.

The horizontal component of the electric field at the ground, which is assumed to be a perfect conductor, is zero, so that the reflection coefficient is $\rho_0 = -1$.

To sustain propagation of a wave in a waveguide, the total phase shift of the component of the wave in the guide's transverse direction must be an integral multiple of 2π. The total phase shift of the reflection coefficient, expression (5.97), is

$$-\frac{2}{3}\frac{2\pi}{\lambda}\frac{\sin^3\theta}{\kappa} + \frac{\pi}{2} - \pi$$

where $\pi/2$ represents the phase shift due to the j operator and π is due to the phase shift on reflection from the perfectly conducting ground. Then, for propagation in the guide to be possible, the condition

$$-\frac{2}{3}\frac{2\pi}{\lambda}\frac{\sin^3\theta}{\kappa} - \frac{\pi}{2} = 2m\pi$$

must be fulfilled. From here,

$$\sin^3\theta = \left(m - \frac{1}{4}\right)\frac{3\lambda\kappa}{2} \tag{5.98}$$

with m an integer denoting the order of the propagation mode that cannot be zero or negative because $\sin\theta = n' \sin\psi$ when $h = 0$; that is, the direction of the wave normal to the ground must be real and positive. For a positive m, $\sin\theta$ is real and positive. The other roots of (5.98) have negative real parts. The possible propagation modes necessitate a positive θ; the modulus of the reflection coefficient is unity.

All propagation modes are locked with most of their energy confined to the region below the virtual height of reflection, which from (5.95) when (5.98) is substituted for $\sin\theta$, becomes

$$h_r = \frac{1}{2\kappa}\left[\left(m - \frac{1}{4}\right)\frac{3\lambda\kappa}{2}\right]^{2/3} \tag{5.99}$$

This height is usually referred to as the *track width* for the mode of order m. It increases with m and λ, however, the number of possible modes is determined by the height over which the anomalous lapse rate prevails. The angle θ for all the possible modes is small; only rays launched near the normal to the earth become trapped.

The wave equation to be satisfied by the horizontal component E_y for the vertical propagation of the wave is the Stokes' equation [Budden, 1961a]:

$$\frac{d^2E_y}{dh^2} + \left(\frac{2\pi}{\lambda}\right)^2 (\sin^2\theta - 2\kappa h)E_y = 0 \tag{5.100}$$

with a solution

$$E_y = Ai\left\{\left[2\kappa\left(\frac{2\pi}{\lambda}\right)^2\right]^{1/3}(h - h_r)\right\} \tag{5.101}$$

where $Ai(\xi)$ is an Airy integral function.

The height dependency of the field components in an atmosphere with a linear decrease of the squared refractive index is shown in Figure 5.21. The three upper curves of modes shown by the figures at their top refer to horizontal polarization or TE-modes, whereas the lower ones for vertical polarization, or TM-modes. The virtual height of reflection is shown by h_r.

Example 5.8

Determine the track width of a 1500 MHz horizontally and vertically polarized wave of the first mode, when $\kappa = 10^{-4}$/km. Explain the meaning of your findings.

Solution: Formula (5.99), with $m = 1$ and κ, the specified modified index of refraction, leads to

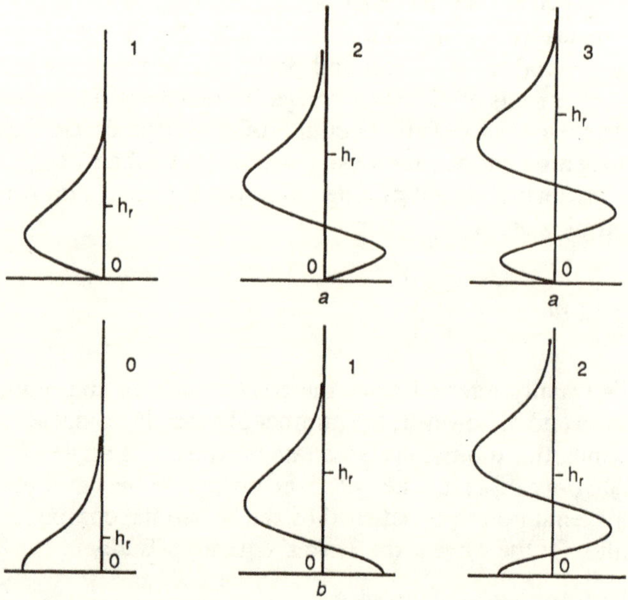

Figure 5.21 The height dependency of the field for a linear decrease of the squared modified refractive index (after Budden, 1961b).

$$h_r = 3.98m$$

for the horizontally polarized wave's track width.

The value for vertical polarization follows from expression (5.99):

$$h_r = 2.71m$$

The computed track widths show the duct heights that contain most of the energy of a propagating trapped wave of mode 1.

In vertical polarization, considering the behavior of the magnetic field that has only a horizontal component, H_y, all modes of propagation are TM-modes. The variation of H_y with height is very small for all values of h and n'^2 of interest and the wave equation, like (5.100), is

$$\frac{d^2H_y}{dh^2} + \left(\frac{2\pi}{\lambda}\right)^2 (\sin^2\theta - 2\kappa h)H_y = 0$$

Considering the ground to be a perfect conductor, its reflection coefficient becomes $\rho = +1$ and the differential $dH_y/dh = 0$ at $h = 0$.

Proceeding then as in the case of horizontal polarization, we arrive at

$$\sin^3\theta = \frac{3\kappa\lambda}{2}\left(m + \frac{1}{4}\right) \tag{5.102}$$

with m as zero or an integer, for the launch angle required to support propagation in mode of order m.

The track width is

$$h_r = \frac{1}{2\kappa}\left[\frac{3\kappa\lambda}{2}\left(m + \frac{1}{4}\right)\right]^{2/3} \tag{5.103}$$

Due to the asymptotic approximation of the Airy integral, both expressions (5.98) and (5.102), for the mode conditions for horizontal and vertical polarization, are only approximate. A more accurate result may be obtained by solving the wave equation

$$H_y = Ai\left\{\left[2\kappa\left(\frac{2\pi}{\lambda}\right)^2\right]^{1/3}(h - h_r)\right\}$$

Because we assumed that the ground is a perfect conductor and expression (5.95) was used for the height of reflection, we can write

$$Ai\left[-\sin^2\theta\left(\frac{\pi}{\kappa\lambda}\right)^{2/3}\right] = 0$$

as the requirement for satisfying the boundary condition $dH_y/dh = 0$ when $h = 0$. The error due to the approximation amounts to about 10% for the lowest propagation mode but it decreases for higher ones.

Figure 5.21 shows that, in the case of horizontal polarization, there are m half-waves in the field profiles and $m + ½$ half-waves for vertical polarization.

Assuming that the lapse rate of the modified index of refraction in the duct is $\kappa = 10^{-4}$/km, the angles of the wave normal elevations at ground level and the track widths for the first four propagation modes for the two considered polarizations at a frequency of 1.3 GHz are as tabulated in Table 5.6.

As mentioned earlier, the hypothetical atmosphere considered here, having a linearly decreasing refractive index profile, is only an artifice easing the complicated theoretical considerations concerning ducting of electromagnetic waves. In the real atmosphere, the square of the modified refractive index n'^2 is some slowly varying

Table 5.6
Wave Normal Elevation Angles at Ground Level and Track Widths for Ducting (from Rohan, 1983)

Mode	*Horizontal Polarization*		*Vertical Polarization*	
m	θ *Degrees*	h_r *Meters*	θ *Degrees*	h_r *Meters*
0	—	—	0.118	21.1
1	0.169	43.8	0.201	61.6
2	0.225	77.1	0.245	91.2
3	0.261	104.2	0.277	116.5

function of height, and the reflection coefficient may be expressed by the phase integral

$$\rho = j \exp\left[-2j\frac{2\pi}{\lambda}\int_0^{h_1} q\, dh\right] \tag{5.104}$$

The variable q is the so-called Booker variable given by

$$q^2 = n'^2 - n_0'^2 \cos^2\theta \tag{5.105}$$

where n' is the real part of the refractive index, and θ is the angle between the wave normal and the horizontal [Budden 1961b].

For waves propagating in a TE-mode, the condition for propagation within the duct having a perfectly conducting ground for one of its boundaries, requires satisfying equation

$$\int_0^{h_r} q\, dh = \frac{\lambda}{2}\left(m - \frac{1}{4}\right) \tag{5.106}$$

The coefficient m is a positive, nonzero integer. At higher frequencies and heights where dn'^2/dh is negative, real values of θ obtained with the aid of (5.106) apply to locked propagation modes, and complex values to leaky ones.

The field of the locked modes within the duct, when $h < h_r$, is the sum of the fields of two crossing waves, or

$$E_h = \frac{2E_0}{\sqrt{q}} \exp\left(j\frac{\pi}{4}\right) \exp\left(-j\frac{2\pi}{\lambda} x \cos\theta\right) \cos\left(\frac{2\pi}{\lambda}\int_h^{h_r} q\, dh - \frac{\pi}{4}\right)$$

with x the direction of propagation, and E_0 determined by the type of radiator launching the wave. This expression breaks down near h_r, where a linear variation of n'^2 with h often may be assumed. The field then, for $h \approx h_r$, is

$$E_h = 2E_0\pi^{2/3}(\kappa\lambda)^{-1/6}Ai\left[-\left(\frac{3}{2}\frac{2\pi}{\lambda}\int_h^{h_r} q\,dh\right)^{2/3}\right]$$

Fok, Vainshtein, and Belkina [1958] studied the attenuation of waves propagating in a surface duct for cases where the transmitter and the receiver antenna are within the duct and when one antenna is high above the inversion layer, and the other is at one-fifth the inversion height. The attenuation in dB increases almost linearly with distance and depends on the antenna heights and the frequency used. Values as high as about 30 dB may be found at a range of 300 km at a frequency of 0.5 GHz, whereas it is about 3 dB at 3 GHz.

Surface reflection, turbulence within the ducting layer, gaseous absorption, and losses associated with propagating waveguide modes within the duct are the possible causes of loss. Generally, the decrease of energy within the duct is inversely proportional to the distance from the transmitter, and inversely proportional to the square of that distance for waves propagating outside it.

Rotheram [1973a] divides the coverage area of a transmitter in a standard atmosphere into a domain on the near side of the radio horizon, where the main mechanism of propagation is reflection and refraction, and one on the far side, where the dominant mechanism of propagation is diffraction. These domains merge into each other near a well-defined radio horizon. When anomalous atmospheric conditions prevail, the domains and hence the radio horizon no longer are clearly defined. On the transmitter's side of the coverage area propagation is due mainly to superrefraction, whereas further on, where the field is dominated by the leaky waveguide modes, he ascribes the dominant mechanism of propagation to superdiffraction. Superrefraction and superdiffraction merge into each other in an intermediate, sometimes quite extensive, domain.

Elevated ducts caused mainly by the compression and consequent warming and drying of subsiding air masses in a high pressure system occur predominantly above clouds at heights of about 1500 m, when the modulus of the modified refractive index increases with height near the earth but decreases over an elevated altitude interval to resume its increasing trend above it. Sketch 3 of Figure 5.20 illustrates the characteristic modified refractivity profile variations for these atmospheric conditions.

Within the region of the negative gradient, n' reaches a maximum value, which decreases for both increasing and decreasing height and causes an upward traveling ray to bend downward and a downward traveling one to bend upward.

Denoting this maximum value at a height h_1 by n_1' and assuming that the wave normal of an upward traveling wave at this height is θ, the Booker variable (5.105) becomes

$$q^2 = n'^2 - n_1'^2\cos^2\theta$$

This is equal to zero at a height h_0 below and height h_{00} above h_1 where

$$n' = n_1' \cos\theta$$

The reflection coefficient ρ_1 referred to height h_1 for an upward traveling wave, as found earlier in (5.104), is

$$\rho_1 = j \exp\left[-2j\frac{2\pi}{\lambda}\int_{h_1}^{h_{00}} q \, dh\right]$$

and, for a downward traveling one, is

$$\rho_2 = j \exp\left[+2j\frac{2\pi}{\lambda}\int_{h_1}^{h} q \, dh\right]$$

The mode conditions applicable to TE- and TM-modes require that a wave traveling in a waveguide, after two reflections, should remain identical with the original wave entering the first reflection (i.e., that the total phase shift should be $2m\pi$), or that the product of the reflection coefficients $\rho_1\rho_2$ should be equal to unity. This, applied to the preceding reflection coefficients, yields

$$\exp\left[-2j\frac{2\pi}{\lambda}\int_{h_0}^{h_{00}} q \, dh\right] = -1$$

from which we have

$$\int_{h_0}^{h_{00}} q \, dh = \frac{\lambda}{2}\left(m - \frac{1}{2}\right)$$

where m is a positive integer. A negative value of m would yield the same series of modes for a wave traveling in the opposite direction; or a positive m applied to a wave traveling upward, and a negative one to a downward traveling wave. The levels h_0 and h_{00} are the bottom and top of the track and their difference is the track width.

The previous approximate formulas, applicable only to elevated ducts, cannot be used where the track width is small; that is, the $q = 0$ levels are close together. If the upper reflection level is close to the top of the track, a leaky mode may develop.

The determination of the heights h_0 and h_{00} requires the knowledge of the variation of the modified refractive index n' across the duct. An approximation of this variation is acceptable for small height differences; accurate knowledge of the variations is required for large height differences and higher order modes.

Figure 5.22 shows some possible ray trajectories in a standard atmosphere and in surface and elevated ducts for rays launched at height h_1 at different angles θ_1 over

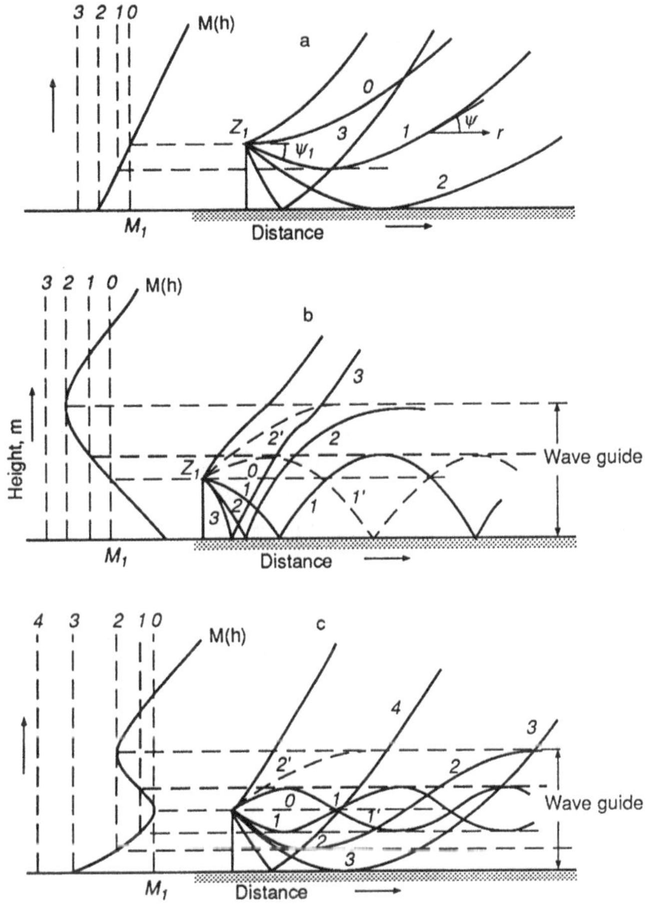

Figure 5.22 Possible ray trajectories in an atmosphere with different modified refractive index profiles: (a) "standard" atmosphere; (b) surface duct; (c) elevated duct (after Al'pert, 1973, © Vsesojuznoje Agentstvo).

an effectively flat earth for three types of modified index M variations. The vertical lines on the left, annotated by values of $M \times 10^{-6} = M_1 \times 10^{-6} - \theta_1^2$, intersect the $M(h)$ profile at heights where $\theta = 0$ for the correspondingly annotated ray paths or where the rays reverse their direction of travel. These heights are the waveguide boundaries.

Waveguide propagation extends only over the range over which the appropriate atmospheric parameters remain constant so that the wave's energy is confined to the duct. If, however, the height-dependent refractive index changes, energy is permitted to leak out of the duct.

The track width required for duct propagation at various wavelengths follows from the formula for λ_{max} (Section 5.4.6.1):

$$d \geq \left[\left(\frac{\lambda}{2.296}\right)^2 \frac{1}{\Delta n/\Delta h}\right]^{1/3} \tag{5.107}$$

Duct propagation has been observed by a number of experimenters. Booker's [1946] observations were mentioned earlier.

Smith and Soden [1974], in Australia, observed radar displays at two operational radars with a view of correlating ducting occurrence with prevailing meteorological conditions. They described the observations at Darwin, Australia, where radar returns from various terrain features beyond the unambiguous radar range were taken to be indicative of the presence of duct propagation. A season-dependent diurnal variation of the occurrence of echoes was established in a geographic location, where the greatest likelihood of ducting at altitudes below about 2.5 km prevails during the tropical dry season and the dry-wet season transition periods. The occurrence of ducting at Darwin is most likely during the early morning and late evening hours, depending on the season.

The observed correlation between anomalous propagation and meteorological data from radiosonde measurements was 0.82 for the wet season (January and February), but it dropped gradually during the wet-dry transition period and dry season to about half that value during the dry-wet transition period, and rose again afterward.

Due to the great variability of the refractive index at microwaves, it is not practicable to predict, on the basis of meteorological data alone, the occurrence of superrefracting conditions in the tropics. Some rough estimates based on collected statistical data may be successful. A reasonably reliable prediction of ducting conditions is possible in more moderate climates.

Generally, oceanic evaporation ducts over all oceanic areas are experienced nearly all the time. Ducting alters the effective earth radius and hence all the phenomena and distances depending on it.

Whereas superrefraction requires that the gradient of the modified index of refraction satisfy the condition

$$0 < \frac{dM}{dh} < 118.11 \ M\text{-units/km}$$

where the right-hand side of the inequality is the accepted gradient of a standard atmosphere, and ducting requires $\Delta N/\Delta h \leq -157$ N-units/km, subrefraction is known to occur for positive gradients algebraically greater than that of the standard atmosphere's value.

Subrefraction is not as frequent as superrefraction. Ground-based subrefractive layers may occur in the same tropical and subtropical regions as superrefraction.

It may develop when the ground temperature in a nocturnal inversion falls below the dewpoint temperature and the water in the lowest layers of the air condenses. The heat of condensation is released directly to the air; under conditions of radiative fog formation, the humidity lapse tends to counteract any temperature inversion. If the humidity inversion is sufficiently strong, substandard refraction having a positive gradient may ensue.

According to Bean *et al.* [1966], subrefractive gradients may occur in the afternoon at locations where nocturnal or early morning superrefraction or ducting prevails. At other locations subrefraction may occur in winter and superrefraction in summer. Surface conditions conducive to the development of subrefractive gradients are either a temperature in excess of 30°C and a relative humidity of less than 40%, or a temperature between 10 and 30°C accompanied by a relative humidity of above 60%.

The first of these environmental conditions occurs at warm, dry continental locations, usually in daytime during months of intense solar heating, when a very nearly homogeneous surface layer, up to several meters thick, may form. Intense convection, which occurs in a layer of moist unstable air, tends to concentrate the available water vapor near the layer's top; a moist adiabatic upper boundary is formed where the superadiabatic lapse rate changes abruptly to a subadiabatic very stable rate, resulting in a stable, dry term but a rather large increase in height through the ground-based layer of the wet term, thus leading to the occurrence of a positive, subrefractive gradient.

In locations where temperature inversions may develop, this subrefractive layer may persist throughout the early evening hours. In the evening, the ground temperature falls fast, therefore the air near the ground becomes more dense. The water vapor trapped between the two stable layers causes the positive wet-term gradient from the ground to the top of the layer to remain large enough to overcome the increasing gradient of the dry term.

The second of the conditions supporting subrefraction may occur during the night and early morning in coastal areas, where differential heating of land and sea causes advection of warmer and more humid air than that of the surface layer. Both the dry and the wet terms may increase with altitude and create an intense subrefracting surface layer.

Waves propagating in substandard surface layers almost always exhibit scintillations in which as high as 15 dB amplitude variations with an irregular, about 30 s frequency may be experienced [Kerr, 1951]. The variability of the signal levels appears to persist at times when the substandard layer is not strong enough to affect the average level appreciably. Occasionally slower variations are superimposed on these rapid fades. When a fairly intense substandard layer exists at the surface, the average signal level is below the value experienced in a standard atmosphere; variations in received power may reach 5 to 10 dB with periods of 5 to 15 min.

The occurrence of subrefraction at various geographic locations depends not only on time of day but also season. Subrefraction, according to Bean and Dutton

[1966], is potentially a very important refractive factor for distances of less than about 40 km. In moderate latitudes, frequently it is offset by the simultaneous occurrence of an adjacent superrefractive layer.

5.4.7 Satellite Communication

Satellites may serve various purposes and their trajectories or orbits depend on them. The satellite's height and hence velocity determines the length of time it may be reached from any location on the earth. The most economical orbit for communication satellites is that used by geosynchronous ones, in which the rotation period of the satellite is 24 hours, the same as that of the earth, thus giving 24-hour continuous access from locations illuminated by the satellite's antenna. Keppler's third law yields the radius R of a circular orbit for any specified rotation period T.

The formula describing the law reads

$$R = \left(\frac{T^2 r^2 g}{4\pi^2}\right)^{1/3} \tag{5.108}$$

where $g = 9.81$ m/s^2 is the average gravitational acceleration, and r is the earth's radius. The rotation period of a satellite in geosynchronous orbit is $T = 86400$ s; that is, the number of seconds in 24 hours yielding a circular orbit radius $R = 42258.33$ km. This is a much longer path than that possible between terrestrial microwave terminals and therefore the propagation losses will be much larger.

Geosynchronous satellites have a number of advantages over satellites with other rotational periods. These include a continuous 24-hour access; the possibility of using fixed antennas with only minor pointing adjustments; no need for tracking adjustments; the adequacy of only three satellites for the global coverage of the most densely populated areas, and negligible doppler shift due to the satellite's motion with respect to the earth. The disadvantage of this type of orbit is the necessarily long distance from the earth with its accompanying great propagation losses and signal delays of about 270 ms for the up and down paths.

5.4.7.1 Choice of Frequency

The choice of frequency for satellite communication depends on a number of factors. The chosen frequency must be high enough to escape the confines of the ionosphere. Because the path attenuation is very large, due to the path's length, the frequencies are restricted to those that suffer no undue attenuation by atmospheric gasses, shown by Figure 5.10. In addition, the chosen frequency must be removed from those used by other services to avoid mutual interference. Further, because the directive gain of an antenna depends on its dimension in terms of the wavelength and because the satellite cannot accommodate physically large and heavy antennas, the chosen frequency should be as high as reasonably possible.

Some of these criteria are contradictory. The highest frequencies would make possible the construction of very high-gain satellite-borne antennas, but these frequencies suffer the largest absorption by uncondensed water vapor (Figure 5.10). The final choice therefore must be a compromise.

The frequencies used by various services are based on international agreements and allocated by the International Telecommunications Union. Those allocated to communication satellites are in the 4–6 GHz band for the earth to satellite path and in the ¾ GHz band in the opposite direction. Some countries use the 7–8, 10–11, and 12–14 GHz bands, and the use of even higher frequencies is being contemplated.

5.4.7.2 Losses

Equipment loss is of no interest in this study; only loss connected with the propagation phenomenon. Loss due to absorption by atmospheric oxygen and uncondensed water vapor have been treated in Section 5.4.2.2, where formulas and the graph in Figure 5.10 have been given to facilitate their computation.

Loss due to diminishing power density with distance, the free-space loss, is directly proportional to the square of the distance and indirectly to the square of the wavelength. The power density on the surface of a sphere of radius R, due to an isotropic radiator in the sphere's center is $P_t/4\pi R^2$, where P_t is the transmitted power. The fraction of this power intercepted by the receiving antenna will be proportional to this antenna's aperture, A_r. Then, the received power is

$$P_r = \frac{P_t}{4\pi R^2} A_r \qquad (5.109)$$

The ratio of an antenna's gain and aperture is [Starr, 1953] $G/A = 4\pi/\lambda^2$, from which, for an isotropic antenna with a unity gain G, the aperture becomes

$$A = \frac{\lambda^2}{4\pi}$$

Inserting this into expression (5.109) leads to

$$P_r = \frac{P_t\lambda^2}{(4\pi R)^2} = P_t/L_0 \qquad (5.110)$$

where

$$L_0 = \left(\frac{4\pi R}{\lambda}\right)^2 = \left(\frac{4\pi Rf}{c}\right)^2 \tag{5.111}$$

is the free-space loss with c the velocity of propagation of electromagnetic waves and f the used frequency.

Including the gains of the transmitting and receiving antennas into expression (5.110), yields the received power as

$$P_r = \frac{P_t G_t G_r \lambda^2}{(4\pi R)^2} = \frac{P_t G_t G_r c^2}{(4\pi Rf)^2} = P_t / L_l$$

where the link loss

$$L_l = \left(\frac{4\pi Rf}{c}\right)^2 \frac{1}{G_t G_r} = \frac{L_0}{G_t G_r} \tag{5.112}$$

Expressing the antenna gains in term of their apertures as

$$G_t = \eta_t \frac{4\pi}{\lambda_t^2} A_t \text{ and } G_r = \eta_r \frac{4\pi}{\lambda_r^2} A_r \tag{5.113}$$

where the subscripts t and r stand for transmitter and receiver and η for the efficiency of the antenna. The link loss becomes

$$L_l = \left(\frac{4\pi R}{c}\right)^2 \frac{f_t}{G_t} \frac{f_r}{G_r}$$

or with (5.113)

$$\begin{aligned} L_l &= \left(\frac{4\pi R}{c}\right)^2 \frac{f_t \lambda_t^2}{\eta_t 4\pi A_t} \frac{f_r \lambda_r^2}{\eta_r 4\pi A_r} \\ &= \left(\frac{4\pi R}{c}\right)^2 \frac{f_t c^2}{\eta_t 4\pi f_t^2 A_t} \frac{f_r c^2}{\eta_r 4\pi f_r^2 A_r} \\ &= \frac{(cR)^2}{\eta_t \eta_r f_t f_r A_t A_r} \end{aligned} \tag{5.114}$$

In addition to these losses, attenuation by rain, fog, and clouds also must be considered. They may be determined by the formulas and graphs given in Section 5.4.2.2. The path lengths in the preceding meteorological phenomena depend on the angle at which the wave enters the atmosphere.

Example 5.9

A satellite in a geosynchronous orbit carries a 4.2 GHz, 65 W transmitter equipped with a 55% efficient 16 dB gain antenna. The aperture of the ground-based antenna is 92.86 m^2 and its efficiency is also 55%. Compute the received signal power and the free-space and link losses.

Solution: The transmitted power is

$$\begin{aligned} P_t &= PG\eta_t \\ &= 65 \times 39.81 \times 0.55 \\ &= 1423.21 \text{ W} \end{aligned}$$

The received power, expression (5.109), is

$$\begin{aligned} P_r &= \frac{P_t}{4\pi R^2} \eta_r A_r \\ &= \frac{1423.21}{4\pi(42258.33 \times 10^3)^2} 0.55 \times 92.86 \\ &= 3.239 \text{ pW} \end{aligned}$$

The free-space loss, expression (5.111), is

$$\begin{aligned} L_0 &= \left(\frac{4\pi R}{\lambda}\right)^2 \\ &= \left(\frac{4\pi 42258.33 \times 10^3}{0.0714}\right)^2 \\ &= 5.532 \times 10^{13} \\ &= 197.43 \text{ dB} \end{aligned}$$

Link loss, expression (5.112), is

$$L_l = \frac{L_0}{G_t G_r}$$

$$G_t = 16 \text{ dB}, \quad \eta_t = 0.55,$$

$$G_r = 0.55 \frac{4\pi 92.86}{0.0714^2} = 125893.94 = 51 \text{ dB}$$

The link loss then is $L_l = 197.43 - 13.4 - 51 = 133.03$ dB where G_t expressed in dB is 13.4 dB.

PROBLEMS

5.1 Determine the field strength of a 10 kW transmitter operating at a frequency of 24 kHz located 12,000 km away when propagation takes place mainly over sea. Compute also the signal-to-noise ratio using noise data of Figure 5.2.

5.2 Compare the field strengths of a 100 kW transmitter operating at a frequency of 20 kHz at a distance of 5000 km obtained from expressions (5.7) and (5.8) when propagation takes place over land. What attenuation factor α is to be chosen in expression (5.8) to obtain the field strength computed from expressions (5.7) when the ionospheric height is 100 km?

5.3 Assuming that reflection of a 30 MHz wave occurs from the F_2-region at a height of 300 km, compute the phase shift at a location 3800 km away from the transmitter when the height of reflection increases by 10 km.

5.4 Compare the vertical component of the electric field of a 100 kW transmitter at a distance of 1500 km and ionospheric height of 120 km for a one- and two-hop propagation path over land far removed from polar regions. Compare them with values obtained from the appropriate graphs.

5.5 Compute the distance covered in a single-hop propagation path when the initial ray elevation angle is $\epsilon = 0°$ and the ionospheric height is 300 km.

5.6 Assuming that both the transmitting and receiving antennas are 10-m high and their separation is 10 km, find the phase difference between the direct and reflected waves at a frequency of 300 MHz.

5.7 A 2 MW radar operating at 1.3 GHz is equipped with a 20-m high horizontally polarized antenna of 30 dB gain in the desired direction and 10 dB in that of the reflected ray's launch angle. Compute the field intensity due to the direct and reflected rays at the target at a height of 500 m and a ground-range distance of 100 km when the coefficient of reflection is 0.65.

5.8 Compute the attenuation of a 20 and 60 GHz radiation at an angle of 12° by atmospheric oxygen and water vapor and the total atmospheric attenuation.

Compare the results with those read from Figure 5.10. The atmospheric temperature is taken to be 20°C and the water vapor density 7.5 g/m^3.

5.9 Using expression (5.53) compute the attenuation of a 3 GHz transmission by rain of 16 mm/h rate and compare it with that read from Figure 5.12.

5.10 Determine the dielectric constant and conductivity of a soil having a 10% moisture content by weight at 20°C.

5.11 Determine the propagation loss in a "standard" atmosphere between two horizontally polarized antennae when
(a) h = 40 m, H = 10 m separated by 35 km;
(b) h = 40 m, H = 500 m separated by 160 km.

5.12 The terrain profile shown in Figure 5.23 is between a transmitter and receiver separated by a distance d = 10 km. The angle α measured at the transmitter site is 47°; and β, the angle at the receiver site, is 21°. Determine the shadow loss relative to smooth earth when the operating frequency is 60 MHz.

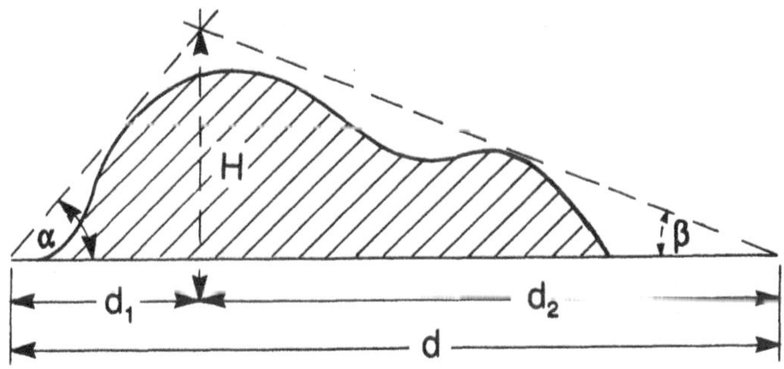

Figure 5.23 Problem 5.12.

5.13 How is a radar's performance affected by superrefractive atmospheric conditions?

5.14 Extend Table 5.2 to show the wave-normal elevation angle and track width for the fourth-order propagation mode.

5.15 Plot the dependency of the wave-normal elevation angle and track width at horizontal and vertical polarization on the frequency for the zeroth and second-order propagation modes over the frequency range 50 MHz to 6 GHz when κ = 10^{-5}/km.

REFERENCES AND RECOMMENDED READING

Adrianov, V.A. (1969) "Applicability Limits of the Concept of the Equivalent Radius of the Earth," *Radio Engineering Electronic Physics,* Vol. 14, No. 2, February, pp. 195–200.

Aitchison, R.F. (1975) "VHF field strength for line of sight reception," *Proc. IREE Australia,* Vol. 36, No. 7, July, pp. 225–231.

Albrecht, H.J. (1962) "Simplified Calculation of Unabsorbed Field Intensity on Overseas Transmission Circuits," *Proc. IRE,* Vol. 50, No. 12, December, pp. 2495–2496.

Albrecht, H.J. (1965) "On the Relationship Between Electrical Ground Parameters," *Proc. IEE,* Vol. 53, No. 5, May, p. 544.

Allenbrock, K., and J.D. Parsons (1977) "Mobile Radio Propagation in British Cities at Frequencies in the v.h.f. and u.h.f. Bands," *Proc. IEE,* Vol. 124, No. 2, February, pp. 95–102.

Aller, L.H. (1953) "Astrophysics, the Atmospheres of the Sun and Stars," Ronald Press Co., New York.

Al'pert, Y.L. (1973) *Radio Wave Propagation and the Ionosphere,* Vol. 2 *Propagation of Electromagnetic Waves near the Earth,* Consultants Bureau, New York and London.

Altshuler, E.E. (1984) "A Simple Expression for Estimating Attenuation by Fog at Millimeter Wavelengths," *IEEE Trans. on Antennas and Propagation,* Vol. AP-32, No. 7, July, pp. 757–758.

Ananasso, F. (1980) "Coping with Rain above 11 GHz," *MSN,* Vol. 10, No. 3, March, pp. 58–72.

Awaka, J. (1984) "A Prediction Method for the Received Power from Rain Scattering," *Radio Science,* Vol. 19, No. 2, March–April, pp. 643–651.

Bailey, V.A., and D.F. Martyn (1934) "Influence of Electric Waves on the Ionosphere," *Phil. Magazine,* Vol. 18, p. 369.

Barghausen, A.F. (1966) "Medium-Frequency Sky Wave Propagation in Middle and Low Latitudes," *IEEE Trans. on Broadcasting,* Vol. BC-12, No. 1, January, pp. 1–14.

Bashir, S.O. (1986) "Microwave Propagation in Dust Storms: A Review," *Proc. IEE,* Pt H, Vol. 133, No. 3, June, pp. 241–247.

Bashkuyev, Y.B. and C.T. Tsydypov (1976) "The Electrical Properties of Soil at Metre Wavelengths," *Telecommunication and Radio Engineering,* Part I, Vol. 30, No. 5, May, pp. 5–8.

Battesti, J. (1983) "Mesures réalisées en France concernant l'effet des hydrométéores sur la propagation," *Annales des Télécom.,* Vol. 38, No. 5–6, pp. 231–243.

Battesti, J., and L. Boithias (1971) "Nouveaux éléments sur la propagation par les hétérogéneités de l'atmosphère," *Annales des Télécom.,* Vol. 26, No. 1–2, pp. 15–30.

Battesti, J., L. Boithias, and P. Misme (1968) "Calcul des affaiblissements en propagation transhorizon à partir des paramètres radiométéorologiques," *Annales des Télécom.,* Vol. 23, No. 5–6, pp. 129–140.

Bean, B.R., and R. Abbott (1957) "Oxygen and Water Vapour Absorption of Radio Waves in the Atmosphere," *Geofisica Pura e Applicata,* Vol. 37, II.

Bean, B.R., B.A. Cahoon, C.A. Samson, and G.D. Thayer (1966) *A World Atlas of Atmospheric Radio Refractivity,* ESSA Monograph 1, U.S. Dept. of Commerce.

Bean, B.R., and E.J. Dutton (1966) *Radio Meteorology,* NBS Monograph No. 92, Washington, DC.

Bean, B.R., E.J. Dutton, and B.D. Warner (1970) "Weather Effects on Radar," *see* Skolnik, 1970.

Benson, R.F. (1964) "A Discussion of the Theory of Ionospheric Cross Modulation," *J. Research N.B.S.,* Vol. 66D, No. 10, October, pp. 1109–1122.

Biggs, A.W. (1970) "Terrain Influences on Effective Ground Conductivity," IEEE Trans. GE-8, No. 2, April, pp. 106–114.

Booker, H.G. (1946) "Elements of Radio Meteorology. How Weather and Climate Cause Unorthodox Radar Vision Beyond the Geometrical Horizon," *J. IEE,* Vol. 93, Pt IIIa, pp. 69–78.

Booker, H.G., and W.E. Gordon (1950) "A Theory of Radio Scattering in the Troposphere," *Proc. IRE,* Vol. 38, No. 4, April.

Bradley, P.A. (1975) "Long-Term h.f. Propagation Prediction for Radio-Circuit Planning," *Radio and Electronic Engineering,* Vol. 45, No. 1–2, January–February, pp. 31–41.

Bremmer, H. (1949) *Terrestrial Radio Waves,* Elsevier Publishing Co., Amsterdam.

Brussard, G. (1981) "Prediction of Attenuation due to Rainfall on Earth-Space Links." *Radio Science,* Vol. 16, No. 5, September–October, pp. 745–760.

Budden, K. (1953) "The Propagation of Radio Waves to Great Distances," *Phil. Magazine,* Vol. 44.

Budden, K. (1961a) *Radio Waves in the Ionosphere,* Cambridge University Press, Cambridge.

Budden, K. (1961b) *The Wave-Guide Mode Theory of Wave Propagation,* Logos Press Ltd., London.

Budden, K.G. "The Influence of the Earth's Magnetic Field on Radio Propagation by Waveguide Modes," *Proc. Royal Society,* Vol. 265A, No. 1323, pp. 538–553.

Bugrov, A.G., V.I. Klyatskin, and B.M. Shevtsov (1985) "Theory of Propagation of Short Radio Waves in a Spherically Layered Atmosphere," *J. Communication Technology and Electronics,* Vol. 30, No. 7, July, pp. 60–67.

Bullington, K. (1947) "Radio Propagation at Frequencies above 30 Megacycles," *Proc. IRE*—Waves and Electrons Section, Vol. 35, No. 10, October, pp. 1122–1136.

Bullington, K. (1972) "Unlocking the Secrets of Microwave Propagation," *Bell Lab. Research,* Vol. 50, No. 1, January, pp. 8–13.

Carlson, A.B., and A.T. Waterman, Jr. (1966) "Microwave Propagation over Mountain Diffraction Paths," *IEEE Trans. on Antennas and Propagation,* Vol. AP-14, No. 4, July, pp. 489–496.

Causebrook, J.H. (1978) "Medium-Wave Propagation in Built-up Areas," *Proc. IEE,* Vol. 125, No. 9, September, pp. 804–808.

C.C.I.R. (1978–1982) "Attenuation by Atmospheric Gases," Report 719-1.

Chang, Hsi-Tien (1971) "The Effect of Tropospheric Layer Structures on Long Range v.h.f. Radio Propagation," *IEEE Trans. on Antennas and Propagation,* Vol. AP-19, No. 6, November, pp. 751–756.

Chu, T.S. (1974) "Rain Induced Cross-Polarization at cm and mm Wavelength," *BSTJ,* Vol. 53, No. 8, October, pp. 1537–1580.

Chu, T.S. (1979) "Effects of Sandstorms on Microwave Propagation," *BSTJ,* Vol. 58, No. 2, February, pp. 549–555.

Clarke, R.H. (1968) "A Statistical Theory of Mobile Radio Reception," *BSTJ,* Vol. 47, No. 6, July–August, pp. 957–1000.

Cox, D.C. (1981) "Depolarization of Radio Waves by Atmospheric Hydrometeors in Earth—Space Paths: a Review." *Radio Science,* Vol. 16, No. 5, September–October, pp. 781–812.

Crane, R.K. (1980) "Prediction of Attenuation by Rain," *IEEE Trans. on Communication,* Vol. COM-28, No. 9, pt 2, September, pp. 1717–1733.

Crombie, D.D. (1960) "On the Mode Theory of Very Low Frequency Propagation in the Presence of a Transverse Magnetic Field," *J. Research NBS,* Vol. 64D, No. 5, May–June, pp. 265–267.

Crombie, D.D. (1963) "Nonreciprocity Propagation of v.l.f. Waves Along the Magnetic Equator," *Proc. IEEE,* Vol. 51, No. 4, April, pp. 617–618.

Crombie, D.D. (1967) "Survey: The Waveguide Mode Propagation of v.l.f. Radio Waves to Great Distances," Conference on M.F., L.F. and V.L.F. Radio Propagation, London, IEE Conference Publication No. 36, pp. 138–158.

David, P., and J. Voge (1969) *Propagation of Waves,* Pergamon Press, London.

Davies, K. (1966) *Ionospheric Radio Propagation,* Dover Publications, New York.

Davis, B.R., and R.E. Bogner (1985) "Propagation at 500 MHz for Mobile Radio," *Proc. IEE,* Vol. 132, Pt F, No. 5, August, pp. 307–320.

Doluchanow, M.P. (1956) *Die Ausbreitung von Funkwellen,* VEB Verlag Technik, Berlin.

Du Castel, P. (1956) "Tropospheric Radiowave Propagation Beyond the Horizon," Pergamon Press, Oxford, England.

Eckersley, T.L. (1932) "Radio Transmission Treated by Phase Integral Methods," *Proc. Royal Society,* Vol. 136.

Eckersley, T.L. (1937) "Ultra-Short-Wave Refraction and Diffraction," *J. IEE,* Vol. 80.

Fedi, F., and U. Bordoni (1981) "Prediction of Attenuation due to Rainfall on Terrestrial Links," *Radio Science,* Vol. 16, No. 5, September–October, pp. 731–743.

Fejer, J.A. (1961) "The Absorption of Short Radio Waves in the Ionospheric D and E Regions," *J. Atmos. Terr. Physics,* Vol. 25, December, pp. 260–274.

Flock, W.L., S.D. Slobin, and E.K. Smith (1982) "Propagation Effects on Radio Range and Noise in Earth-Space Telecommunications," *Radio Science,* Vol. 17, No. 6, November–December, pp. 1411–1424.

Fok, V.A., L.A. Vainshtein, and M.G. Belkina (1958) "Radiowave Propagation in Surface Tropospheric Ducts," *Radiotechnika i elektronika,* Vol. 3, No. 12, December, pp. 1411–1429.

Ford, L.H., and R. Oliver (1946) "An Experimental Investigation of the Reflection and Absorption of Radiation of 9-cm Wavelength," *Proc. Physics Society,* Vol. 58, pt 3, No. 327, May, pp. 265–280.

Gager, D.R. (1971) "Signal Fading Characteristics due to Reflection from the Sea," *Marconi Rev.,* Vol. 34, No. 180, pp. 27–49.

Galejs, J. (1971) "On the Propagation of Long Waves to Large Distances," *Proc. IEEE,* Vol. 59, No. 11, November, pp. 1635–1636.

Galejs, J. (1972) "Ionospheric Interaction of vlf Radio Waves," *J. Atmos. Terr. Physics,* Vol. 34, No. 3, March, pp. 421–436.

Gibbins, C.J. (1973) "Nomograms for Estimating Clear-Sky Zenith Atmospheric Attenuation in Range 80–130 GHz," *Electronic Letters,* Vol. 9, No. 26, December 27, pp. 605–607.

Heeran, M.P., and E.H. Carman (1973) "Transequatorial v.h.f. Transmission and Solar-Related Phenomena," *Australia J. Physics,* Vol. 26, No. 6, December, pp. 797–804.

Hogg, D.C. (1969) "Statistics on Attenuation of Microwaves by Intense Rain," *BSTJ,* Vol. 48, No. 9, November, pp. 2949–2962.

Howell, T.F., and J.R. Shakeshaft (1967) "Attenuation of Radio Waves by the Troposphere over the Frequency Range 0.4–10 GHz," *J. Atm. Terr. Physics,* Vol. 29, No. 12, December, pp. 1559–1571.

Hunsucker, R. (1983) "Anomalous Propagation Behavior of Radio Signals at High Latitudes," AGARD Conf. Proc., No. 332, Propagation Aspects of Frequency Sharing, Interference and System Diversity. 31st Symposium of the Electromagnetic Wave Propagation Panel, Issy-les-Moulineaux, France, October 8–22 1982. AGARD, pp. 11/1–15.

Ibrahim, M.F., and J.D. Parsons (1983) "Signal Strength Prediction in Built-up Areas. I. Medium Signal Strength," *Proc. IEE,* Pt F, Vol. 130, No. 5, August, pp. 377–384.

Ippolito, L.J. (1981) "Radio Propagation for Space Communication Systems," *Proc. IEEE,* Vol. 69, No. 6, June, pp. 697–727.

Ivanov, V.V. (1970) "Long-Range Tropospheric Propagation of Radio Waves," *Radio Engineering Electronic Physics,* Vol. 15, No. 6, June, pp. 1123–1130.

Jakes, W.C., Jr., (1974) *Microwave Mobile Communications,* John Wiley & Sons, New York.

Jeske, H., and K. Brooks (1986) "Comparison of Experiments on Duct Propagation Above the Sea with the Mode Theory of Booker and Walkinshaw." *Radio Science,* Vol. 1 (new series), No. 8, April, pp. 891–895.

Joos, G. (1944) *Theoretical Physics,* Blackie & Son Limited, London and Glasgow.

Jordan, E.C. (1953) *Electromagnetic Waves and Radiating Systems,* Constable & Co. Ltd., London.

Josephson, B., and A. Blomquist (1958) "The Influence of Moisture in the Ground, Temperature and Terrain on Ground Wave Propagation in the vhf Band," *IRE Trans. on Antennas and Propagation,* Vol. AP-6, No. 2, April, pp. 169–172.

Kamal, A.K., and P.F. Christofer (1981) "Communication at Millimeter Waves (Atmospheric Attenuation)," IEEE Int. Conf. on Communication, Denver, June 14–18, Vol. 4, p. 68.2/1–8.

Kerr, D.E. (1951) *Propagation of Short Radio Waves,* M.I.T. Radiation Lab. Series, Vol. 13, McGraw-Hill, New York.

Kline, M., and I.W. Kay (1965) *Electromagnetic Theory and Geometrical Optics,* Interscience Publishers, New York.

Kucherov, L.V. (1968) "Calculation of the Field Strength of Ultra Short Waves in the Presence of Superrefraction," *Radio Engineering Electronic Physics,* Vol. 11, No. 4, April, pp. 537–541.

Macchiarella, G. (1982) "Analysis of Rain Attenuation Models for Short Path," *Electronic Letters,* Vol. 18, No. 11, May 27, pp. 486–487.

Martin, H.G. (1961) "Wave-Guide Propagation of Very Low Frequency Radio Waves," *J. Atm. Terr. Physics,* Vol. 20, No. 2–3, March, pp. 206–209.

Matthews, P.A. (1965) *Radio Wave Propagation—UHF and Above,* Chapman & Hall, London.

Migliora, C.G., and L.B. Felsen (1982) "High-Frequency Propagation in an Elevated Tropospheric Duct," *IEEE Trans. on Antennas and Propagation,* Vol. AP-30, No. 8, November, pp. 1107–1120.

Mondloch, A.J. (1969) "Overwater Propagation of Millimeter Waves," *IEEE Trans. on Antennas and Propagation,* Vol. AP-17, No. 1, January, pp. 82–85.

Morecroft, J.H. (1933) *Principles of Radio Communication,* John Wiley & Sons, New York.

Muggleton, L.M. (1966) "Dependence of Ionospheric Absorption on Solar Zenith Angle," *Proc. IEE,* Vol. 113, No. 12, December, pp. 1909–1912.

N.B.S. (1948) "Ionospheric Radio Propagation," Circular 462.

Oguchi, T. (1960) "Attenuation of Electromagnetic Waves due to Rain with Distorted Raindrops," *J. Radio Research Lab.* [Tokyo], Vol. 7, No. 33, September, pp. 467–485.

Oguchi, T. (1983) "Electromagnetic Wave Propagation and Scattering in Rain and Other Hydrometeors," *Proc. IEEE,* Vol. 71, No. 9, September, pp. 1029–1078.

Ossanna, J.V., Jr. (1964) "A Model for Mobile Radio Fading due to Building Reflections: Theoretical and Experimental Fading Waveform Power Spectra," *BSTJ,* Vol. 43, No. 6, November, pp. 2935–2971.

Pappert, R.A., and C.L. Goodhart (1977) "Case Studies of Beyond-the-Horizon Propagation in Tropospheric Ducting Environments." *Radio Science,* Vol. 12, No. 1, January–February, pp. 75–87.

Pappert, R.A., and C.L. Goodhart (1979) "A Numerical Study of Tropospheric Ducting at HF," *Radio Science,* Vol. 14, No. 5, September–October, pp. 803–813.

Parsons, J.D., and M.F. Ibrahim (1983) "Signal Strength Prediction in Built-up Areas. II. Signal Variability," *Proc. IEE,* Pt F, Vol. 130, No. 5, August, pp. 385–391.

Pedersen, P.O. (1927) *The Propagation of Radio Waves,* Denmarks Naturvidenskabelige Samfung, Copenhagen.

Piggott, W.R. (1967) "Medium Wave Propagation at Night," Conference on M.F., L.F. and V.L.F. Radio Propagation, London.

Picquenard, A. (1974) *Radio Wave Propagation,* Macmillan Press Ltd, London.

Pressey, B.G., G.E. Ashwell, and J. K. Hargreaves (1961) "The Phase Variation of Very Low Frequency Waves Propagated over Long Distances," *Proc. IEE,* Vol. 108, Pt B, No. 38, March, pp. 214–226.

Reed, H.R., and C.M. Russell (1966) "Ultra-High Frequency Propagation," Chapman and Hall, London.

Rohan, P. (1948a) "Umísťování ultrakrátkovlnných stanic v terénu vzhledem k šíření vln," *Vestník Spoj. Vojska,* Vol. 1, No. 2.

Rohan, P. (1948b) "Nejnovější theorie o šíření ultrakrátkých vln," *Krátké vlny,* Vol. 7, No. 5, May.

Rohan, P. (1960) "The Reception and Exploitation of Very Low Frequency Standard Frequency Transmissions," W.R.E., Tech. Memo., E.I.D. 77, December.

Rohan, P. (1962a) "Frequency and Time Standardization." M.E. Thesis, The University of Adelaide, May.

Rohan, P. (1962b) "Radio Wave Propagation from the Frequency and Time Standardization Point of View," W.R.E., Tech. Memo., E.I.D. 127, August.

Rohan, P. (1963a) "Maximum Usable Frequencies and Propagation Delays over the Canberra—Salisbury and Salisbury—Range E Paths During 1962," W.R.E., Tech. Memo., E.I.D. 145, February.

Rohan, P. (1963b) "Formulae for the Duration of Direct Daylight and Their Application to vlf Propagation Research," W.R.E., Tech. Memo., P.A.D. No. 160, June.

Rohan, P. (1981) "Surveillance Radar Performance Assessment by Mathematical Modelling," PhD Thesis, University of Adelaide, May.

Rohan, P. (1983) *Surveillance Radar Performance Prediction,* IEE Electromagnetic Waves Series, Vol. 17, Peter Peregrinus Ltd., London.

Rohan, P., L.L. Anderson, and D.J. Cooke (1963a) "The Effect of the U.S. High Altitude Nuclear Explosion on Radiowave Propagation," *Nature,* Vol. 197, No. 4870, 2/3.

Rohan, P., L.L. Anderson, and D.J. Cooke (1963b) "Sudden Changes in Amplitude and Phase of the vlf Signal from Station GBR at Salisbury, South Australia," *Nature,* Vol. 197, No. 4869, February 23.

Rohan, P., L.L. Anderson, and D.J. Cooke (1963c) "Interfering Radio Signals on 18 kc/s Received in Australia and New Zealand," W.R.E., Tech. Note, P.A.D. No. 76, September.

Rotheram, S. (1973a) "Propagation Theory for the Evaporation Duct," paper presented at NATO Advanced Study Institute, Sorento.

Rotheram, S. (1973b) "Theory of Transhorizon Propagation in the Evaporation Duct," Conference on Propagation of Radio Waves at Frequencies above 10 GHz., London, April 10–13, pp. 70–75.

Rotheram, S. (1983) "Microwave Duct Propagation over the Sea," Third Int. Conf. on Antennas and Propagation, ICAP 83, Norwich, England, April 12–15, pp. 9–13, pt 2.

Rugg, D.E. (1967) "Theoretical Investigation of the Diurnal Phase and Amplitude Variations of vlf Signals," *Radio Science,* Vol. 2, No. 6, June, pp. 551–556.

Ryde, J.W. (1946) "The Attenuation and Radar Echoes Produced at Centimetre Wave-Lengths by Various Meteorological Phenomena," report of Conf. on Meteorological Factors in Radio-Wave Propagation, The Physical Society, London.

Saxton, J.A. (1970) "Radio and Weather," *Proc. IEE,* Vol. 117, No. 1, January, pp 9–15.

Schiavone, J.A. (1981) "Meteorological Parameters Affecting Line-of-Sight Microwave Fading," IEEE Int. Conf. on Communication, Denver, June 14–18, pp. 68.5/1–3.

Sengupta, N. (1983) "Sunrise and Sunset Effects as Observed on Microwave Propagation over Line-of-Sight Path," *IEEE Trans. on Antennas and Propagation,* Vol. AP-33, No. 4, April, pp. 441–445.

Sengupta, N., and M.J. Das Gupta (1984) "Effect of Thunderstorm on Microwave Line-of-Sight Propagation," *IEEE Trans. on Antennas and Propagation,* Vol. AP-32, No. 2, February, pp. 196–200.

Sengupta, P.R. (1964) "Ionospheric Effects of High-Altitude Nuclear Explosions," *J. Atm. Terr. Physics,* Vol. 26, No. 9, September.

Shlionskiy, A.G. (1985) "Forecasting of Duct Radiowave Propagation," *Telecommunication and Radio Engineering,* Pt 2, Vol. 40, No. 4, April, pp. 112–114.

Skolnik, M.I. (1970) *Radar Handbook,* McGraw-Hill, New York.

Smith, A.A., and L.B. Soden (1974) "Field Study of Ducting and Tropospheric Radar Propagation Around Darwin, Australia," *Proc. IREE* [Australia], Vol. 35, No. 12, December.

Sofaer, E., and C.P. Bell (1966) "Factors Affecting the Propagation and Reception of Broadcasting Signals in the u.h.f. Bands," *Proc. IEE,* Vol. 113, No. 7, July, pp. 1133–1140.

Starr, A.T. (1953) *Radio and Radar Techniques,* Sir Isaac Pitman & Sons, Ltd., London.

Stratton, J.A. (1941) *Electromagnetic Theory,* McGraw-Hill, New York.

Suzuki, H. (1977) "A Statistical Model for Urban Radio Propagation," *IEEE Trans. on Communication,* Vol. COM-25, No. 7, July, pp. 673–680.

Taylor, W.L. (1960) "Daytime Attenuation Rates in the Very Low Frequency Band Using Atmospherics." *J. of Research, N.B.S.,* Vol. 64D, No. 4, July–August, pp. 349–355.

Van Vleck, J. H. (1947a) "The Absorption of Microwaves by Oxygen," *Physics Rev.,* Vol. 71, No. 7, April, 413–424.

Van Vleck, J.H. (1947b) "The Absorption of Microwave by Uncondensed Water Vapor," *Physics Rev.* Vol. 71, No. 7, April, pp. 425–433.

Volland, H. (1961) "Comparison Between Mode Theory and Ray Theory of vlf Propagation," *J. Research N.B.S.,* Vol. 65D, No. 4, July–August, pp. 357–381.

Wait, J.R. (1959) "Diurnal Change of Ionospheric Heights Deduced from Phase Velocity Measurements at vlf," *Proc. IRE,* Vol. 47, No. 5, May, p. 998.

Wait, J.R. (1960) "Terrestrial Propagation of Very-Low Frequency Radio Waves," *J. of Research N.B.S.,* Vol. 64D, No. 2, March–April, pp. 153–204.

Wait, J.R. (1961) "A New Approach to the Mode Theory of v.l.f. Propagation," *J. Research N.B.S.,* Vol. 65D, No. 1, January–February, pp. 37–46.

Wait, J.R. (1969) "Reflection from Subrefractive Layers," *Electronic Letters,* Vol. 5, No. 4, February, pp. 64–65.

Wait, J.R., and H.H. Howe "The Waveguide Mode Theory of vlf Ionospheric Propagation, *Proc. IEE,* Vol. 45, No. 1, January, p. 95.

Wang, J.C.R. (1977) "Prediction of Medium Frequency Skywave Field Strength in North America," *IEEE Trans. on Broadcasting,* Vol. BC-23, No. 2, June, pp. 43–49; also in *Proc. IEEE,* Vol. 65, No. 7, July 1976, pp. 1085–1086.

Weiss, P. (1936) "Poruchy při příjmu krátkých vln," *Radioamater,* Vol. 15, No. 1, January.

Westfall, W.D. (1967) "Diurnal Changes of Phase and Group Velocity of vlf Radio Waves," *Radio Science,* Vol. 2, No. 1, January, pp. 119–125.

Yeh, K.C., and C.H. Liu (1982) "Radio Wave Scintillation in the Ionosphere," *Proc. IEEE,* Vol. 70, No. 4, April, pp. 324–360.

Young, R., Jr. (1952) "Comparison of Mobile Radio Transmission at 150, 450, 900 and 3700 Mc," *BSTJ,* Vol. 31, No. 6, November, pp. 1068–1085.

Ziem, G. (1960) "Empfang und Peilung elektromagnetischer Wellen in Seewasser," *Telefunken Zeitung,* Vol. 33, No. 128, June, pp. 141–150.

Chapter 6
NOISE AND THE SIGNAL-TO-NOISE RATIO

6.1 INTRODUCTION

The preceding chapters dealt with the behavior of electromagnetic waves of the various frequencies used in practice and considered the field intensity variations with distance. Although the knowledge of this quantity is important, it is not the only quantity that determines the usability of a signal because a receiver also receives unwanted signals originating either in transmitters operating at frequencies adjacent to the wanted one or in atmospherics and electrical equipment installed at sites neighboring the receiver's location. The received unwanted signals, designated by the cumulative name of *noise,* add to the noise generated within the receiver.

It is well known that we have to raise our voices to conduct a conversation in a noisy environment; by doing this, we increase the signal-to-noise ratio. Similarly, in radio communication the wanted signal level, generally, must exceed the total interfering signal's level, or noise, if reliable communication is to be maintained. The total noise (i.e., that generated within the receiver and the received one) determines the lowest useful signal level required for achieving the signal-to-noise ratio dictated by the requirements of the various services for reliable error-free operation.

The signal-to-noise ratio, a numerical criterion of circuit performance, then is a very important concept that deserves some attention. Because signals can be amplified to the required level, determined by the level of the ambient noise, nature of the signal, and the required signal-to-noise ratio, noise will be discussed to some extent in the following text.

Noise is a cumulative name for unwanted disturbances within a frequency band. It may originate in any of various sources. Radio frequency noise, generally, is random in character and contains uniform energy components extending over the whole frequency range when it is termed *white noise.* Because communication networks are restricted in bandwidth, the noise in communication circuits is band-limited white noise or colored noise.

Although there are other types of noise, for example, impulse noise consisting of nonoverlapping switching transients due to spark discharges in switches, commutators, or motorcar ignitions, mainly radio frequency noise is of interest here. Such noise may originate in the communication equipment itself or it may be of external origin. This noise may be artificial or due to natural phenomena, like lightning discharges.

6.2 EQUIPMENT NOISE

Random motion of electrons in the resistive components of receiver or amplifier input circuits causes noise to appear at the output without any external influence. The generated noise voltage depends on the value of the resistance, the temperature, and the bandwidth of the network following it. Its mean square value may be expressed by [Rohan, 1964]

$$\bar{e}^2 = 4kT_0BR \tag{6.1}$$

where

k = Boltzmann's constant (1.38×10^{-23} J/K),
T_0 = the absolute temperature of the resistor (usually taken to be 290 K or about 17°C);
B = the equivalent noise bandwidth of the receiver or amplifier, that is, the width of an ideal receiver having a rectangular bandpass characteristic of the same area and peak value as the power-gain-frequency response of the used receiver; and
R = the value of the resistor in Ω.

For theoretical noise considerations we can represent the resistor by a noiseless resistor of the same value and a zero impedance noise generator supplying the above noise voltage.

Maximum noise power will be delivered to a load resistor matched in value to the noise generator's output impedance, or

$$N_{max} = N_{in} = \bar{e}^2/4R = kT_0B \tag{6.2}$$

This maximum available input noise power appears at the output multiplied by the gain G of the receiver or amplifier augmented by the noise contribution of the network of the receiver or amplifier; that is,

$$N_{out} = kT_0BG + \text{network noise } N$$

Maximum signal input power is delivered to a matched load by a signal source of power S_{in}. The output signal power is GS_{in}, so that the ratio of the input to the output signal-to-noise ratio becomes

$$F = \frac{S_{in}/N_{in}}{GS_{in}/N_{out}} = \frac{N_{out}}{N_{in}}\frac{S_{in}}{GS_{in}} = \frac{N_{out}}{kT_0B}\frac{1}{G} \tag{6.3}$$

from which the output noise power

$$N_{out} = FGkT_0B \tag{6.4}$$

where F stands for the receiver's noise factor as defined by (6.3).

The receiver or amplifier's noise contribution is

$$N = N_{out} - kT_0BG = (F - 1)kT_0BG \tag{6.5}$$

The noise factor is usually expressed in dB when it is called the *noise figure.* Then, on the basis of (6.4),

$$F_{dB} = 10 \log \frac{N_{out}}{kT_0BG} \tag{6.6}$$

Example 6.1

A microwave receiver having an overall gain of 100 dB and a 10-dB noise figure is to receive an 8-MHz-wide television channel and a 1.2-MHz-wide telephony transmission. Determine the minimum input power required for a 40-dB output signal-to-noise ratio.

Solution: The input signal-to-noise ratio, in terms of the noise factor, is given by expression (6.3), as

$$\frac{S_{in}}{N_{in}} = F\frac{S_{out}}{N_{out}}$$

or, given the quantities in dB,

$$\frac{S_{in}}{N_{in}} = 10 + 40 = 50 \text{ dB}$$

which corresponds to a 10^5 power ratio.

The input noise power when the required minimum bandwidth, the sum of the television and telephony bandwidths, is 9.2 MHz, becomes

$$kTB = 1.38 \times 10^{-23} \times 290 \times 9.2 \times 10^{6}$$

or

$$N_{\text{in}} = 3.648 \times 10^{-14} \text{ W}$$

The minimum input power then becomes

$$S_{\text{in}} = N_{\text{in}}FS_{\text{out}}/N_{\text{out}} = 3.648 \times 10^{-9} \text{ W}$$

When a single stage amplifier of gain G_1 and noise factor F_1 is followed by a second stage having a gain G_2 and noise factor F_2, then the available noise power output of this stage is the noise power output of the first stage; that is, the input of the second stage, given by (6.4), amplified by the second stage, together with its noise contribution. This, according to (6.5), is

$$(F_2 - 1)kT_0BG_2$$

so that the noise output power is

$$FG_1G_2kT_0B = F_1G_1G_2kT_0B + (F_2 - 1)kT_0BG_2 \tag{6.7}$$

and the noise factor of the two amplifiers in series is

$$F = F_1 + \frac{F_2 - 1}{G_1} \tag{6.8}$$

A receiver or amplifier consists, usually, of several stages in cascade, each with its noise factor and gain. The total noise power output of such a cascade of stages having a noise factor F, on the pattern of (6.7), is

$$FG_1G_2 \cdots G_nkT_0B = F_1G_1G_2 \cdots G_nkT_0B + (F_2 - 1)G_2G_3 \cdots G_nkT_0B + \cdots (F_n - 1)G_nkT_0B$$

from which the noise factor of the cascade

$$F = F_1 + \frac{F_2 - 1}{G_1} + \frac{F_3 - 1}{G_1G_2} + \cdots + \frac{F_n - 1}{G_1G_2 \cdots G_{n-1}} \tag{6.9}$$

shows that the main contributor to the noise output of the cascade of stages is the first stage of the chain. In a receiver the first stage has to be designed carefully to reduce, as far as possible, the receiver's noise contribution.

Example 6.2

Two amplifiers of equal bandwidths, having gains of 30 or 20 dB and noise figures of 4 and 6 dB, are to be connected in series. What will be the contribution of the second amplifier to the total noise figure?

Solution: According to expression (6.8) the total noise factor is

$$F = F_1 + (F_2 - 1)/G_1$$

when F_1 and F_2 are the noise factors of the amplifiers and G_1 is the gain of the first one.

Inserting the given quantities leads to

$$F = 2.51 + (3.98 - 1)/1000 = 2.51298$$

or

$$F = 4.00189 \text{ dB}$$

showing that the contribution of the second amplifier to the total noise figure is negligible.

6.3 EXTERNAL NOISE

As indicated earlier, noise of origin external to the receiver may be either natural or artificial.

6.3.1 Natural Noise

6.3.1.1 Thermal Noise and Noise Temperature

The noise power delivered by the antenna to the receiver is thermal in origin, as has been proven experimentally. It depends on the thermal radiation of the surrounding objects; and, with highly directional antennas, it depends on the elevation angle of the antenna's beam. The noise power is highest when the antenna points toward the horizon because of noise of terrestrial origin. The antenna's noise contribution may be defined by its equivalent noise temperature.

The relationship between temperature and noise may be derived by considering the emissive power of a black body for unpolarized radiation in a bandwidth B, as given by Planck's radiation law [Joos, 1934]

$$\frac{E^2}{2Z_0} = \frac{P}{A} = \frac{2kf^3B}{c^2\,(e^{hf/kT} - 1)}$$

which for small values of hf/kT, for which

$$e^{hf/kT} - 1 = 1 + \frac{hf/kT}{1!} + \frac{(hf/kT)^2}{2!} + \cdots - 1 \approx \frac{hf}{kT}$$

and due to $c = \lambda f$ reads

$$\frac{E^2}{2Z_0} = \frac{P}{A} = \frac{2k^2TB}{\lambda^2 h}$$

from which we have

$$T = \frac{hP\lambda^2}{2k^2AB}$$

where

T = the noise temperature in kelvins;
P = radiation density in W/m^2;
λ = wavelength in m;
f = frequency in Hz;
A = solid radiation angle or the directivity of the antenna in steradians (4π for an isotropic radiator);
B = bandwidth in Hz;
h = Planck's constant (6.625×10^{-34} joule second);
k = Boltzmann's constant;
E = electric field intensity in V/m; and
Z_0 = characteristic impedance of free space ($120\pi\Omega$).

Considering the antenna to be a generator matched to the input impedance of the receiver of bandwidth B, the noise power delivered by the antenna to the receiver, on the basis of expression (6.2), is

$$N_{ext} = kT_aB \tag{6.10}$$

so that the antenna's noise temperature becomes

$$T_a = \frac{N_{ext}}{kB}$$

The total noise power of a receiver referred to its *input* as the sum of the external noise power, as outlined, and the internal noise power contribution (i.e., the noise power generated by the receiver referred to its *input terminals*), given by

$$N_{int} = (F - 1)kT_0B \tag{6.11}$$

The total input noise power then is

$$N_{tot} = N_{ext} + N_{int}$$

where the right-hand side terms were given by (6.10) and (6.11). Hence,

$$\begin{aligned} N_{tot} &= kT_aB + (F - 1)kT_0B \\ &= kT_0B\left(\frac{T_a}{T_0} + F - 1\right) \end{aligned} \tag{6.12}$$

T_0 usually is taken to be 290 K corresponding to room temperature or 17°C. The expression in parentheses often is referred to as the *operating noise factor* or *operating noise figure,* if expressed in dB.

A network connected to a generator at a reference temperature T_0 has a noise factor F and gain G; the available noise power at its output is FGT_0, which referred to its input is FT_0. The total noise power at the input terminals consists of T_0 due to the noise temperature T_0 supplied by the generator and $T_e = (F - 1)T_0$, the effective noise temperature contribution of the network.

On the basis of expression (6.9), the effective input noise temperature of a cascade of n networks is

$$T_{e_{1,2,\cdots,n}} = T_{e_1} + \frac{T_{e_2}}{G_1} + \frac{T_{e_3}}{G_1G_2} + \cdots + \frac{T_{e_n}}{G_1G_2 \cdots G_{n-1}} \tag{6.13}$$

Because the sky's temperature differs from the ambient temperature of the atmosphere near the receiving antenna, an antenna pointing toward the sky introduces into the receiver's input noise proportional to the ratio of the sky's temperature and that of the prevailing ambient one.

The minimum signal power required for a specified *signal-to-noise ratio* (S/N), based on expression (6.12), is [Rohan, 1964]

$$S_{\min} = kT_0B\left(\frac{T_a}{T_0} + F - 1\right)\frac{S}{N} \tag{6.14}$$

with T_a the noise temperature of the antenna and hence of that part of the sky at which the antenna is pointed.

The approximate limits of the sky temperature as a function of frequency are shown by Figure 6.1, and the distribution of celestial noise temperatures by Figure 6.2. Formulas for the conversion of sky noise temperatures from those at the measurement frequency to those at other frequencies, as given by Menzel [1960], are

$$dB_f = dB_{f_1} - k_1 \log \frac{200}{f_1} + k \log \frac{200}{f}$$

with $dB_f = 10 \log T_f$; f_1 and k_1 refer to the frequency of the background measurements given in the relevant illustrations, and f and k to the frequencies of interest.

The values of the coefficients are

$$k, k_1 = 2 \quad \text{when } f < 200 \text{ MHz}$$

$$k, k_1 = 2.7\left[35 - dB_{f_1} + k_1 \log \frac{200}{f_1}\right] \quad \text{for } f > 200 \text{ MHz}$$

The intensity of thermal noise caused by the sun varies to a great extent, especially at frequencies above about 30 MHz and during years of high solar activity. The contribution of the moon and planets is negligible compared to that of the sun.

At frequencies above about 0.1 GHz, the ratio $T_a/T_0 \approx 1$, so that expression (6.14) reduces to

$$S_{\min} = kT_0BFS/N \tag{6.15}$$

Assuming then that in space, at the center of a sphere of radius R, is an isotropic source of radiation of power P, the power density on the surface of the sphere is $P/4\pi R^2$; and at the area toward which an antenna of directional gain G_T is pointed it is $PG_T/4\pi R^2$.

An antenna of an effective area A located at the illuminated part of the sphere's surface will intercept the power:

$$P_R = \frac{PG_TA}{4\pi R^2} \tag{6.16}$$

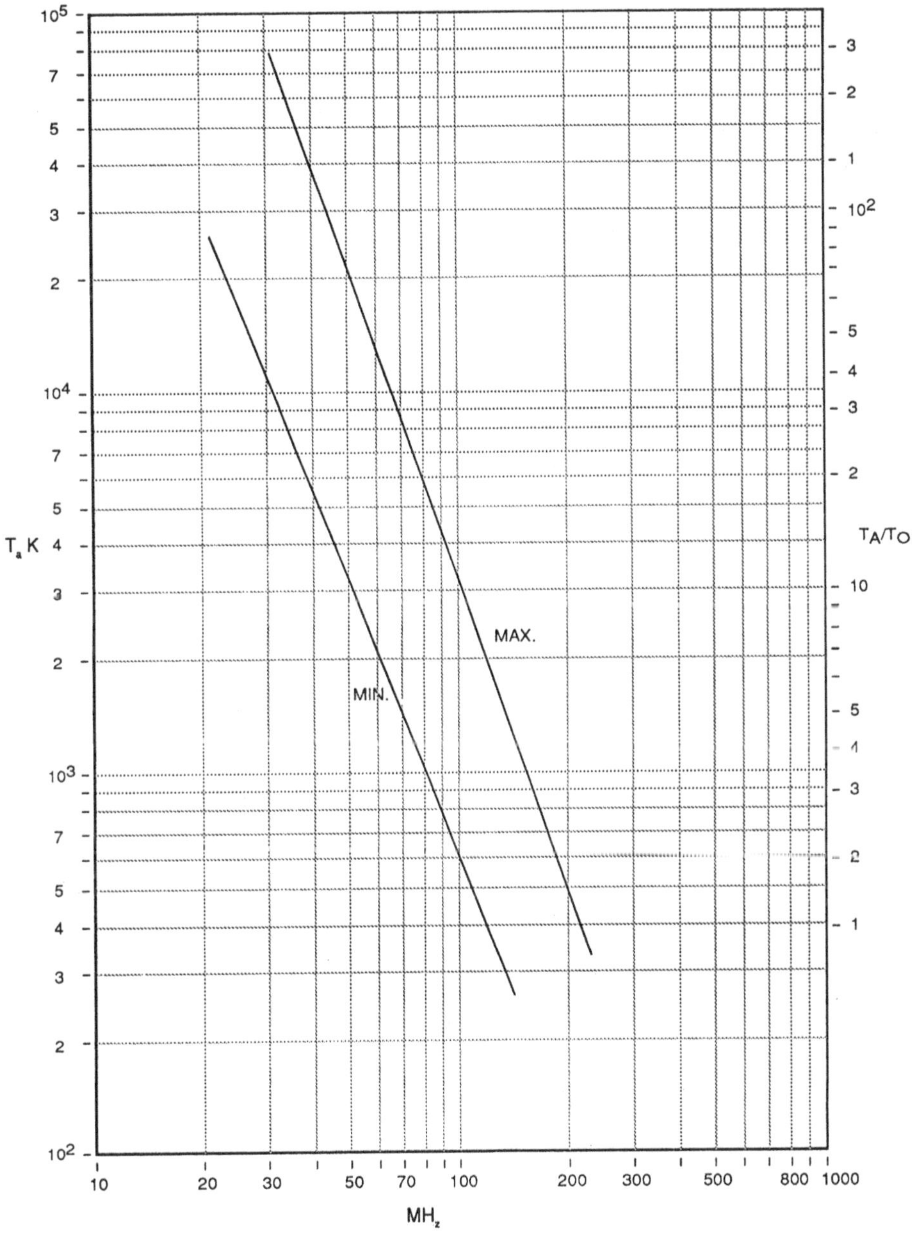

Figure 6.1 Range of noise temperatures T_a of an antenna pointing to the sky as a function of frequency (T_0 = 290 K) (after Rohan, 1964, © DRCS).

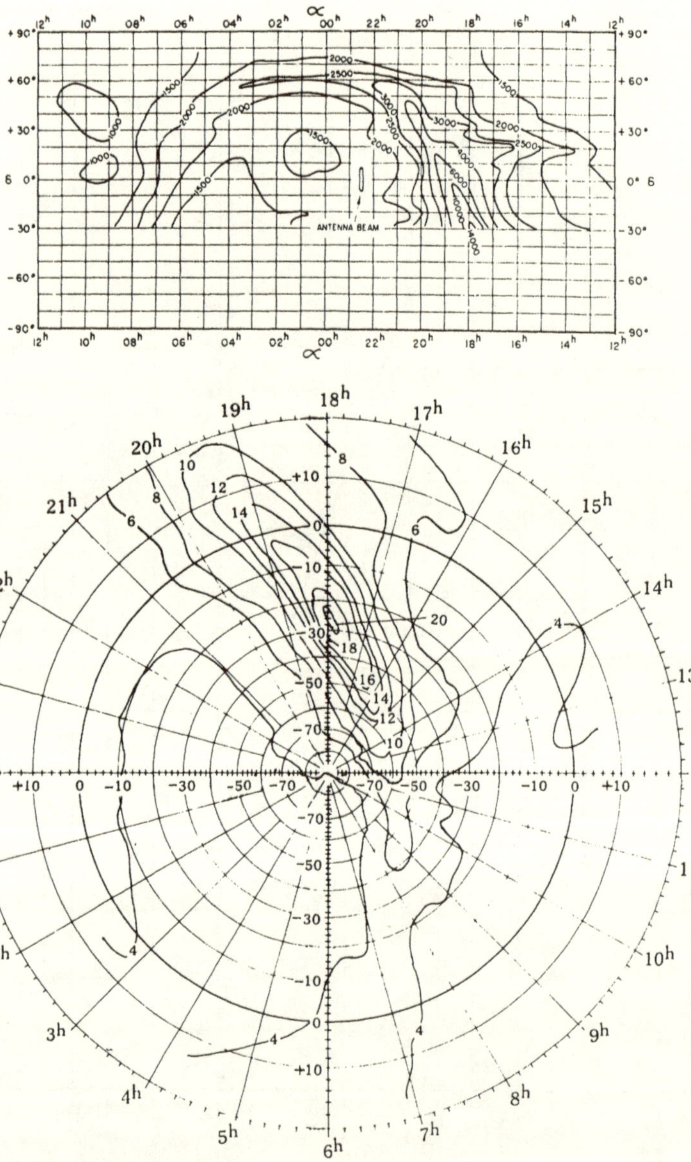

Figure 6.2 Isophotal maps: (a) northern sky at 81 MHz (after Ko, 1958, © 1958 IRE); (b) southern sky at 55 MHz, contours labelled in 1000 K units (after Rohan and Soden, 1970, © *Aust. J. Physics*

which with the area in terms of the receiving antenna's gain:

$$A = \frac{G_R \lambda^2}{4\pi}$$

becomes

$$P_R = \frac{P G_T G_R \lambda^2}{(4\pi R)^2} \tag{6.17}$$

Equating this to (6.14) yields the transmitted power required to produce a signal power for reception at levels specified by the required signal-to-noise ratio as

$$P = \frac{(4\pi R)^2 k T_0 B (T_a/T_0 + F - 1) S/N}{G_T G_R \lambda^2} \tag{6.18}$$

or

$$P = \frac{(4\pi R)^2 k T_0 B F \; S/N}{G_T G_R \lambda^2} \tag{6.19}$$

if the transmission is at a frequency above 0.1 GHz.

We must realize, however, that the last two expressions do not consider losses. The computed powers therefore must be increased to account for these.

Under certain conditions very high-frequency radiations may be absorbed by the ionosphere, which acts then as a source of thermal radiation. The ionospheric absorption is inversely proportional to the square of the frequency and decreases rapidly with rising frequencies.

Example 6.3

Modify expression (6.19) to make it applicable to radar.

Solution: The given formula was developed from expression (6.17) derived on the basis of power density considerations. In the radar case, we consider a target of radar echoing area σ m^2 in the location of the considered receiving antenna. The target intercepted power σP is reradiated toward the radar (receiving) antenna at a distance R where the power density, due to the reradiated power, is $\sigma P/4\pi R^2$.

Expression (6.17) then becomes

$$P_R = \frac{\sigma P G_T G_R \lambda^2}{(4\pi)^3 R^4}$$

which, in agreement with the considerations outlined in the text, leads to

$$P = \frac{(4\pi)^3 R^4 k T_0 B F S/N}{\sigma G_T G_R \lambda^2}$$

6.3.1.2 *Atmospheric Noise*

Noise experienced at frequencies below about 30 MHz is predominantly atmospheric noise caused by lightning discharges during thunderstorms. We generally refer to this noise as *atmospherics.*

About 50,000 thunderstorms occur daily in the world, causing, on the average, 100 lightning strokes per second. A typical lightning discharge radiates an energy of about 5×10^4 J at peak currents around 50,000 A [Picquenard, 1974]. A large part of the energy at very low frequencies peaks at a frequency near 10 kHz. It reaches the receiver, like other radio frequency radiation, by all the previously described mechanisms of propagation. Electromagnetic waves generated by lightning discharges are vertically polarized with field strengths of several V/m at short distances. The discharges are pulses of 0.1 to 3 ms duration, which produce continuous frequency spectra. The amplitude of the frequency components above about 10 to 20 kHz decreases with approximately the square of a rising frequency.

Statistical data collected over long periods of time show that lightning discharges are present continuously within the ionosphere-ground resonator. The low-frequency components of the generated disturbances, which are in the extreme and very low-frequency range, travel within the *D*-region–ground waveguide very long distances with very little attenuation. Hence, atmospherics, being most pronounced in the extreme low-frequency range extending to about 3 kHz, make this unsuitable for communication.

Atmospherics at higher frequencies, for example, in the broadcast band, are caused mainly by local storms. At the receiver, the energy level due to atmospherics is greatest between sunset and midnight and tapers off to a broad minimum at sunrise.

Several investigators studied the electromagnetic energy spectrum radiated by lightning discharges. Their findings were summarized in graphic form by Oh [1969]. The curve in Figure 6.3, based on Oh's curve, shows the estimated average level of spectral amplitude distribution of lightning discharges or atomspherics in dB above 1 μV/m at a 1 kHz receiver *bandwidth* (BW) and a distance of 1 km. The field-strength variation is assumed to obey the inverse distance law. The peak response of the receiver at less than 10 kHz bandwidth is taken to be directly proportional to the bandwidth, whereas at a greater than 10 kHz bandwidth it is assumed to be directly proportional to its square root.

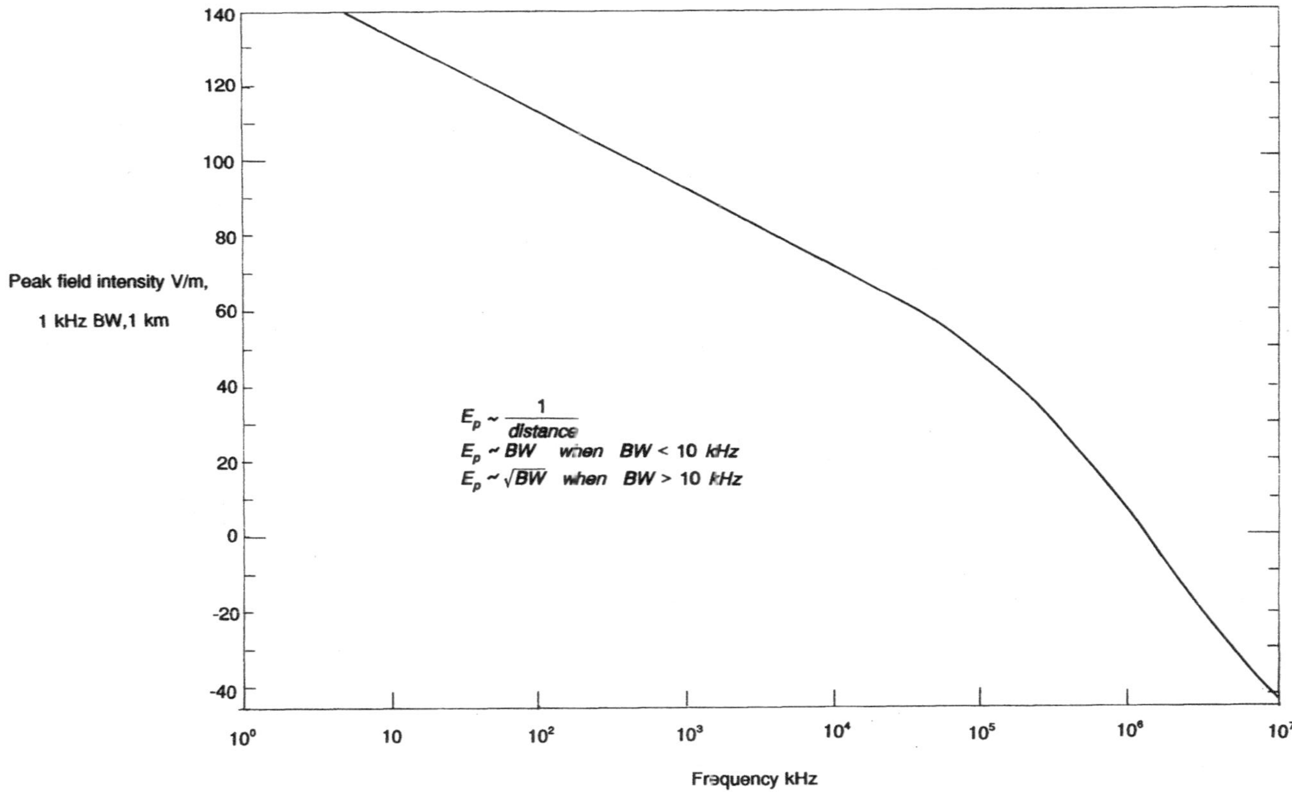

Figure 6.3 Spectral amplitude distribution of a lightning discharge (after Oh, 1969, © IEEE).

Figure 6.3 intimates that the spectral amplitude distribution of the energy radiated by lightning discharges at frequencies between 1 kHz and 50 MHz is inversely proportional to the frequency and to approximately $\sqrt{f^5}$ between 50 and 500 MHz. At 10 kHz, the peak field intensity, normalized to lightning to observer distance of 1 km when the receiver bandwidth is 1 kHz, is about 135 dB above 1 μV/m. The peak field intensity in dB above 1 μV/m is approximately 95 dB at 1 MHz, 45 dB at 100 MHz, 10 dB at 1 GHz, and -40 dB at 10 GHz. The given field intensity data are to be taken to be only approximate because it is very difficult to obtain the precise distance to the location of the lightning discharge.

The frequency of thunderstorm activity depends on the time of day, the season, and the geographic location. Most thunderstorms occur in the equatorial belt. Maps of the expected values of atmospheric radio noise in dB above kT_0B at a frequency of 1 MHz, together with graphs of the frequency dependence of radio noise and data on noise variability and character for six 4-hr intervals of a day during the four seasons of a year have been published by the CCIR.

The curves, samples of which are shown in Figures 6.4(a) and 6.5(a), permit the determination of the noise level at any geographic location on the earth. This in turn allows us to compute the required signal level for a specified signal-to-noise ratio.

The curves in Figure 6.4(b) and 6.5(b), showing the noise level variation with frequency, extend to only about 10 MHz because, at higher frequencies at many of the data gathering stations, noise of galactic origin exceeded that of atmospherics and the derivation of estimates on variability of atmospheric noise alone was not practicable.

The nomogram in Figure 6.6 converts the median effective antenna noise factor E_n resulting from external noise power available in a loss-free antenna into rms noise field strength along the antenna as a function of frequency. It solves the equation [CCIR, 1964]:

$$E_n = F_a + 20 \log f_{\mathrm{MHz}} \cdot 65.5 \quad \mu\mathrm{V/m} \tag{6.20}$$

Atmospheric noise, a random quantity, is treated in theoretical investigations on a statistical basis. It obeys a log-normal distribution [Picquenard, 1974], the same as the signal's median value. However, in atmospheric propagation, the instantaneous signal value varies around an hourly median value according to a Rayleigh distribution. The change in signal level required to maintain the signal-to-noise ratio above a level specified by the grade of service for a given percentage of time dictated by the time availability is the signal fluctuation margin. This is determined on the basis of data given in Figures 6.4(c) and 6.5(c).

The standard deviation of the ratio of the hourly median values of signal-to-noise is

$$\sigma = \frac{(D_u^2 + D_s^2)^{1/2}}{1.3} \tag{6.21}$$

where D_u, read from the preceding figures, is the maximum deviation of F_{am}, the median value of hourly values of the equivalent antenna noise factor due to the external noise power F_a available from a loss-free antenna over 90% time availability. D_s is the value of the received signal power exceeded for 90% of the time, in dB below the median value of the day-to-day variations in the hourly median. It is zero for a ground wave and 7 dB for an ionospheric one.

The fluctuation margin C, a protection factor necessary to ensure the required carrier-to-noise ratio for a given percentage of time within the time block for this computed standard deviation, is obtained either from Figure 6.7, when a single receiver is used, or Figure 6.8, for diversity reception, by interpolating the value of σ in the given curves and entering the complement to unity of the time availability on the horizontal axis.

Because the computed values of F_{am} and C are mean values, another protection factor to provide the required carrier-to-noise ratio for 90% of the time block has to be introduced if the service probability is to exceed 50%. This factor, known as the *uncertainty margin* C_u, is obtained from

$$\sigma_{C_u} = (\sigma_{D_u}^2 + \sigma_{D_s}^2)^{1/2} \tag{6.22}$$

where σ_{D_u}, from Figure 6.4(c) or 6.5(c), is the standard deviation of D_u, the ratio of the upper decile to median value of F_{am}, and σ_{D_s} is zero for a ground wave and 1.5 dB for an ionospheric one.

The nomogram of Figure 6.9 yields the value of σ_C, the standard deviation of C and the uncertainty margin for known values of the percentage of time availability p, on the left-hand scale, and σ_{C_u} or σ_T, the total standard deviation or total uncertainty. The total uncertainty σ_T, assuming that the errors are not correlated, is

$$\sigma_T = (\sigma_C^2 + \sigma_{F_{am}}^2 + \sigma_p^2 + \cdots)^{1/2} \tag{6.23}$$

with σ_p, in the range of 2 to 5 dB, expressing the quadratic average error in the calculation of the received power. The application of this data is demonstrated by Example 6.4.

In addition, noise at frequencies below about 40 kHz may be caused by whistlers, that is, lightning energy guided by the earth's magnetic lines of force half-way around the earth with an energy spectrum extending from a few Hz to tens of MHz.

a

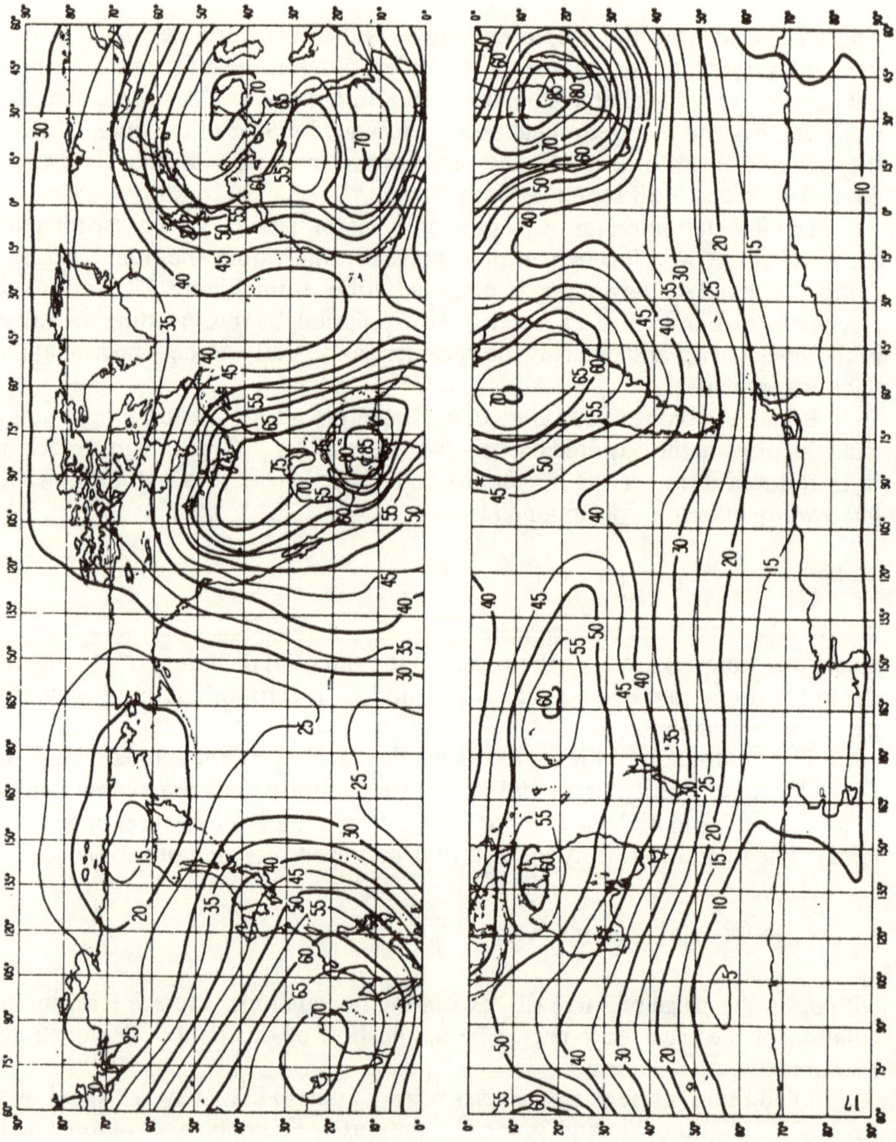

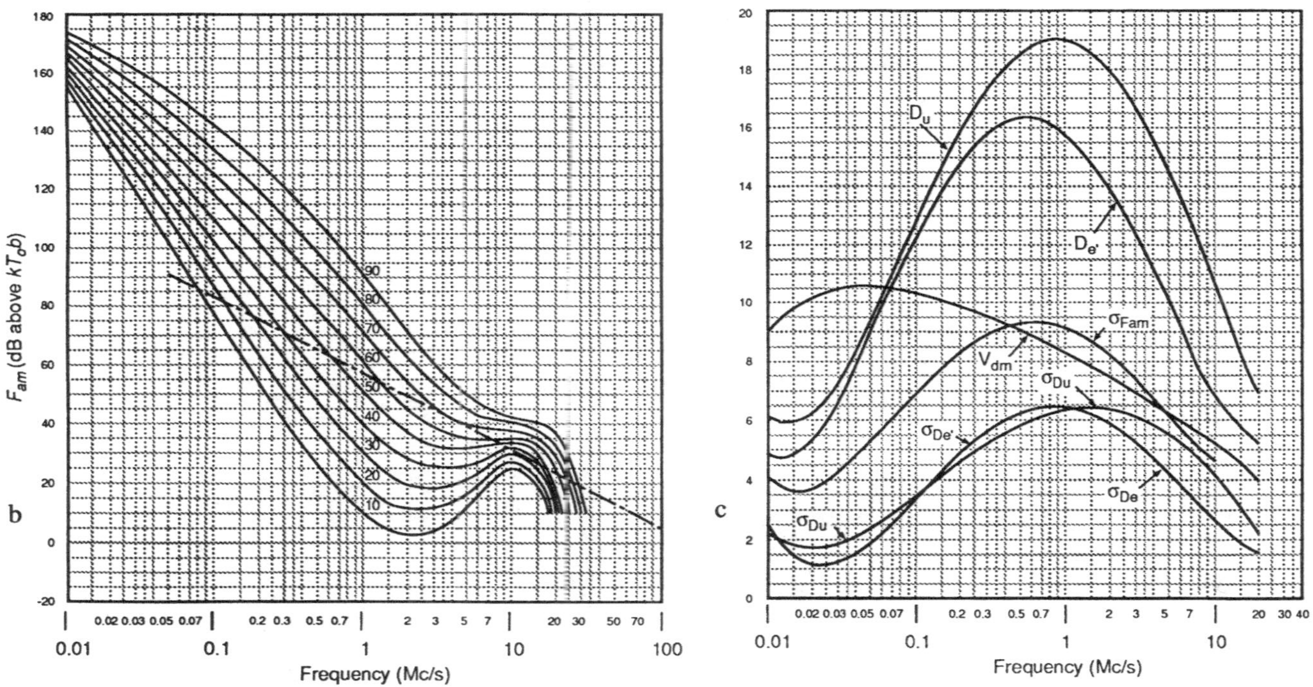

b c

Figure 6.4 Atmospheric noise between noon and 4 P.M. during a northern summer: (a) expected values of atmospheric radio noise F_{am} in dB above kT_0B at 1 MHz; (b) frequency dependence of radio noise; (c) data on noise variability and character (after CCIR, 1964).

a

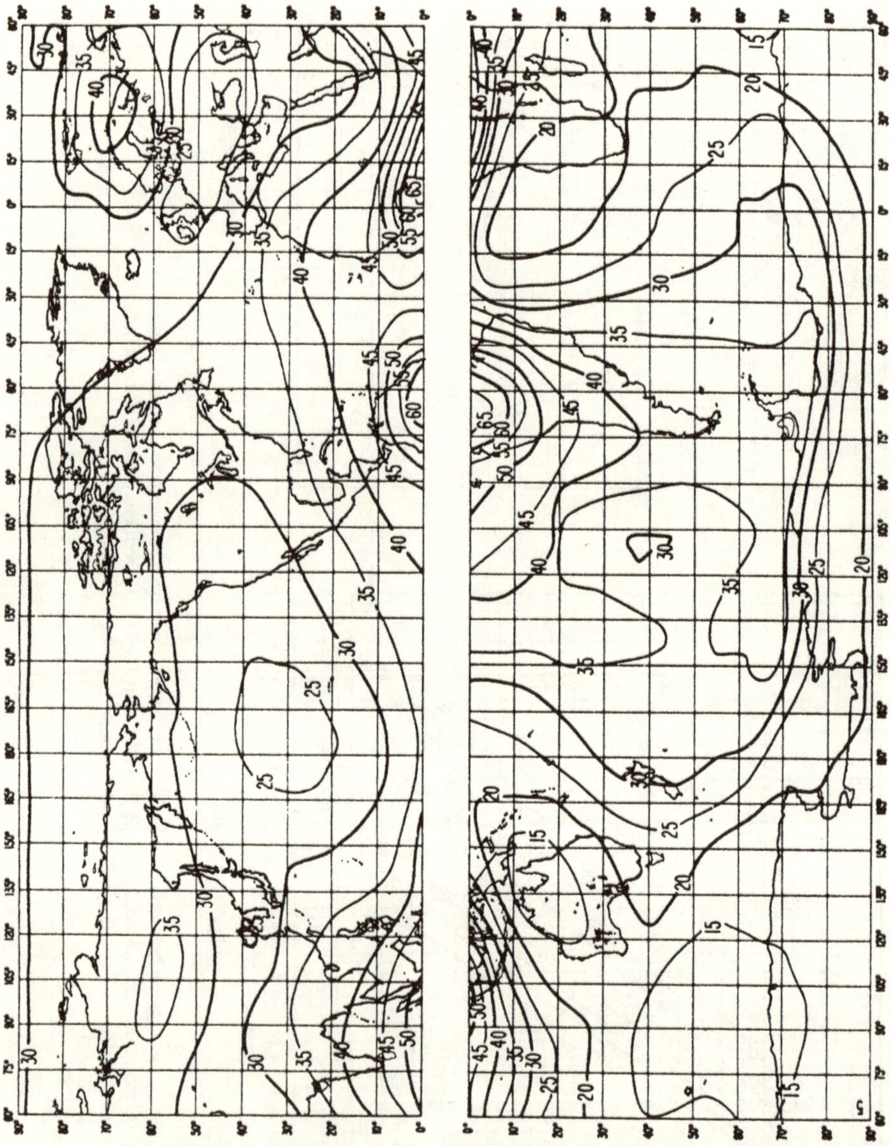

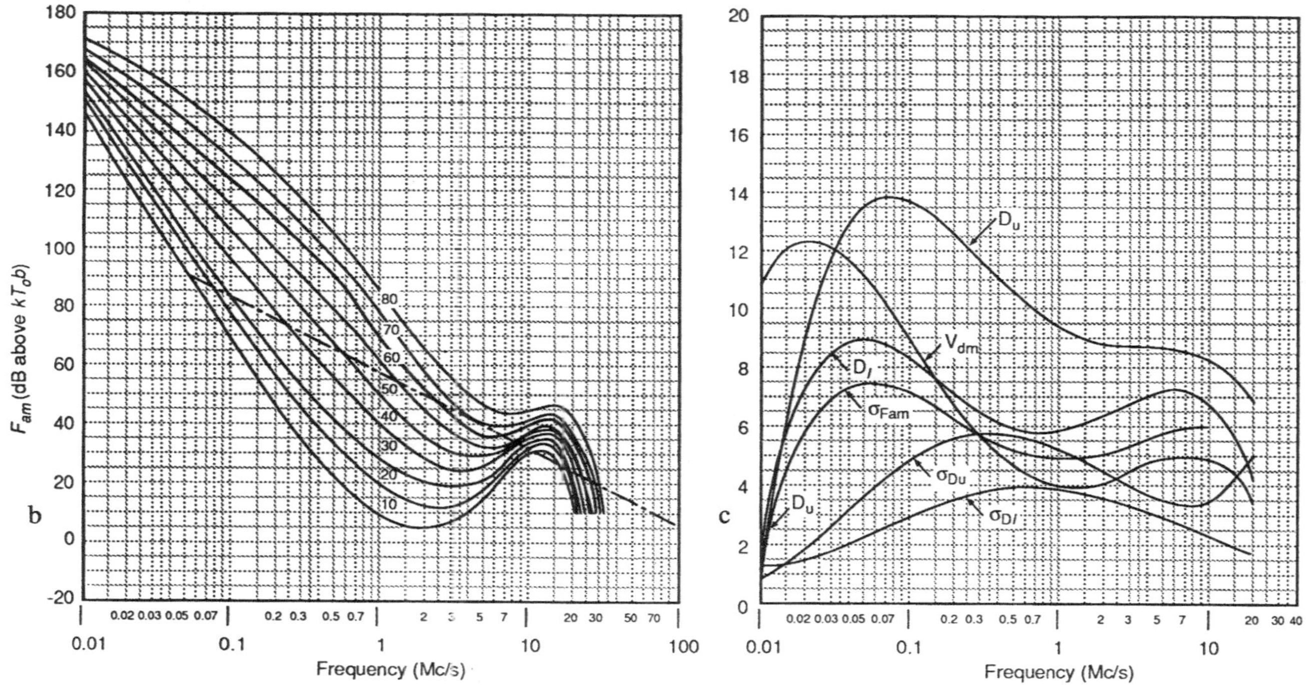

Figure 6.5 Atmospheric noise between noon and 4 P.M. during a northern winter: (a) expected values of atmospheric radio noise F_{am} in dB above kT_0B at 1 MHz; (b) frequency dependence of radio noise; (c) data on noise variability and character (after CCIR, 1964).

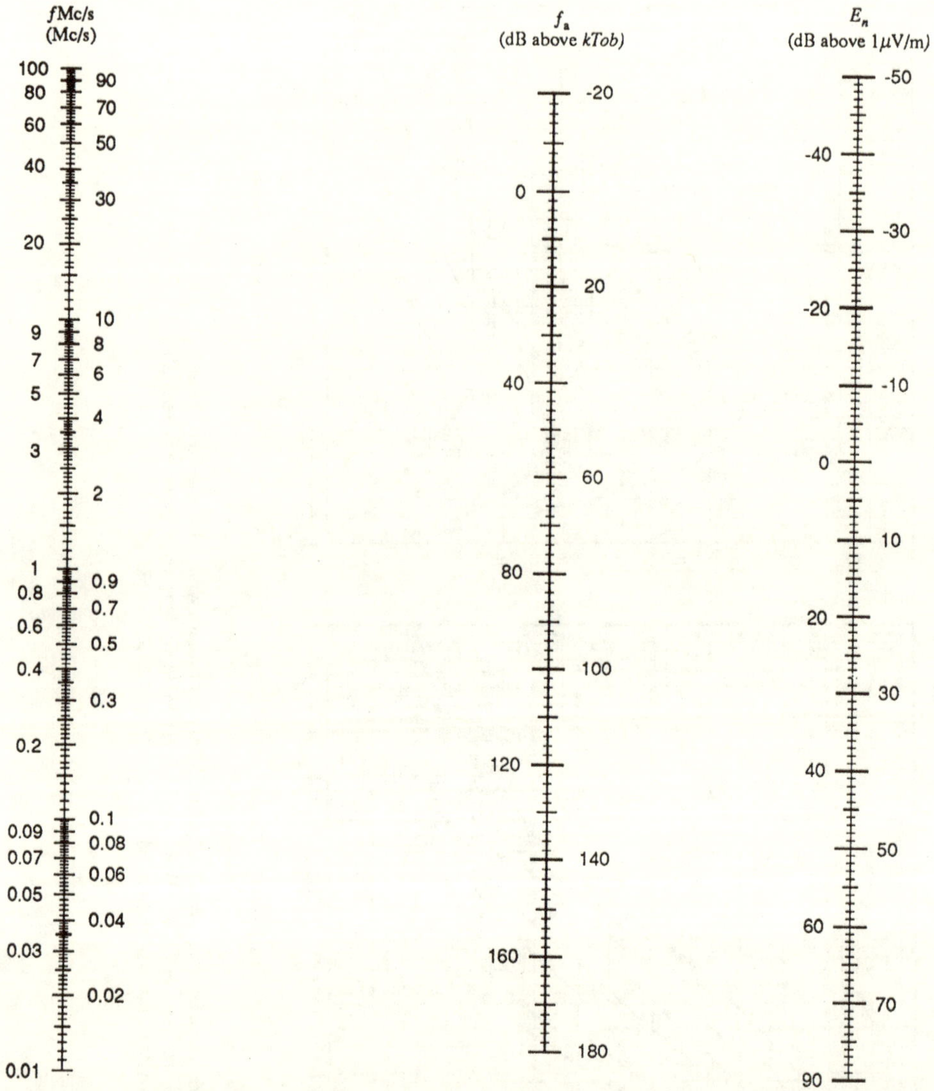

Figure 6.6 Nomogram for transforming effective antenna noise factor to noise field strength as a function of frequency; expression (6.20) (after CCIR, 1964).

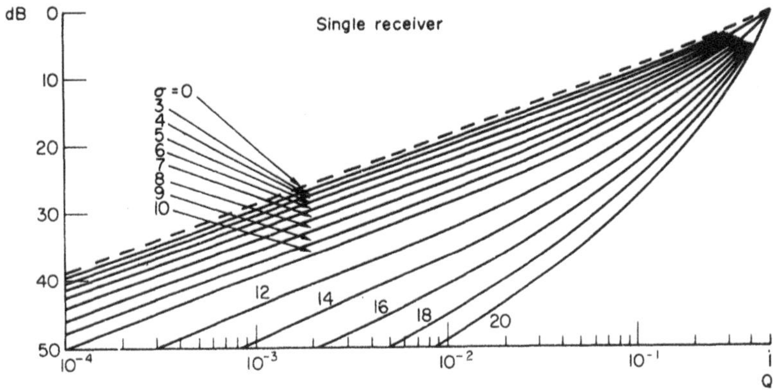

Figure 6.7 Convolution of the Rayleigh and log-normal distributions (after Picquenard, 1974, Macmillan Press).

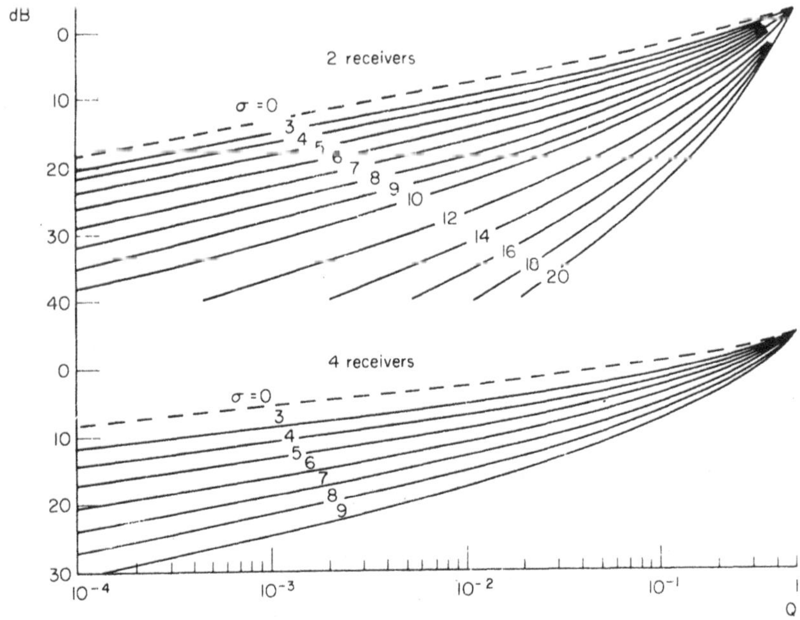

Figure 6.8 Fading distribution for diversity reception (switch type) (after Picquenard, 1974, Macmillan Press).

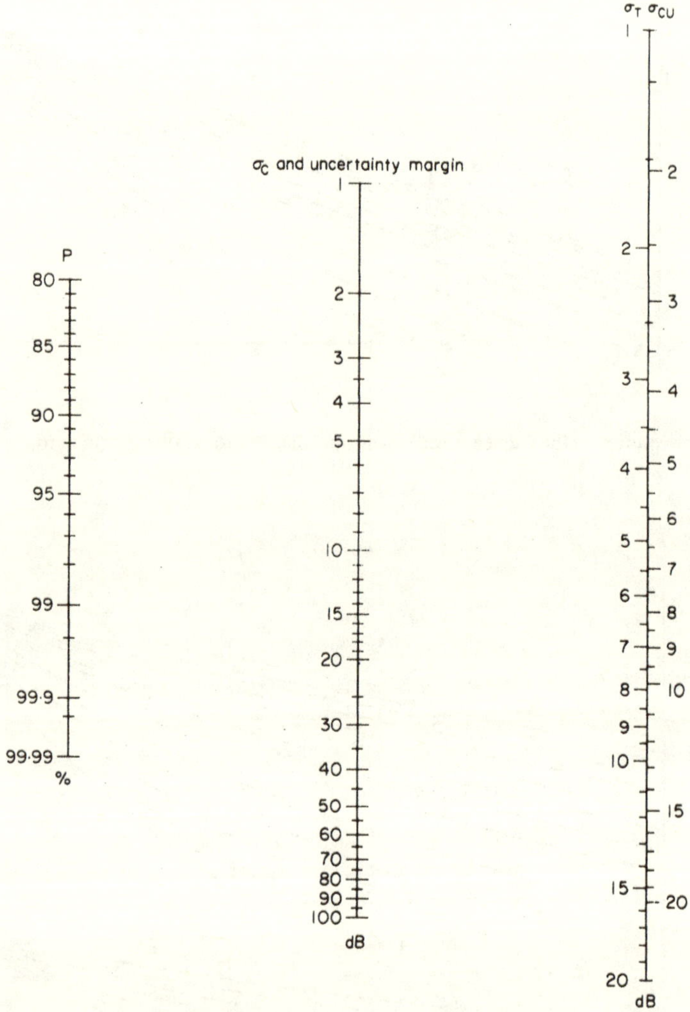

Figure 6.9 The uncertainty margin (after Picquenard, 1974, Macmillan Press).

6.3.1.3 Other Sources of Natural Noise

Other natural phenomena may cause noise and affect reception. Raindrops, snowflakes, dust, and sand particles moving through the air become charged and, on reaching an antenna, transfer their charge to it. Low, highly charged clouds passing over an antenna may cause corona discharges at the tip of an antenna. Little is known about the magnitude of the charges causing the observed noise; however, it was found that they may lift the noise level by more than 50 dB at all frequencies up to about 20 MHz.

6.3.2 Artificial Noise

Artificial disturbances are caused by electrical equipment of various sorts. Sparks are a source of wideband electromagnetic radiation, hence automobile interference to high frequency reception. Fluorescent light tubes cause another often experienced interference. Even power lines, especially at atmospheric conditions supporting the possibility of corona discharges, are known to interfere with the reception of electromagnetic waves.

Artificial noise propagates mainly over power lines and by ground waves, so it is relatively unaffected by diurnal and seasonal variations of the ionosphere, although some experimental evidence shows ionospheric propagation of artificial noise.

Artificial interference must be suppressed, where possible, at its origin; it may be reduced by proper earthing, shielding, and capacitive bypassing. It is higher in built-up areas, like cities and especially industrial suburbs, than in areas remote from human habitation, industry, and other sources of radiation. Such interference may be deliberate. Its aim may be to prevent the reception of foreign broadcasts or the disruption of military communication or radar operation. This type of interference is difficult, and sometimes impossible, to suppress.

6.4 SIGNAL-TO-NOISE RATIO REQUIRED BY VARIOUS COMMUNICATION SERVICES

As was pointed out earlier, the reliability of a communication circuit depends on the prevailing signal-to-noise ratio, because noise may distort signals and thus cause errors in the received messages. Different types of services, depending on the type of modulation used for the radio frequency carrier wave and detection, require different signal-to-noise ratios. The derivation of formulas for signal-to-noise ratios appropriate to various modulating systems is beyond the scope of this book, and the reader is referred to the specialized literature on the subject.

There is no accurate information on the required signal-to-noise ratio for a given degree of reliability of the various communication systems. Figures in Table 6.1, recommended by the Plenary Assembly of the CCIR, are to be used only as guidelines and great care is to be taken in their interpretation, as the given data are functions of many possible variables; for example, bandwidths in various parts of a communication system, type of modulation and demodulation used, and propagation conditions that vary during a day.

Table 6.1
Required Signal-to-Noise Ratios (Stable Conditions, Note 6)*

Type of Service	*Receiver Audio Bandwidth, kHz*	*Audio Signal-to-Noise Ratio, dB*	*Receiver Bandwidth, kHz*	*Ratio of Peak RF Signal-Noise in 6 kHz Band (Note 1), dB*
A1 telegraphy:				
8-baud low grade	1.5	−4	3	−7
24-baud	1.5	11	3	8
120-baud recorder	0.6	10	0.6	0
50-baud printer	0.25	16	0.25	2
A2 telegraphy:				
8-baud low grade	1.5	−4	3	−3 (Note 2)
24-baud	1.5	11	3	12 (Note 2)
F1 frequency-shift telegraphy:				
120-baud recorder	0.25	4	1.5	2
50-baud printer	0.10	10	1.5	−2
Phototelegraphy F4:				
Subcarrier frequency-modulation single-sideband emission	3	15	3	12
Hellschreiber, frequency-shift	1.5	6	3	3
A3 telephony:		(Note 3)		
Double-sideband, just usable quality, operator to operator (Note 4)	3	6	6	18
Double-standard, marginally commercial (Note 5)	3	15	6	27
Double-sideband, good commercial quality (Note 5)	3	33	6	35†
Single-sideband:				
1 channel	3	33	3	26†
2 channels	3	33	3‡	28†
3 channels	3	33	3‡	29†
4 channels	3	33	3‡	30†
Broadcasting	5	33	10	47

**CCIR Document of the IXth Plenary Assembly,* Los Angeles, 1959, vol. VI, *Recommendation 161,* p. 120, Geneva, 1959.

†Assuming 10-dB improvement due to the use of noise reducers.

‡Per channel.

Note 1: Measured as the ratio of r.m.s. signal corresponding to peak output of the transmitter and r.m.s. noise in a 6 kHz band, assuming stable conditions.

Note 2: Carrier keyed. Beat-frequency oscillator used.

Note 3: For A3 telephony the figures in this column represent the ratio of the audio signal as measured on a standard vu meter, to rms noise for a bandwidth of 3 kHz. (The corresponding peak signal power, i.e., when the transmitter is 100 percent tone-modulated, is assumed to be 6 dB higher.)

Note 4: For 90 percent intelligibility of unrelated words.

Note 5: When connected to the public service network.

Note 6: These values are based on Washington Doc. No. 138, Doc. No. 112 of Geneva, 1951, and The Hague Doc. No. 11.

Example 6.4

A 3-MHz double-sideband A3 telephony signal is to be received 95% of the time with good quality, at 3 P.M. local time during the northern winter, in Rome, Italy. Determine the needed signal level for the required signal-to-noise ratio in the presence of atmospheric noise.

Solution: The required signal-to-noise ratio from Table 6.1 is 35 dB. Proceeding according to Section 6.3.1.2, we read the expected value of atmospheric radio noise at 1 MHz at Rome's coordinates in Figure 6.5(a) to be about 26 dB. Entered into Figure 6.5(b) at a frequency of 1 MHz, this changes the value of F_{am} to about 23 dB above kT_0B at 3 MHz.

The standard deviation of F_{am}, as obtained from Figure 6.5(c) at 3 MHz, is 5.2 dB, so that

$$F_{am} = 23 + 5.2 = 28.2 \text{ dB}$$

The nomogram of Figure 6.6 permits the conversion of this value to the noise field strength of 28 dB below 1 μV/m.

This value, added to the required signal-to-noise ratio, yields 35 − 28 = 7 dB. The fluctuation margin, computed from expression (6.21) as

$$\sigma = (D_u^2 + D_s^2)^{1/2}/1.3$$

where D_u, from Figure 6.5(c), is 8.8 dB. Because ionospheric propagation is assumed, D_s = 7 dB, so that

$$\sigma = 8.65 \text{ dB}$$

Entered into Figure 6.7 at the curve for σ = 8.65 and 1 − 0.95 = 0.05, where 0.95 is the required time availability, on the horizontal axis, this leads to the fluctuation margin of about 32 dB.

The uncertainty margin, as given by expression (6.22), requires reading the value of σ_{D_u} at 3 MHz from Figure 6.5(c). The obtained value is 4 dB, and as ionospheric propagation has been assumed, the value of σ_{D_s} = 1.5, as given in the text. Then,

$$\sigma_{C_u} = 4.27 \text{ dB}$$

The standard deviation of the fluctuation margin C, as obtained from the nomogram in Figure 6.9 when entered at the percentage of time availability p = 95% and the preceding value of σ_{C_u}, is σ_C = 5.5 dB. The uncertainty margin for this standard deviation and a chosen $\sigma_P \approx 4$ dB, from expression (6.23), is σ_T = 9.59 dB.

The uncertainty margin, from Figure 6.9, is 16 dB and the required signal field strength is the sum

$$7 + 32 + 5.5 + 16 = 60.5 \text{ dB above } 1\ \mu\text{V/m}$$

This signal level is required to overcome atmospheric noise. Application of the formulas given in the previous chapter allows the determination of the transmitter power to yield this signal level.

PROBLEMS

6.1 The measured noise figure of two amplifiers in cascade is 5.5 dB. If the noise figure of the second amplifier is 10 dB and the first amplifier's gain is 30 dB, determine the second stage's contribution to the total noise figure.

6.2 Determine the total input noise power of a receiver at a frequency of 100 MHz when its bandwidth is 0.5 MHz and the antenna, at a location 30° south of the equator, points vertically up at 8 P.M. local time. The receiver's noise figure is 9 dB.

6.3 The external noise power reaching the input terminals of a 0.5-MHz-wide receiver is 4.14×10^{-14} W. The receiver's equivalent noise temperature is 2600 K. Determine the total input noise power of the receiving station.

REFERENCES AND RECOMMENDED READING

Aiya, S.V.C. (1979) "Atmospheric Noise Data for Tropical Regions," *Radio Science,* Vol. 14, No. 6, November–December, pp. 1001–1010.

CCIR (1959) Documents of the Ninth Plenary Assembly, Vol. VI, *Recommendation 161,* Los Angeles.

CCIR (1964) "World Distribution and Characteristics of Atmospheric Radio Noise," *Report 332,* International Telecommunication Union, Geneva.

Flock, W.L., S.D. Slobin, and E.K. Smith (1982) "Propagation Effects on Radio Range and Noise in Earth-Space Telecommunications," *Radio Science,* Vol. 5, No. 6, November–December, pp. 1411–1424.

Hamsher, D.H. (1967) *Communication System Engineering Handbook,* McGraw Hill, New York.

Joglekar, P.J., and R.A. Agarwala (1978) "Variation of Atmospheric Noise with Sunspot Number during the 1600–2000 Hours Time Period at Delhi," *IEEE Trans. on Electromagnetic Compatibility,* Vol. EMC-20, No. 1, Part 2, February, pp. 202–206.

Joos, G. (1934) *Theoretical Physics,* Blackie & Son Ltd., London.

Ko, H.C. (1958) "The Distribution of Cosmic Radio Background Radiation," *Proc. IRE,* Vol. 48, No. 1, January.

Menzel, D.H., ed. (1960) *The Radio Noise Spectrum,* Harvard University Press, Cambridge, MA.

Oh, L.L. (1969) "Measured and Calculated Spectral Amplitude Distribution of Lightning Sferics," *IEEE Trans. on Electromagnetic Compatibility,* Vol. EMC-11, No. 4, November, pp. 125–130.

Picquenard, A. (1974) *Radio Wave Propagation,* Macmillan Press Ltd, London.

Rohan, P. (1964) "Noise Factor, Noise Figure and Noise Temperature," W.R.E. Tech. Memo. PAD 176.

Rohan, P., and L.B. Soden (1970) "A Survey of the Southern Sky at 55 MHz," *Australian J. of Physics,* Vol. 23, pp. 223–225.

Chapter 7
RADIO LINK DESIGN CONSIDERATIONS

The purpose of radio links is the provision of reliable radio communication between two or more sites. This communication may be affected by various factors described in the earlier chapters, all of which must be considered in the initial studies to establish a link. Parameters of the equipment and location of the link terminals must be chosen to reduce the various effects of disruptive phenomena to an acceptable minimum.

Because many of the phenomena are random, long-term observations are required to gain adequate statistical data for their prediction. The phenomena of interest are ionospheric characteristics, atmospheric noise, and meteorological conditions, depending on the frequencies used. Ionospheric observatories publish data on which predictions of ionospheric behavior may be based. Atmospherics, as mentioned in the preceding chapter, are always present. However, statistical data on noise levels due to atmospherics at various geographic locations are available and may be used to determine the probable noise levels at the considered locations. Most meteorological stations have adequate data on the occurrence of meteorological phenomena that may affect the propagation of electromagnetic waves and hence the reliability of communication.

Bear in mind that we cannot rely entirely on computations because some local conditions, especially at higher frequencies, may exert additional influences on the performance of the terminal equipment. Although the required equipment parameters may be determined on the basis of computations, the final siting of the equipment should be confirmed by experiment.

7.1 THE CHOICE OF FREQUENCY

The separation of the link's terminals is the basic criterion affecting the choice of frequency. Communication over long distances necessitates the use of frequencies below about 30 MHz. Communication over distances of several thousand km is possible at high frequencies, whereas communication over distances within the radio horizon will be at very or ultra high frequencies. The chosen frequency determines the mechanism of propagation and hence the phenomena that may affect it.

It should be mentioned that the whole range of frequencies usable for communication is divided into bands allocated by international agreement to various services, and the chosen frequency and the used transmitted power and type of service must comply with these agreements. Further, due to the diurnal and seasonal changes in the behavior of the ionosphere, several frequencies are required in the high-frequency range for uninterrupted communication over 24 hours; all these frequencies must be considered when planning a radio link. The internationally agreed upon allocation of frequencies to the various services is available from the International Telecommunication Union, Geneva, Switzerland. The various services may be grouped into fixed services, aeronautical and maritime services, mobile services, and broadcasting.

The frequencies in the low-frequency range, below about 200 kHz, require very large antenna installations; in addition, due to the low antenna efficiency, they require rather high-power transmitters. The inherent equipment characteristics permit only slow signaling speeds. Due to these properties, such frequencies seldom are used in new installations, but still are used occasionally between fixed points to back up high-frequency ones in times of severe ionospheric disturbances. Frequencies between about 500 and 1600 kHz are used for broadcasting. The 1.6 to 4 MHz range of frequencies can be used over distances of about 150 to 400 km. At low sunspot activity, they may be used at night for long-distance communication.

Long-distance communication circuits generally use frequencies between 4 and 30 MHz; however, as indicated above, several frequencies are required for uninterrupted 24-hour traffic. Due to the great demand for frequencies within this range, strict adherence to the allocated frequency is obligatory to avoid interference with services on adjacent frequencies.

Frequencies in the 30 to 60 MHz range have been used for ionospheric scatter communication links, where the interference caused by the high powers and antenna configurations is tolerable.

Point-to-point links over short distances mainly use frequencies above 60 MHz. Longer distances are covered in several sections with repeater stations at distances equal to or preferably shorter than the combined radio horizons of the two subterminals. Due to the line-of-sight limitations of transmission at these frequencies, the frequencies used can be reused by several sections of the link. Links of this kind, in addition to higher reliability, have lower atmospheric and manufactured noise susceptibility and lower power requirements over those using lower frequencies. Tropospheric scatter links operating at these frequencies and using high-power transmitters are potential sources of interference.

Microwave links use frequencies above 1 GHz. Frequencies for satellite communication usually are chosen between those at which high attenuation by atmospheric gasses is experienced (see Figure 5.10).

7.2 MODULATION AND BANDWIDTH

The required bandwidth of communication systems is determined by the type of modulation used, which, in turn, depends on the type of service used; that is, the information to be transmitted. The overcrowded frequency bands necessitate bandwidth restrictions. Radio telegraphy, in which the carrier wave is keyed to convey information as a sequence of short and long symbol elements, requires the narrowest bandwidth if only an indication of the presence of the symbol elements suffices. A restricted bandwidth distorts signals. A larger bandwidth is required if the shape of the transmitted waveform is to be preserved. Nyquist [1924] has shown that the signaling speed is proportional to the bandwidth of a circuit or signal.

The transmission of messages by Morse code, teletype, or data transmission can be considered to be telegraphic communication. Radio circuits used for the transmission of telegraphic signals must be carefully designed to reduce the possibility of signal distortion and hence message alteration. Modern telegraphic systems have some mode of checking the accuracy of the transmitted message, for example, some long-distance radio telegraphy uses some form of error detecting code and others employ an automatic request for repeating characters considered to have been wrongly received [Hamsher, 1967]. Telegraphy also is used to transmit large volumes of alpha-numeric data for which no immediate answers are expected.

Facsimile systems are employed for transmitting photographs, weather maps, and similar information on subcarrier amplitude or frequency modulated by signals representing individual picture elements.

Data, for example, digital computer input or output data, usually are transmitted as binary signals using one frequency for a mark and another one for a space; a method referred to as *frequency shift keying*. Error checking, as a rule, is incorporated in data transmission circuits.

Voice communication requires modulating a high-frequency carrier wave by the information to be conveyed. Various modulation processes are used in communication. The continuous wave carrier's amplitude, phase, or frequency may be altered in accordance with the information to be transmitted or the information may be conveyed by the amplitude, width, or position variations of a sequence of pulses.

In the first of these cases, we speak about amplitude modulation that may be used as a single-sideband system, where only one sideband containing the modulating frequencies is transmitted with or without the radio frequency carrier wave, or a double-sideband system, adopted generally for broadcasting purposes where the additional cost of the more complicated single-sideband receivers forces authorities to retain the older double-sideband transmissions. Single-sideband circuits require half the double-sideband bandwidth and generally are used for fixed and mobile services at frequencies below about 25 MHz. The sideband width depends on the highest modulating frequency to be transmitted.

Phase and frequency modulation are special cases of angle modulation. In phase modulation, the phase of the carrier varies with the modulating signal. In frequency modulation, it varies with the integral of this signal.

Because frequency modulation is a nonlinear process, new frequencies are generated due to which the transmission of frequency modulated waves requires a much wider bandwidth than that necessary for the transmission of amplitude modulated ones. For this reason, frequency modulation is restricted to very and ultra high frequencies, where it is used for high-fidelity music and television sound transmission.

A periodic sequence of pulses represents the carrier in all these pulse modulation systems. Their amplitude, width, or position may vary in sympathy with the information to be transmitted. We speak then about pulse amplitude modulation in the first case, pulse width or pulse duration modulation in the second one, and pulse position in the last one.

The applicability of pulse modulation is based on the sampling theorem's finding that a bandlimited signal may be specified exactly by sufficiently frequent samples even if it varies continuously with time. More sophisticated pulse modulation systems have been developed to increase the reliability of communication. These include various types of delta modulation and pulse code modulation. Pulse code modulation quantizes samples of the signal by allowing them to assume only certain preassigned amplitude values and transmitting a binary number representing the sample closest to the signal level. These more sophisticated modulation systems improve the signal-to-noise ratio at the expense of bandwidth requirement. The modulated pulse sequence usually frequency modulates a radio frequency carrier.

Multiplexing of signals to be transmitted over a common channel is possible due to the finding of the sampling theorem. The signals then share the sampling cycle in a so-called time multiplexed system.

Amplitude modulation requires a bandwidth on the order of the highest modulating frequency to be transmitted, or twice that in case of double-sideband transmission. Wideband frequency modulation uses a bandwidth approximately equal to twice the maximum frequency deviation determined by the maximum amplitude of the modulating signal.

The bandwidth requirement of the various pulse modulation systems depends on what pulse parameter is to convey the encoded information. A pulse-amplitude modulation system must provide a sequence of pulses with amplitude variations corresponding to those of the transmitted signal. It may be shown that a bandwidth equal to about the reciprocal of the pulse width will satisfy this requirement. On the other hand, the variation of the position of the pulses with respect to a regular sequence of pulses carries the information in a pulse-position modulation system and the pulse rise time is of great importance. The transmission of steep rise times requires a very wide bandwidth.

The required bandwidth of a pulse position modulated system is inversely proportional to twice the pulse rise time. A pulse-code modulated system's bandwidth is proportional to the number of pulses in a sample slot multiplied by the signal bandwidth [Schwartz, 1959]. The required RF bandwidth is twice the modulation bandwidth.

7.3 POWER CONSIDERATIONS

Up to frequencies at which daytime propagation is predominantly by ground waves (i.e., frequencies below about 3 MHz), high-power transmitters are required to obtain the sought coverage. At high frequencies to about 30 MHz, low-power transmitters of about 100 to 1000 W are adequate for worldwide communication under favorable propagation conditions. It is well known that radio amateurs using transmitting powers on the order of tens of Watts and operating in the allocated high-frequency bands can establish and maintain communication with others at great distances. Even less power input is adequate at very and ultra high frequencies for communication over distances shorter than those to the radio horizon.

The choice of transmitter power depends then, in addition to the frequency to be used, on the required field strength to achieve the wanted signal-to-noise ratio at the remote link terminal. This is further conditioned by the permissible weight and volume of the transmitter and economic factors.

Electron tubes capable of delivering several hundred kW of continuous power at the lower radio frequencies used for broadcasting are readily available. The availability of high-frequency transistors capable of delivering about 1 to 100 W of continuous power, depending on the frequency, and about 1 kW of pulse power at frequencies in the GHz region permits the combination of such devices in high-power applications [Rohan *et al.*, 1968].

High-power transmitters require high primary power and therefore are restricted to fixed applications. Mobile stations usually operate at lower powers. The chosen transmitter power then depends on, in addition to factors dictated by the type of service and propagation effects:

(a) the availability of suitable devices for generating and amplifying signals at the chosen frequency;
(b) the availability of primary power and an adequate cooling water supply at the chosen site;
(c) the acceptable volume;
(d) the acceptable weight.

Although, as indicated earlier, devices are available for the generation and amplification of powers required for most current applications, the size of efficient radiators determines the required real estate and contributes to the economic considerations to be fulfilled. Volume and weight restrictions are particularly rigid for missile- and satellite-borne applications, where small batteries or solar cells serve as primary energy sources.

The minimum acceptable radiated power is computed considering the maximum range to be covered. A safety factor on the order of 2 usually is applied to the specification of the power to be provided.

The power fed to the antenna determines the field strength across the antenna's terminals. The break-down strength of the gap across the terminals is reduced by the environment's ionization. The ionization of the higher atmospheric layers at ionospheric levels may initiate glow discharges at fairly low voltages across the antenna terminals.

The ratio of the maximum available transmitted power and the minimum required signal power yields the acceptable propagation loss. This may be used to compute the required antenna gains at the link's terminals. High-gain antennas have a small beamwidth; this increases the difficulty of exact pointing of the antenna's beam toward the corresponding terminal. Exact antenna alignment is particularly critical in airborne applications and requires rigid antenna stabilization, especially for satellite communication.

7.4 COMMUNICATION LINK DESIGN PROCEDURE

The aim of communication link designs is the specification of equipment characteristics capable of fulfilling the requirements of reliable communication. The design procedure must consider all losses (i.e., those due to electromagnetic wave propagation and equipment design or deterioration) and determine the transmitter power capable of producing, at the receiver's input terminals for most of the time, signal levels satisfying the appropriate signal-to-noise ratio requirements for the type of service in question.

As was mentioned earlier, frequencies up to about 2 MHz are used either for special services, like time-signal transmission; communication with ships at sea and submerged submarines; or broadcasting where the widest coverage usually is desirable. At these frequencies, we cannot speak about radio links in the narrow sense of the meaning and therefore choose the highest economically acceptable transmitter powers and, of necessity, omnidirectional antennas. This consideration may be extended to frequencies up to about 30 MHz where, however, directional antennas often are used to increase radiation in a sought direction. We speak then about links only when communication at frequencies above 30 MHz is considered.

Propagation at these frequencies is mainly over line-of-sight distances; nevertheless, even here bands of frequencies are restricted by international agreements to specific applications. The frequency chosen for a microwave link must be within the appropriate frequency band; for example, the bands of

1.850–1.990 GHz,

2.130–2.150 GHz,

2.180–2.200 GHz,
6.575–6.875 GHz, and
12.2–12.7 GHz

are allocated to operational fixed services. And those at

1.990–2.110 GHz,
6.875–7.125 GHz, and
12.95–13.2 GHz

are for studio-transmitter links [Freeman, 1981].

A number of frequency bands are allocated to military applications and, although they might not be used in certain geographic locations at a given time, their use at some time in the future is possible. Some thought should be given to the correct frequency choice, therefore.

The pathlength, as well as other factors, determines the path loss. The desired application determines the required signal-to-noise ratio for a reliable service. This, together with the path loss, influences the required transmitter power. The initial design procedure is roughly the same at all frequencies, when considering the factors relevant to these. The sequence of design steps is outlined below.

7.4.1 Link Design Procedure

Various environmental conditions exert influence on the propagation of electromagnetic waves of the various frequencies. Whereas the curvature of the earth is of great importance in microwaves, it is of no consequence at low frequencies. Similarly, the ionosphere manifests itself at frequencies below about 30 MHz. Therefore the link designer must decide, on the basis of the material in the preceding chapters, which environmental influences to consider. The following considerations, although applying strictly to microwaves, may be taken as illustrative of similar ones at other frequencies.

7.4.1.1 Propagation

Propagation at lower frequencies is either by ground waves or sky waves, as described earlier. Propagation by microwaves is to a line-of-sight distance. This depends on the antenna height and atmospheric conditions, as described earlier in Section 5.4.2.2. It may be determined by expression (5.39):

$$d = \sqrt{2kr}\,(\sqrt{h} + \sqrt{H})$$

where h and H are the transmitting and receiving antenna heights, r is the true earth radius, and k is an earth radius correction factor given by expression (5.44) or (5.45). All lengths and heights are to be inserted in the same units.

As mentioned in Section 3.5, propagation between two elevated antennas may occur simultaneously over several paths. The resulting field exhibits various phenomena, the lobing pattern of the vertical coverage diagram being but one. Other frequently observed effects due to multipath propagation are "ghosting" of television pictures and signal strength variations of received signals by a mobile receiver, like a car radio. Multipath propagation may prove rather troublesome in pulse-modulated systems, where it may lengthen the received pulses or their repetition, depending on the delay between the pulses propagating over the direct and reflected paths.

In radar, multipath propagation may occur at propagation in both directions; namely, from the radar to the target and back. This is rather objectionable, because the target's range is determined by the delay between the transmitted pulse and the received echo. Narrowbeam antennas, permitting the complete or at least partial suppression of rays propagating toward the ground, may reduce or even eliminate the effects of multipath propagation.

Terrain features, including the earth's hump between the transmitting and receiving antennas, may restrict the mutual visibility of the antennas and hence affect propagation. The approximate maximum height of the bulge of a spherical earth half-way between the terminals separated by a distance d is

$$b \approx \frac{d^2}{8kr} \tag{7.1}$$

The bulge of the earth, in meters, at the position of a terrain feature at a distance $d_1 \leq d/2$ km from the nearer link terminal when $d_2 = d - d_1$ and $r \approx 6378$ km may be computed from

$$b = 0.0784 d_1 d_2 / k \tag{7.2}$$

the approximate height of the segment of a circle subtending a small angle at the circle's center when d_1 differs but little from d_2. This is to be added to the feature's height when antenna heights are determined to yield direct mutual visibility between them.

The optimum clearance between a straight line joining the antennas and the terrain feature's tip, in meters, according to Freeman [1981] is

$$b' = 10.38 \sqrt{\frac{d_1 d_2}{f(d_1 + d_2)}} \tag{7.3}$$

when the used frequency is expressed in GHz and d_1 and d_2, the feature's distances from the two link terminals, are in km.

Reading terrain feature heights and distances along a straight line connecting the antenna sites from appropriate maps and increasing them by the determined terrain bulges at their locations and the required clearances lets us draw path profiles on the basis of which suitable antenna heights over selected pathlengths may be determined.

Keep in mind that the earth's radius correction factor k, which depends on meteorological conditions, has a value of approximately 4/3 only for an artificial (the so-called standard) atmosphere, and it changes with atmospheric conditions. The "standard" atmospheric value may be used in the first approach to link design. Antenna heights should be chosen so that the terrain at which reflection occurs supports dispersion of the energy to minimize multipath effects.

Example 7.1

Determine the required antenna heights for a microwave link operating in a "standard" atmosphere at a frequency of 6880 MHz over a distance of 40 km: (a) over a spherical earth; (b) when a 25-m-high obstacle is at a distance of 5 km from the transmitter antenna site.

Solution: The earth's bulge is accounted for by considering a spherical earth of a finite radius kr.

(a) Optimum ray clearance at center of the path, from (7.3), is

$$b' = 10.38 \sqrt{\frac{20 \times 20}{6.88 \times 40}} = 12.51 \text{ m}$$

Assuming that both antennas will be of the same height, the ray connecting them will be tangential at the midpoint to an earth of a radius $kr + b'$, where $r = 6378.228$ km is the average earth radius. Then (see Figure 7.1),

$$(kr + h)^2 = (kr + 12.51 \times 10^{-3})^2 + X^2$$

with X the distance between the antenna and the point of contact between the ray and the earth of the extended radius, and h is the sought antenna height.

We can thus write,

$$\theta = \frac{20}{kr} = 2.352 \times 10^{-3} \text{ rad}$$

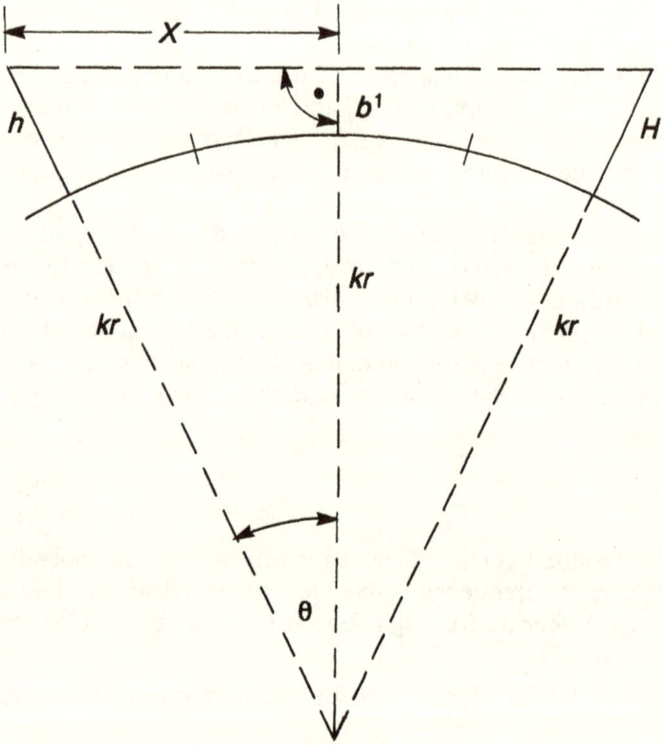

Figure 7.1 Example 7.1 for a spherical earth.

$$\tan\theta = \frac{X}{kr + 12.51 \times 10^{-3}}$$

from which $X = 20.00$ km and the antenna height $h = 36$ m.

(b) Because mutual visibility of the antennas must be maintained, the ray connecting the transmitter's antenna, the point at optimum clearance above the obstacle, and the receiving antenna must be on a straight line. This comes closest to the earth halfway between the obstacle and the receiving antenna; that is, at a distance of $35/2 = 17.5$ km from the obstacle (see Figure 7.2).

The optimum clearance there is

$$b_1' = 10.38 \sqrt{\frac{17.5^2}{6.88 \times 35}} = 11.71 \text{ m}$$

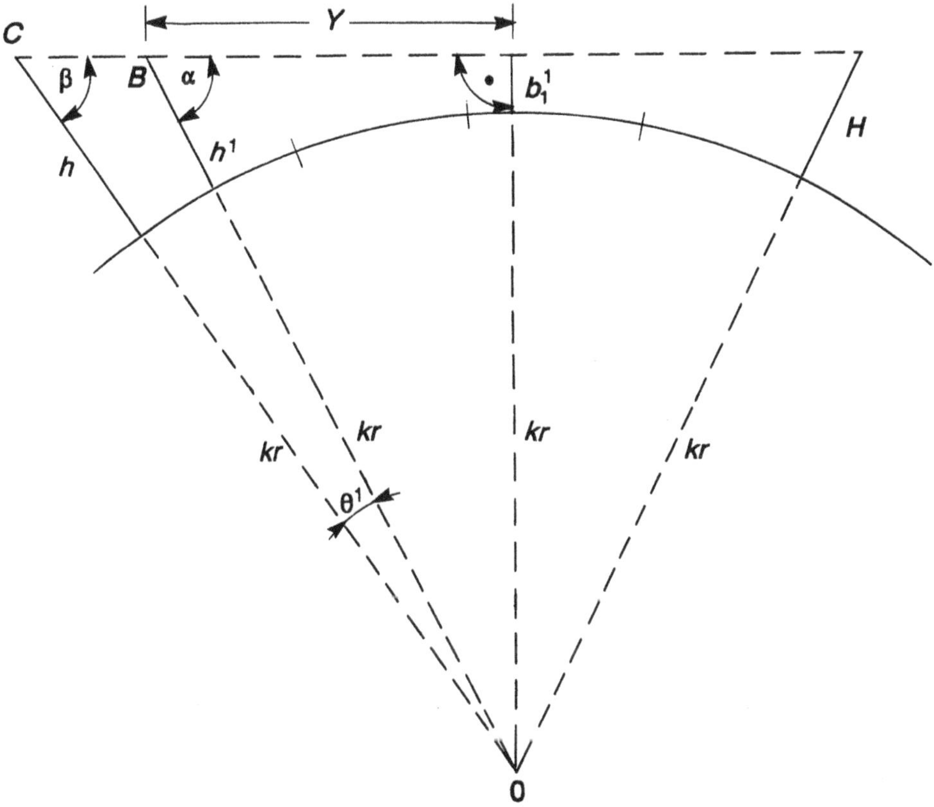

Figure 7.2 Example 7.1 with an obstacle between the antennas.

and the optimum clearance above the obstacle is

$$b' = 10.38 \sqrt{\frac{5 \times 35}{6.88 \times 40}} = 8.28 \text{ m}$$

The length of the line Y joining the point of optimum clearance above the obstacle at a height of $25 + b' = 33.28$ m and the one halfway from there to the receiving antenna is computed as follows. Assuming that the ray is orthogonal to the extended earth radius at point A, we can write

$$Y = \sqrt{(kr + 33.28 \times 10^{-3})^2 - (kr + 11.71 \times 10^{-3})^2}$$
$$Y = 19.15 \text{ km}$$

Further,

$$\tan \alpha = \frac{kr + 11.71 \times 10^{-3}}{Y}$$

$$\alpha = 89.871°$$

The complementary angle at B is 90.13°, so that

$$\beta = 180° - \theta' - 90.13°$$

with

$$\theta = \frac{5}{kr} = 0.0337°$$

leading to

$$\beta = 89.84°$$

From the law of sines applied to the triangle COB,

$$\frac{kr + h}{\sin(180° - \alpha)} = \frac{kr + 33.28 \times 10^{-3}}{\sin\beta}$$

yielding the transmitting antenna height $h = 44$ m.

Due to symmetry about the extended radius OA, the receiving antenna height H will be 33.28 m; that is, the sum of the obstacle height and the optimum clearance above it. In more complicated situations, a path profile drawing, as mentioned earlier, leads to the required solution.

7.4.1.2 *Losses*

Losses may be caused by free-space propagation or lack of optimal performance by equipment. The free-space propagation loss, according to expression (5.111), is

$$L_0 = \frac{P_t}{P_r} = \left(\frac{4\pi Rf}{c}\right)^2$$

or, expressed in decibel notation,

$$L_{0_{dB}} = 32.44 + 20 \log R_{km} + 20 \log f_{MHz} \tag{7.4}$$

where R is the straight-line distance between the isotropic antennas in km, and f is the frequency in MHz.

Most radio-link terminals use antennas with parabolic reflectors having a directional gain due to which the computed losses must be reduced by

$$G_{dB} = -42.3 + 20 \log f_{MHz} + 20 \log d_m \tag{7.5}$$

where d, in meters, is the diameter of the antenna's aperture. This expression considers a 54% surface efficiency.

Even the most carefully designed equipment experiences unavoidable losses. These include losses in transmission lines and waveguides, connectors, circulators, directional couplers, *et cetera,* and those due to the deterioration in equipment performance; for example, detuning of circuits or mismatches. Adding then, in decibel notation, all losses yields the figure to be considered in the link design computations.

7.4.1.3 Fading

Section 5.4.5.2 intimates that fading (i.e., variation in signal level with time), due to changing atmospheric conditions, may occur at very and ultra high frequencies. The severity of fading increases with an increasing frequency and pathlength. Deep fading probability follows an approximate $fd^{3.5}$ law [Freeman, 1981], where f is the carrier frequency and d the distance between the antennas.

Variations in atmospheric conditions cause variations in the radius of curvature of the ray paths in the vertical plane, resulting in variations in the phase differences between the direct and reflected rays and hence signal strength. This may be particularly pronounced when narrowbeam antennas are used because a large part of the arriving energy may be outside the receiving antenna's beamwidth. Signal level variations caused by this effect are small compared with fading due to multipath propagation.

It is good engineering practice to allow for possible signal strength variations due to fading and use a fading margin depending on the sought reliability of communication. Recommended margin values for a microwave link of about 50 km path length are

Link Availability (%)	*Fading Margin (dB)*
90	10
99	20
99.9	30
99.99	40

Freeman [1981], quoting the work of Barnett and Vigants, gives the formula

$$M_{\mathrm{dB}} = -10 \log \frac{1.67 \times 10^{9}(1 - p_r)}{abf_{\mathrm{MHz}} d_{\mathrm{km}}^{3}} \tag{7.6}$$

for the computation of the fading margin. Here p_r is the percentage of time availability of the radio link expressed as a fraction, f is the carrier frequency in MHz, d is the pathlength in km, and the parameters

a = 4 for very smooth terrain, over water, or flat desert;
= 1 for average terrain with some roughness;
= 0.25 for mountains, very rough, or very dry terrain;
b = 0.5 for hot, humid coastal areas;
= 0.25 for normal, interior temperate or subarctic areas;
= 0.125 for mountainous or very dry but nonreflective areas.

The obtained fading loss margin usually is increased by at least 6 dB to account for minor antenna misalignment and system degradation, like waveguide corrosion or transmitter output and receiver noise figure degradation due to aging.

Although fading cannot be predicted accurately, the percentage of time during which fading occurs may be ascertained on a statistical basis.

7.4.1.4 *Receiver Noise*

Because the signal-to-noise ratio at the input terminals of the receiver is of interest, the noise contribution of the receiver must be found. Using expression (6.4) for the receiver's output noise level and dividing it by the receiver's gain G leads to the input noise level:

$$N_{\mathrm{in}} = kT_0BF \tag{7.7}$$

Knowing the value of Boltzmann's constant (1.38×10^{-23} Ws/K) and that of the generally used ambient temperature (290 K) permits the expression of the input noise level in dB:

$$\begin{aligned} N_{\mathrm{indB}} &= 10 \log kT + 60 + 10 \log B_{\mathrm{MHz}} + F_{\mathrm{dB}} \\ &= -144 + 10 \log B_{\mathrm{MHz}} + F_{\mathrm{dB}} \end{aligned} \tag{7.8}$$

where the equivalent noise bandwidth, or the bandwidth of the intermediate frequency amplifier, is to be inserted in MHz and F_{dB} is the receiver's noise figure defined by expression (6.3).

Some modulation systems provide signal-to-noise improvement, as mentioned earlier. The improvement threshold of a wideband frequency modulation system occurs at an about 11 dB input signal-to-noise ratio [Schwartz, 1959] when a "capture effect" manifests itself by a jump in the output signal-to-noise ratio. The input signal must be of a level to yield the required input signal-to-noise ratio permitting the exploitation of the advantages of the more sophisticated modulation systems. Designating this improvement, expressed in dB, by I, the required input signal level S becomes

$$S_{\mathrm{dBW}} = N_{\mathrm{in}_{\mathrm{dB}}} - I + \mathrm{Losses(dB)} \tag{7.9}$$

7.4.1.5 *Required Transmitter Power*

Based on the derived input signal level, we can determine the required transmitter power needed. Expression (7.9) considers neither the ambient-noise power level at the receiver's site, nor the fading margin, nor does it consider the input signal-to-noise ratio required for the system's operation above the improvement threshold.

Adding the ambient noise power N_{amb} at the receiver's site, together with the fading margin, required signal-to-noise ratio appropriate for the type of service, and threshold power level T_p (about 11 dB in wideband frequency modulation) required for the system's operation above the improvement threshold, expression (7.9) yields the required transmitter power as

$$P_{t_{\mathrm{dBW}}} = S_{\mathrm{dBW}} + N_{\mathrm{amb}_{\mathrm{dB}}} + M_{\mathrm{dB}} + S/N + T_{p_{\mathrm{dB}}} \tag{7.10}$$

Example 7.2

Determine the transmitter power required for a 16-km-long frequency modulated studio-to-transmitter link operating at a frequency of 6880 MHz over an average terrain in a humid coastal area. The receiver, having an 8-dB noise figure, has a 240 kHz i.f. amplifier bandwidth. Both antennas use a 1 m diameter parabolic reflector with a 54% surface efficiency. The link is to operate for 99.99% of the time above the F.M. improvement threshold with a 30 dB input signal-to-noise ratio. The used modulation index will be $\beta = 5$.

Solution: From expression (7.4), the free-space loss is

$$\begin{aligned} L_{0_{\mathrm{dB}}} &= 32.44 + 20 \log 16 + 20 \log 6880 \\ &= 133.27 \mathrm{~dB} \end{aligned}$$

The antenna gain factor from expression (7.5) is

$$G = -42.3 + 20 \log 6880 = 34.45 \text{ dB}$$

The free-space loss then becomes

$$L_{0_{\mathrm{dB}}} = 133.27 - 34.45 - 34.45 = 64.37 \text{ dB}$$

The fading margin for $p_r = 0.9999$ is

$$\begin{aligned} M_{\mathrm{dB}} &= -10 \log \frac{1.67 \times 10^9 \times 10^{-4}}{0.5 \times 6880 \times 16^3} \\ &= 19.26 \text{ dB} \end{aligned}$$

Attenuation by atmospheric gasses is about $0.005 \times 16 = 0.08$ dB. Waveguide and other losses are estimated to be 0.5 dB. The sum of losses is 64.97 dB. FM improvement is $I = 3 \times \beta^2 = 75$, or 18 dB. The input noise level is

$$\begin{aligned} N_{\mathrm{in}_{\mathrm{dB}}} &= -144 + 10 \log 0.24 + 8 \\ &= -142.20 \text{ dB} \end{aligned}$$

The ambient noise at these frequencies is negligible.

By using an FM improvement threshold of $T_p = 11$ dB; S/N, the required signal-to-noise ratio, of 30 dB; and a safety margin of 6 dB, the required power, from expression (7.10), is

$$\begin{aligned} P_{t_{\mathrm{dBW}}} &= -142.20 - 18 + 64.97 + 30 + 11 + 6 + 19.26 \\ &= -28.97 \text{ dB} \end{aligned}$$

The required transmitter power is then 1.28 mW.

PROBLEMS

7.1 Determine the one-way path loss of a synchronous satellite link operating at a frequency of 4 GHz. The distance D of the satellite from the center of the earth may be computed from

$$D = \sqrt[3]{\left(\frac{Tr}{2\pi}\right)^2 g_0}$$

obtained from Newton's second law and Kepler's third law. Here T is the satellite's period of orbit in seconds ($T = 86400$ s for a day), r is the earth's radius in meters (use the average radius $r = 6378228$ m), and $g_0 = 9.81$ m/s^2 is the gravitational acceleration at the earth's surface.

7.2 A radio link operating at a frequency of 5 GHz consists of an 80-km-long path between a city location and a passive reflector of 20 m^2 aperture at the top of a 720-m-high cliff and an 800-m-long path to the corresponding station at the bottom of the cliff. Determine the total path loss when the passive reflector gain in both directions is

$$G = \left(\frac{4\pi A f^2}{c^2} \cos\alpha\right)^2$$

where A is the aperture and α is half of the angle between the two paths.

7.3 A power of -60 dBW is required at the receiving antenna. The gain of each (i.e., the transmitting and the receiving) antenna is 40 dB. Find the required transmitter power when operating at a frequency of 5 GHz over a 50-km-long path.

7.4 A microwave television transmitter operating at a frequency of 6500 MHz over a 60-km-long path uses a parabolic reflector antenna of 60 cm diameter and 54% aperture efficiency. The receiver uses a half-wave dipole. The receiver bandwidth is 4 MHz and its noise figure is 12 dB.

(a) What transmitter power is required to assure a 30 dB signal-to-noise ratio for 99% of the time when propagation is over a rough terrain in a temperate climate?

(b) A repeater at part a's receiver location has an overall gain of 120 dB. Both of its antennas are identical with part a's transmitter antenna. What transmitter power will satisfy the preceding requirements when the given receiver is 18 km from the repeater and propagation for the second part of the path is over water in a temperate climate?

REFERENCES AND RECOMMENDED READING

Barsis, A.P., K.A. Norton, and P.L. Rice (1962) "Predicting the Performance of Tropospheric Communication Links, Singly and in Tandem," *IRE Trans. on Communication Systems,* Vol. CS-10, No. 1, March, pp. 2–22.

Barnett, W.T., and A. Vigants (1975) "Engineering Considerations for Microwave Communication Systems," Lenkurt Electric Corp., San Carlos, CA.

Bradley, P.A., and C. Bedford (1976) "Prediction of h.f. Circuit Availability," *Electronics Letters,* Vol. 12, No. 1, 8 January, pp. 32–33.

Brussaard, G. (1981) "Prediction of Attenuation Due to Rainfall on Earth-Space Links." *Radio Science,* Vol. 16, No. 5, September–October, pp. 745–760.

Chu, T.S. (1979) "Effects of Sandstorms in Microwave Propagation," *BSTJ,* Vol. 58, No. 2, February, pp. 549–555.

Crane, R.K. (1977) "Prediction of the Effects of Rain on Satellite Communication Systems," *Proc. IEEE,* Vol. 65, No. 3, March, pp. 456–474.

Daniel, L.D., and R.A. Reinman (1976) "Performance Prediction for Short-Range Troposcatter Links," *IEEE Trans. on Communication Technology,* Vol. COM-24, No. 6, June, pp. 670–672.

Daniel, L.D., and R.A. Reinman (1976) "Modification of the Radio Model for Performance Prediction of Short Range Troposcatter Links," 1976 International Conference on Communication, Part III, Philadelphia, June 14–16, pp. 46/24–26.

East, F.R. (1967) "Planning an h.f. Transmission Circuit," *Proc. IREE* [Australia], Vol. 28, No. 6, January, pp. 207–212.

Fedi, F., and U. Bordoni (1981) "Prediction of Attenuation Due to Rainfall on Terrestrial Links," *Radio Science,* Vol. 16, No. 5, September–October, pp. 731–743.

Freeman, R.L. (1981) *Telecommunication Transmission Handbook,* 2d Ed., J. Wiley & Sons, New York.

Fujita, M., and K. Nakamura (1979) "Seasonal Variations of Attenuation Statistics in Millimetre-Wave Earth-Satellite Link Due to Bright-Band Height," *Electronics Letters,* Vol. 15, No. 20, September 27, pp. 654–655.

Hearson, L.T. (1967) "Unusual Propagation Factors in Point-to-Point Microwave System Performance," *IEEE Trans. On Communication Technology,* Vol. COM-15, No. 4, August, pp. 615–625.

Hamsher, D.H. (1967) *Communication System Engineering Handbook,* McGraw-Hill, New York.

Lane, J.A., A.C. Gordon-Smith, and A.M. Zavody (1967) "Absorption and Scintillation Effects at 3 mm Wavelength on a Short Line-of-Sight Radio Link," *Electronics Letters,* Vol. 3, No. 5, May, pp. 185–186.

May, G. (1977) "Determining the Point of Reflection on MW Radio Links," *Microwave J.,* Vol. 20, No. 9, September, pp. 74–76.

Nyquist, H. (1924) "Certain Factors Affecting Telegraph Speed," *BSTJ,* Vol. 3, No. 2, April, pp. 324–346.

Panter, P.F. (1972) *Communication System Design,* McGraw-Hill, New York.

Petrie, H.E., and R.D. Hunsucker (1967) "Short-Term Forecasting for a High-Frequency Long-Distance Point-to-Point Communication Path," *IEEE Trans. on Antennas and Propagation,* Vol. AP-15, No. 3, May, pp. 390–394.

Reference Data for Radio Engineers, 4th Ed. (1956) International Telephone and Telegraph Corp., New York.

Rohan, P., M. Viola, E.C. Templar, and J.B. Wilcox (1968) "A High Power Radio Frequency Transistor Amplifier," *Electronics Letters,* Vol. 4, No. 21, October; reprinted *Proc. IREE* [Australia], Vol. 32, No. 2, February.

Schwartz, M. (1959) *Information Transmission, Modulation and Noise,* McGraw-Hill, New York.

Shur, A.A. (1976) "The Effect of Climate on the Long-Range Propagation of Ultrashort Radio Waves," *Telecommunication and Radio Engineering,* Part I, Vol. 30, No. 5, May, pp. 8–10.

Suzuki, H. (1977) "A Statistical Model for Urban Radio Propagation," *IEEE Trans. on Communication Technology,* Vol. COM-25, No. 7, July, pp. 673–680.

Weibel, G.E., and H.O. Dressel (1967) "Propagation Studies in Millimeter-Wave Link Systems," *Proc. IEE,* Vol. 35, No. 4, April, pp. 497–513.

INDEX

The Artech House Remote Sensing Library

Fawwaz T. Ulaby, *Series Editor*

Handbook of Radar Scattering Statistics for Terrain by F.T. Ulaby and M.C. Dobson

Introduction to Electromagnetic Wave Propagation by P. Rohan

Microwave Radiometric Systems: Design and Analysis by N. Skou

Microwave Remote Sensing Fundamentals and Radiometry, Volume I by F.T. Ulaby, R.K. Moore, and A.K. Fung

Microwave Remote Sensing: Radar Remote Sensing and Surface Scattering and Emission Theory, Volume II by F.T. Ulaby, R.K. Moore, and A.K. Fung.

Microwave Remote Sensing: From Theory to Applications, Volume III by F.T. Ulaby, R.K. Moore, and A.K. Fung

Radargrammetric Image Processing by F.W. Leberl

Radarpolarimetry for Geoscience Applications, F.T. Ulaby and C. Elachi, eds.

Radar Scattering Statistics for Terrain: Software and User's Manual by F.T. Ulaby and M.C. Dobson

www.ingramcontent.com/pod-product-compliance
Lightning Source LLC
LaVergne TN
LVHW101457200926
848483LV00001B/149
* 9 7 8 0 8 9 0 0 6 5 4 5 7 *